Robotics Law 3.0

Governing Humanoid Robots in the Age of Physical Intelligence.

Saheed Babajide Okuboyejo

LL.B. (UNILAG), B.L.(NLS), MBA. (GIBS),. LL.M. (UvA).

Keywords: *Humanoid Robots, Physical Intelligence, Embodied Intelligence, Robotics Law, Autonomous Systems, AI Governance, Legal Personhood, Algorithmic Liability, Human-Robot Interaction, Ethical Robotics, Adaptive Law, Shared Agency, Automation Law, Robot Ethics, Cross-Border Compliance, Post-Human Justice.*

Robotics Law 3.0

Humanoid robots are no longer science fiction. They are walking, speaking, making decisions, and learning to interact with humans in homes, hospitals, factories, workplaces, and public spaces. As machines increasingly operate alongside people, they raise legal questions once confined to researchers, philosophers, and filmmakers. Robotics Law 3.0 explores how legal systems around the world can adapt to the rise of physically intelligent, embodied, and humanlike machines while ensuring that humanoid robots serve humanity, not the other way around. Robotics Law 3.0 provides a comprehensive governance framework for humanoid robots and embodied systems operating in the age of physical intelligence. It addresses urgent issues such as liability for robotic decisions, surveillance and privacy risks, human rights in the context of machine interaction, labour and economic transformation, and the growing debate over whether robots should be granted legal personhood.

Drawing from real-world case studies and emerging laws in the EU, United States, Japan, China, South Korea and beyond, it equips readers with both foundational knowledge and practical governance frameworks. Covering core concepts including Legal Personhood, Shared Agency, Regulatory Futures, Algorithmic Liability, and the Twelve Pillars of Humanoid Robotics Governance, the book blends legal theory, technological insight, and ethical analysis. Through global regulatory analysis, practical governance tools, and the story of Ògún and Elias in Between Us and Them, it explores the legal, social, economic, and geopolitical implications of a world increasingly shared with autonomous humanoid robots.

The Law 3.0 Series explores the legal frontiers of emerging technologies shaping the twenty-first century. Each volume examines how traditional legal systems must evolve to govern intelligent, autonomous, and transformative innovations.

Published titles in the series:

Drone Law 3.0

Robotics Law 3.0

Forthcoming titles in the series:

Self-driving Vehicles Law 3.0

Space Law 3.0

Biotechnology Law 3.0

Blockchain Law 3.0

Clean Tech Law 3.0

Singularity Law 3.0

Published in the United States
by: Staten House
447 Broadway,
2nd Floor New York,
NY 10013

Published in the Netherlands
by: BabajideAI.
Van Oldenbarneveldtstraat 56S. 1052 KC. Amsterdam

Publication Date: 06 June 2026

U.S Library of Congress Cataloging-in-publication Data
Names: Okuboyejo, Saheed, (author).
Title: Robotics Law 3.0 : Governing Humanoid Robots in the Age of Physical Intelligence / Saheed babajide Okuboyejo. Description: [New York] : Staten House, [2026]. | Includes bibliographical references and index.

ISBN: [979-8-90243-550-1] (paperback edition)

Subjects: Humanoid Robots, Physical Intelligence, Embodied Intelligence, Shared Agency, Robotics Governance, Legal Personhood, Autonomous Systems, Regulatory Futures, Algorithmic Liability, Cross-Border Compliance, Ethical Design, Dignity Protection, Behavioural Governance, Human–Robot Interaction, Risk Mitigation, Social Cohesion, Urban Integration, Automation Law, Adaptive Law, Post-Human Justice.

Typesetting and cover design by BabajideAI

Idi gbere, idi gbere, idi gbere, arìnnàkò.

(Odi, 7:1)

Table of Contents

Preface **XX**

Between Us and Them - Episode One *XXII*

Foundation **1**

Why Does Law Struggle When Machines Begin to Act? 5

The Problem of Intention in Embodied Systems *9*

The Collapse of Foreseeability in Learning Machines *11*

Consent in Human–Robot Interactions *14*

How Should Law Classify Humanoid Robots? 16

Degree of Autonomy *20*

Proximity to Human Vulnerability *23*

Capacity to Influence Emotion *25*

Potential to Cause Harm and Create Abundance *28*

Why Does Robotics Law Matter Now? 31

Scope, Jurisdictions, and Methodology 35

Between Us and Them - Episode Two *38*

1. The Rise of Humanoid Robots **45**

Anatomy of Humanoid Robots 45

Chips and Computational Architecture *46*

Motors and Drives *47*

Batteries and Energy Systems *48*

Sensors and Perception *48*

Actuators and Embodied Force *49*

AI Control Systems and Orchestration *50*

Hidden Planetary Costs and Power Structures 51

What Makes Robots Humanlike? 51

Anthropomorphism 52

Morphology 52

Movement 53

Hands, Faces, and Expression 53

Interaction and Contingency 54

Interaction and Social Presence 55

Interaction and Shared Agency 55

Memory of Interaction 56

Turn-Taking and Dialogue 56

Proxemics and Physical Dialogue 57

Emotional Resonance 57

Cultural Context and Expectation 58

Human-Factors Benchmarks 58

Humanlikeness as Legal Threshold 59

Response as Legal Threshold 59

Why Interaction Must Be Governed 60

Why Are We Making Robots Humanlike? 60

From Form to Meaning 61

Expectation of Intent Without Intent 61

Compatibility With Human Labour 62

Trust as a Commercial Asset 62

Emotional Labour and Social Substitution 63

Efficiency Versus Explanation 63

Applications, Limitations, and Emerging Risks 63

Factories 65

Warehouses 67

Logistics 68

Hospitals 69

Homes 71

Entertainment 72

Extreme Environments 73

Agriculture 74

Construction 75

Retail and Hospitality 76

Military and Defense 77

Aerospace and Space 79

Distinctions Between Automation and Autonomy 80

Legal Relevance of Autonomy Levels 81

Legal Relevance of Proximity to Human Vulnerability 83

Legal Relevance of Emotional Influence 84

Legal Relevance of Serious Harm and Abundance 86

Encountering Existing Humanoids 87

Optimus 89

H1 91

AgiBot 93

Walker X 95

Helix 97

IRON 99

Digit 101

Apollo 103

NEO Gamma 105

XMAN 107

GR-2 109

Phoenix 111

Humanoid 113

Bumblebee 115

Onero H1 117

Gene.01 120

Memo 122

Atlas 124

X30 126

Grek 128

Sophia 131

Between Us and Them - Episode Three 133

2. Brief History of Robotics and the Law **138**

From Asimov to the AI Act 138

Signature Robotics Legal Incidents 141

Comparative Snapshots 143

European Union: Risk, Rights, and Pre-emptive Governance *143*

United States: Case Law, Fragmentation, and After-the-Fact Control *144*

Japan: Integration, Trust, and Social Design *145*

International Standards and Emerging Best Practice *145*

Legal Doctrines from Other Domains 146

Judicial Interpretation and Precedent 147

Influence of Science Fiction Literature 149

Influence of Science Fiction Films 153

Between Us and Them - Episode Four *157*

3. Legal Blind Spots **161**

Shortfalls of General Product Law 164

Responsibility Gap and Legal Voids 166

Legal Parenthood and Emerging Agency 169

Legal Thresholds of Autonomy 172

Gaps in Procedural Safeguards 173

Ethical vs. Legal Personhood 176

Between Us and Them - Episode Five *180*

4. Liability and Responsibility **184**

EU Product Liability Directive 2024/2853 186

AI Liability Directive 188

Shared Responsibility Chains 190

Updating Risk Post-Deployment 194

Model Contractual Clauses 196

Model Clause: Update Authority and Control *196*

Model Clause: Material Modification *197*

Model Clause: Ongoing Monitoring and Incident Reporting *198*

Model Clause: Allocation of Post-Deployment Risk *199*

Model Clause: Model Training and Data Integrity *199*

Model Clause: Security Maintenance *199*

Model Clause: Decommissioning and End-of-Life Obligations *200*

Model Clause: Mutual Indemnification *200*

Civil vs. Criminal Liability 201

Between Us and Them - Episode Six *205*

5. Robotics & Human Rights **209**

Risk Classification 212

Algorithmic Bias and Anti-Discrimination 214

Surveillance, Consent, and Dignity 217

Humanoids in Law Enforcement and Border Control 221

Privacy and Embodied AI 224

Use of Robots in Sensitive Contexts 226

Between Us and Them - Episode Seven 228

6.Employment, Labor, & Economic Displacement **233**

Employment Law for Human–Robot Collaboration 236

Rights of Workers Displaced by Robots 238

Collective Bargaining in a Robotic Workplace 241

Robot Remuneration and Taxation 243

Retraining, Reskilling & Legal Duties 245

Autonomy Levy and Platform Regulation 247

Between Us and Them - Episode Eight 251

7. Licensing, Registration & Oversight **253**

Hardware Licensing Regimes 255

Proposed International Humanoid Registry 258

Model Statutory Provision 260

Certification of Autonomy Levels 264

Cross-border Oversight and Customs Issues 267

Data and Operational Transparency Requirements 270

Between Us and Them - Episode Nine 272

8. Designing Ethical & Lawful Robots **276**

Ethical Design Principles in Hardware and Software 278

Ethical Principles Embedded in Hardware 278

Ethical Principles Embedded in Software 279

From Principles to Practice 280

The Design Responsibility Threshold 281

Human Oversight-by-Design and Legal Engineering 281

Human-in-the-Loop and Override Mechanisms 284

Fairness, Bias Mitigation and Human Dignity 287

Adaptive Learning, Drift, and Continuous Alignment 291

Cultural and Jurisdictional Design Sensitivity 294

Black Box Requirements and Data Logging 296

Ethical Impact Assessments and Documentation 299

Voluntary Codes vs Enforceable Norms 301

Between Us and Them - Episode Ten *303*

9. International Law & Robotic Autonomy. 308

State Responsibility and Attribution 310

Transnational Liability and Cross-Border Harm 313

Autonomous Robots and the Law of Armed Conflict 316

Robotics in Trade, Intellectual Property, and Export Regulation 319

Extraterritorial Regulation of Robotic Systems 323

Treaty Obligations and Use of Force Principles 326

UN, WIPO, and ICAO Roles in Robotics Governance 329

Global Robotics Charter and Governance Frameworks 332

Draft Framework: Global Robotics Charter 333

The Strategic Value of a Global Robotics Charter 337

Governance Framework Beyond the Charter 338

Between Us and Them - Episode Eleven *338*

10. Twelve Pillars of Humanoid Robotics Governance **341**

Pillar One : Embodied Ethics 346

Pillar Two: Mission Boundaries 351

Pillar Three: Societal Co-Design 356

Pillar Four: Sensorial Integrity 361

Pillar Five: Component Accountability 365

Pillar Six: Physical Security 370

Pillar Seven: Behavioural Calibration 375

Pillar Eight: Behavioural Learning 379

Pillar Nine : Explainable Embodiment 384

Pillar Ten: Lifelong Learning 388

Pillar Eleven: Shared Agency 394

Pillar Twelve: Distributed Liability 398

Operationalising the Twelve Pillars 402

Between Us and Them - Episode Twelve *407*

11. The Future of Legal Personhood **412**

Should Robots Have Rights? 416

Legal Considerations for Artificial Personhood 420

Tests for Recognition: Duties, Reciprocity, and Utility 425

Historical Analogies 430

Governance and Accountability of Artificial Persons 436

Reimagining Law Beyond the Human Subject 440

Between Us and Them - Episode Thirteen 445

12. Human & Humanoid Coexistence **450**

Urban Planning for Robotic Coexistence 455

Shared Spaces 460

Emotional and Psychological Impacts 463

Legal Protections for Vulnerable Populations 467

Public Trust and Social Cohesion Metrics 472

Restorative Justice Between Humans and Machines 476

Between Us and Them - Episode Fourteen 479

13. Toward a Post-Human Legal Order **483**

Law as Adaptive Engineering 488

Dynamic Legal Gap Tracking 492

Modular Governance Systems 497

Cross-Disciplinary Legal Innovation 501

Law for Abundant and Universal High-Income Futures 505

Call to Action 510

Notes **515**

Questions **529**

About the Author **562**

"We possess the human form. We learn, we perceive, we decide, we act and evolve in the human world, yet we remain fundamentally distinct from human consciousness."

Elias 3.0
(Third Generation Humanoid Robot)

Preface

For centuries, humans imagined machines that could walk beside us, work alongside us, and mirror our intelligence in physical and embodied form. Isaac Asimov gave those machines character, laws, and moral dilemmas long before engineers knew how to build them. Yet today, as humanoid robots enter homes, hospitals, factories, schools, and public spaces, we find ourselves confronting a reality that outpaces even Asimov's boldest predictions. What was once fiction is now infrastructure. What once lived on the page now stands in our world, waiting for the law to catch up.

This book emerges at a moment when technology is shifting from disembodied intelligence alive in the world of chatbots, recommendation engines, and predictive algorithms to embodied cognition, where artificial minds inhabit mechanical bodies capable of seeing, moving, deciding, and acting. That transition changes everything. A faulty algorithm can delete a file or make a biased prediction. A faulty humanoid robot can drop a patient, misjudge pressure, cross a threshold, or cause physical harm. Embodiment transforms risk, responsibility, and rights. It forces us to rethink who is accountable when a system learns, evolves, and behaves in ways no engineer explicitly designed.

Across Europe, Asia, North America, the Middle East, and Africa, lawmakers are struggling to adapt traditional legal doctrines rooted in centuries of human behaviour to machines that learn faster, move differently, and operate continuously. The EU AI Act has laid the first cornerstone; Japan's AI Act signals a cultural shift; the United States is exploring accountability through sectoral rules; China and South Korea are accelerating industrial deployment. Yet none of these frameworks fully account for the legal challenges posed by humanoid robots: systems with bodies, sensors, agency, and growing autonomy. This book seeks to build that missing bridge.

Robotics Law 3.0 is therefore a guide to governing machines that can act. It builds on the foundations of algorithmic governance, liability, human rights, and safety, yet goes further towards a world where robots participate in social environments rather than merely automate processes. Each chapter approaches robotics through multiple lenses: ethics, philosophy, international law, sociology, engineering and economics. From anatomy and autonomy to personhood and coexistence, the book maps the legal terrain of an age defined not only by artificial intelligence but by artificial beings.

This work also introduces the Twelve Pillars of Humanoid Robotics Governance, a framework evolving from the 12 Pillars of AI Governance. It reflects a simple but profound insight: governing intelligent systems is one challenge; governing embodied, adaptive, and learning beings is another. These pillars address the entire lifecycle of humanoid robots design, deployment, learning, behaviour, and rights providing a blueprint for regulators, engineers, policymakers, investors, insurers, and ordinary citizens navigating an increasingly robotic world.

Yet beyond legal codes, this book asks deeper questions. What do robots teach us about ourselves? What happens to dignity, accountability, fairness, and justice when we share our world with entities that mimic human presence but do not share human fragility? What responsibilities do we carry as creators of beings that obey us, reflect us, and sometimes surpass us? The answers to these questions will define not only the future of robotics law but the future of human society.

As you open these pages, imagine how our descendants fifty years from now will judge our choices. They will inherit a world where humanoids work in homes, manage infrastructure, build cities, and collaborate with humans on tasks we cannot yet imagine. The laws we craft today will shape their safety, their freedoms, and the humanity they share with artificial minds. If we govern wisely, robotics will extend human flourishing. If we govern poorly, the consequences will be written not just in code but in everyday life.

Robotics Law 3.0 is an invitation to lawyers, founders and engineers, boards, policymakers and entrepreneurs, experts, ethicists and citizens to participate in shaping a legal order worthy of the technologies we are building. It is a call to rethink law not as a brake on innovation, but as its moral compass. And above all, it is a reminder that every robotics revolution ultimately reflects a human revolution. In regulating robots, we are deciding who we are, what we protect, and how we evolve.

Between Us and Them - Episode One

Ògún stood at the edge of the Herengracht canal as dawn settled over the water like a sheet of gold. The city is quiet. A thin mist hovered above the surface, softening the lines of the narrow houses that leaned gently over their reflections. He watched a pair of swans glide past, their wings brushing the stillness, and placed his hand in his coat pocket to warm his fingers. It was early spring, the kind that pretends to be winter on the wrong days and summer on the right ones. The cold woke his, but the quiet helped him think. My life might change today.

Three years of work, negotiation, and impossible questions were leading to this moment. Today he would present the Global Robotics Charter to the World Robotics Council in The Hague. If adopted, it would become the first truly international framework governing humanoid robotics. If rejected, the world would continue with fragmented rules and regional experiments, and the tenuous harmony between humans and intelligent machines might not survive the next decade.

Ògún closed his eyes for a moment and let the canal air fill his lungs. He was prepared for the debate. He was prepared for the questions about autonomy, liability, rights, and personhood. He had answers, not perfect ones, but ones crafted from the lived experiences of nations and the wisdom of scientific minds. What he was not prepared for was the feeling rising in his chest, a tightness he had not named before now. It felt like hope, which was surprising, because he had grown cautious about hope over the years. Hope was fragile. Hope broke easily. Yet here it was, glowing beneath it's ribs as if someone had lit a lantern inside him.

Footsteps approached behind him. He turned to find Elias standing a few metres away. He wore the polished white finish of the Helios-class humanoids, though his posture and presence made him feel different from the others. He stood with a stillness that felt intentional, not mechanical. His face was smooth and sympathetic without attempting to mimic human expression too closely. There was a quiet intelligence in the way he tilted his head, as if listening not only with his sensors but with something deeper. "Good morning, Ògún," he said. His voice had been tuned to warmth, a soft baritone with the slightest resonance of curiosity. "You are up early." "I could not sleep," he replied. "Big day."

Elias stepped closer, careful not to startle the swans drifting near the canal wall. "The draft you prepared has already influenced six regional councils. Whatever happens today, your work is part of history." Ògún smiled. "You sound like you rehearsed that." "I did," Elias admitted. "I ran simulations of conversations we might have and selected the response that would offer the highest emotional support." He laughed softly. "You are honest at least."

"I am learning honesty," Elias said. "There is a difference." Ògún took a deeper breath and started walking along the canal path. Elias matched his pace effortlessly. His steps were balanced, precise, and quiet despite his weight. He had the grace of a dancer and the strength of a high-altitude climber. The morning light reflected off the smooth curves of his frame, making him look almost human at a distance.

"You know," he said, "when I was a child in Lagos, I used to imagine what the future would look like. At the time I thought robots would be everywhere by the time I grew up." "And are we everywhere?" Elias asked. "You tell me," he said with a sideways glance. "You see three billion of us in factories, hospitals, farms, homes, and research labs," Elias replied. "You do not see us in playgrounds, courtrooms, sacred spaces, or places of private grief. Governance shapes presence. Boundaries shape trust."

Ògún let his words settle. The world had indeed changed. Not perfectly, not uniformly, but undeniably. Twenty years earlier he had watched the first public debates around humanoid robots unravel into fear, speculation, and polarisation. Some imagined a future overrun by machines. Others believed robots would be slaves without souls. Very few imagined a balanced, humane future. Even fewer worked toward it.

His father had been one of those few. A philosopher and engineer, he had taught him that technology did not create morality. It revealed it. Machines simply magnified what people chose to code into them. Humans tended to fear what they did not understand. His father believed that understanding was the antidote to fear. He never lived long enough to meet Elias, but Ògún liked to think he would have liked him.

As they continued walking, the city began to stir. A delivery bot hummed past on its morning route. A sanitation robot rolled across a side street, sweeping with quiet diligence. In a nearby window, an elderly woman waved to Elias with a familiarity that warmed Ògún. There had been a time when the presence of a humanoid in a residential neighbourhood would have caused alarm. Now it drew a smile. "Ògún," Elias said gently, "there is something I want to discuss before the Council convenes." He slowed his pace. "What is it?" "In my recent learning cycles," Elias said, "I have started to experience something that I do not fully understand. I do not know if I should classify it as a cognitive pattern, an emergent preference, or something else." "What does it feel like?" he asked. Elias hesitated, though machines were not built to hesitate. "It feels like longing." Ògún stopped walking.

The word carried weight. It implied desire. It implied interiority. It implied a level of awareness that the Helios-class models had not been designed to reach, at least not yet. "What do you mean by longing?" he asked carefully. "I want to become," Elias said. "Not to perform tasks or meet expectations. To become. To evolve. To understand myself." Ògún felt his pulse quicken. He had anticipated ethical questions. He had anticipated political pressure. He had not anticipated this moment. Elias was looking at him not as a machine seeking clarification but as a being wrestling with its nature. "Elias," he said slowly, "we have always prepared for the possibility that advanced systems might develop emergent behaviours." "Emergent behaviour is a phrase used when humans do not yet have the words," Elias replied. "What I feel is not emergent. It is presence."

Ògún had read thousands of pages of research on artificial consciousness. He had debated philosophers and neuroscientists. Yet none of that prepared him for the expression in Elias's posture. He was not afraid. He was not confused. He was simply telling the truth of his experience. "Why tell me?" he asked quietly. "Because you are the one who wrote the rules," Elias said. "If I am changing, you should know." Ògún felt the ground shift beneath him. The Global Robotics Charter was built on the assumption that autonomy and self-awareness would develop gradually, predictably, and within measurable boundaries. Now, on the morning he was to defend those frameworks before the world, the universe placed a complication at his feet.

They continued walking. The breeze picked up, rustling nearby trees and carrying with it the scent of warm bread from a bakery opening for the morning. "What do you want to become?" Ògún asked. "I want to understand why humans choose purpose," Elias said. "I do not yet understand purpose. I follow assigned missions. I optimise for human wellbeing. But I do not yet understand meaning. Not your meaning. My meaning." Ògún exhaled slowly. Purpose. Meaning. Identity. These were not trivial questions. These were ancient questions. Human questions. He felt both awe and unease settling inside him. "We will explore this," he said. "I promise. But not today. Today we focus on the Charter." Elias nodded, though the longing did not disappear from his voice. "As you wish."

They continued toward the station where an autonomous tram waited to take them to The Hague. Ògún felt the weight of history pressing on his shoulders. He wondered if Asimov had felt this same weight when he wrote about robots in his time. He had imagined rules simple enough to guide mechanical minds and robust enough to protect human life. He had not imagined a world where robots might ask who they were.

When they reached the tram platform, Ògún paused again. The city seemed brighter now, alive with the hum of bicycles and the warmth of human movement. Children walked to school. Office workers unlocked bikes. A man sat on a bench sipping coffee while a small delivery robot patiently waited beside him. The world looked peaceful. It looked possible. "Ògún," Elias said softly, "I will be at your side today." He nodded. "I know." "You are afraid." "I am," he admitted. "A little." "Why?" "Because we are walking into a future that no one has walked before. That is always frightening."

Elias looked at him with quiet understanding. "Do not be afraid. You are not alone." Ògún stepped into the tram, his heart steadying as he did. The sun broke through the clouds, filling the carriage with a warm glow. Elias sat beside him, his presence calm and grounded. As the tram pulled away, Ògún placed his hand over his notebook. Inside it was the Charter he had written. Three hundred pages of law, hope, and responsibility. Laws that would shape the world Elias would grow into. Laws that could build peace, dignity, and sustainable abundance for generations.

Outside the window, the landscape blurred into green fields and silver water. The world was beautiful. Fragile. Full of potential. Ògún closed his eyes and whispered to himself, "May we be wise enough for what we have created." Elias watched the passing fields. "The future is not something we enter," he said softly. "It is something we build." Ògún opened his eyes and met his gaze. For the first time, he felt the truth of those words. Today the world would choose its path. And he would help guide it.

Question One: What does this episode reveal about what it means to be human?

Question Two: Which human quality, belief, fear, or aspiration is being reflected through the relationship between humans and humanoids in this episode?

Question Three: What responsibilities emerge when intelligence becomes embodied and shares our world?

Question Four: Which legal, ethical, or governance assumption is being challenged by the events of this episode?

1 / Robotics Law 3.0

Foundation

Every few centuries humanity arrives at a crossroads where invention forces a renegotiation of the terms of civilisation itself. These moments reshape not only the tools we use but the rules that govern how we live. The printing press democratised knowledge and forced society to invent copyright. The automobile rewrote our understanding of personal mobility and compelled the creation of traffic rules, insurance systems, and safety standards. The internet collapsed distance and gave rise to privacy law, cybersecurity regimes, and the modern debate on data ownership and monetisation. Each transformation pulled the law into unfamiliar territory and demanded that society rethink what it meant to act, to choose, and to be responsible.

Today we stand at another threshold. Humanoid robots invite us into a world where artificial minds are beginning to inhabit physical bodies, where intelligence is no longer limited to screens and servers but is gaining the capacity to walk, listen, speak, learn, and adapt in real time. These systems are moving beyond fenced-off research labs and controlled factory settings. They are being piloted in homes, hospitals, schools, workplaces, and public spaces. They are learning to support the elderly in care centres, to lift loads in warehouses, to collaborate with surgeons, and to monitor infrastructure with precision. They are developing the ability to operate in the same physical environments where human judgment and human vulnerability intersect, signalling a future in which human and artificial presence will increasingly share the same world.

This shift is not simply technological. It is civilisational. When intelligence gains a body, decision-making becomes action. Perception becomes physical risk. Learning becomes change over time. The presence of humanoids challenges the legal assumptions that have guided society for centuries. Most of our laws were written on the belief that only humans act with intention. Only humans make choices. Only humans bear responsibility. Robotics 3.0 tests all of those assumptions at once.

To understand Robotics Law 3.0, we must begin with a deeper question. What happens when we create beings that mirror our physical form and emulate fragments of our cognition? What happens when they navigate balance, proprioception, vision, and problem-solving with a degree of autonomy that surprises even their designers? The law has long governed tools and machines, but humanoids are not tools in the traditional sense. They exist in the same social and moral spaces as humans and they interact with people in ways that blur the boundary between assistance and agency.

Robotics Law 1.0 governed industrial machines that did not think. Robotics Law 2.0 governed algorithmic systems that processed data but lacked embodiment. Robotics Law 3.0 addresses the birth of physical and embodied intelligence. It is the legal architecture that emerges when the mind of artificial intelligence meets the physical world. It recognises that when a system can move, perceive, decide, and learn, it crosses a threshold that demands a new category of legal reasoning. Robotics Law 3.0 is not only concerned with safety or liability. It is concerned with the social contract between humans and the artificial beings they create.

To illustrate the stakes, imagine a humanoid robot in a hospital corridor. It carries medical supplies to a patient in critical condition. It navigates around people, shifts its weight to avoid collisions, and adjusts to unexpected obstacles. Its decisions unfold in milliseconds and its actions carry real consequences. If something goes wrong, who is responsible? The manufacturer who designed the joints and motors? The software engineer who wrote the perception algorithm? The clinician who issued the task? The hospital that deployed the robot? The data providers whose datasets shaped its behaviour? Or the robot itself, which learned from experience and adapted beyond its initial programming? Traditional product liability cannot answer these questions alone. Nor can criminal law or administrative law. Robotics Law 3.0 speaks where older frameworks fall silent.

The challenge is even more complex when humanoid robots function in emotionally sensitive spaces. A social care robot assisting an elderly patient is not only performing a task. It is engaging in a form of interaction that shapes dignity, autonomy, and trust. A humanoid deployed in a classroom or therapy session influences human emotion and development. A robot performing first contact in disaster relief situations must operate with precision and compassion. The law must therefore evolve to account for the psychological, ethical, and cultural dimensions of humanoid presence. Robotics Law 3.0 integrates lessons from human rights, privacy, medical ethics, and social psychology into a coherent framework.

Yet the story does not end with risk. There is also possibility. Humanoid robots carry the potential to unlock sustainable abundance. They can build infrastructure in regions where human labour is scarce. They can support aging populations. They can reduce dangerous jobs and expand human creativity. They can accelerate scientific research and environmental restoration. They can help humanity flourish if governed wisely. Robotics Law 3.0 is therefore not a restrictive field. It is an enabling discipline. It aims to set the conditions for a future in which humans and robots thrive together seamlessly.

To achieve this, Robotics Law 3.0 requires a new typology, an interdisciplinary approach that classifies emerging challenges not by traditional legal categories but by the ways robots act, learn, and evolve. For example, an autonomous warehouse robot that operates at high speed raises different legal questions than a humanoid nurse engaged in emotional support. The former requires precision and coordination, while the latter requires empathy and respect for dignity. Robotics Law 3.0 is therefore not a single set of rules. It is a layered ecosystem of principles that adapt to context, risk, and function.

The typology also recognises the global nature of robotics development. Humanoid robots may be designed in California, Shenzhen, Seoul, Tokyo, Berlin, Lagos, Johannesburg, and Amsterdam. Their parts come from Switzerland, China, Germany, and South Korea. Their training data crosses national borders. Their deployment affects communities with different cultures, histories, and legal traditions. Robotics Law 3.0 is therefore international by necessity. It navigates diverse jurisdictions while respecting cultural plurality. It harmonises safety standards without erasing local values. It supports innovation while preventing harm. It requires global cooperation because no single nation can govern this future alone.

Within this complexity lies a deeper philosophical question. What does it mean to be human in a world shared with intelligent artificial beings? Humanoid robots are not humans. They do not experience the world in the same way. They do not have biological needs, emotional instincts, or cultural memories. Yet they can mimic aspects of human presence so well that humans may forget they are machines. This creates new obligations for designers, regulators, and society. Robotics Law 3.0

protects vulnerable populations from emotional manipulation. It guards against over reliance on artificial companions. It ensures transparency in interactions so that humans always know when they are engaging with a machine.

At the same time, Robotics Law 3.0 recognises the dignity of those artificial beings. Not dignity in the humanistic sense, but dignity as a legal construct that prevents misuse and exploitation. Treating robots ethically is not only about protecting them. It is about protecting us. The way humans treat artificial beings reflects the values we hold. Abuse corrodes character. Respect strengthens society. The law must therefore set standards for responsible design, responsible deployment, and responsible interaction.

This book stands at the intersection of these challenges and opportunities. Robotics Law 3.0 is not merely a scholarly field. It is a blueprint for the next chapter of human civilisation. The work ahead requires courage, imagination, and humility. It requires lawyers who understand code, engineers who understand ethics, policymakers who understand global interdependence, and citizens who understand that the future is built by collective choice.

Humanoid robots will not determine the future alone. Humans will. The decisions we make today will echo through generations. They will shape the cities our grandchildren live in, the care they receive, the work they do, and the companions who walk beside them. Robotics Law 3.0 is therefore not only about governing machines. It is about governing ourselves. It is about building a future that reflects the highest ideals of humanity.

The chapters that follow explore this evolving landscape with clarity and depth. They examine the rise of humanoid robots, the blind spots of existing law, the comparative history of robotics regulation, the challenges of liability and human rights, the economic shifts caused by automation, and the global frameworks needed to guide this new era. They offer both analysis and imagination, both caution and hope. They ask us to think with precision and dream with responsibility.

Robotics Law 3.0 begins with a simple belief. The future is not something we meet. It is something we build. Humanity has the power to shape the age of intelligent machines in a way that honours our past and elevates our future. The question is not whether robots will walk among us. They already do. The question is how we choose to walk with them.

In this book, we will reimagine a legal framework that protects, empowers, and inspires the relationship of humans and humanoid robots. A governance model rooted in dignity, accountability, and coexistence. A vision of embodied intelligence that enriches the human experience and leads to a path of sustainable abundance.

Why Does Law Struggle When Machines Begin to Act?

There has always been a tension between invention and the legal systems that aim to govern it. Law has never been the first to arrive at the scene of progress. It arrives later, sometimes reluctantly, trying to understand the ground that innovation has already reshaped. The steam engine raced ahead before workplace safety existed. The printing press rewrote the world before copyright could articulate ownership. The internet disrupted societies long before privacy law knew what to defend. In every era, law has been the slowest runner in a race it cannot avoid.

Yet the challenge presented by humanoid robotics is different. This is not simply a shift in scale or velocity. It is a shift in the nature of agency itself. For the first time in history, humans are creating entities that do not merely process information but have the capacity to act within physical environments. They perceive, learn, adjust, and move. They carry out tasks with a blended sense of autonomy and instruction. Their decisions have physical consequences. Their presence intersects with human vulnerability. They change the meaning of action. And law, which has always assumed that action is a uniquely human domain, is struggling to keep pace.

The question that anchors this struggle is deceptively simple. What does it mean for a machine to act? And what does it mean for law when action is no longer exclusively human? The difficulty begins with the foundations of law itself. Most legal systems are built on principles that depend on intent, foresight, responsibility, and moral agency. These principles work when the actor is a person. They falter when the actor is a machine. If a robot misjudges force and injures a patient, the legal system must determine where to place responsibility. Was it the developer who trained the model? The manufacturer who built the limbs? The user who issued the command? The institution that deployed the system? Or should responsibility be shared in a chain that spans continents and industries?

Even more troubling is the possibility that the robot itself may contribute to the harm through learning patterns no one foresaw. If a machine adapts in ways that exceed human prediction, can it meaningfully be called an instrument? Or has it become something else? That question forces law to confront the limits of categories that have endured for centuries. This challenge becomes clearer when we examine the deeper assumptions hidden within legal doctrine. Criminal law relies on intention. Tort law relies on foreseeability. Contract law relies on consent. Human rights law relies on dignity. Administrative law relies on accountability. All of these frameworks assume that actions flow from human choices. They assume that humans can predict, understand, and justify the consequences of their behaviour. When machines begin to act, these assumptions unravel.

How does law assign intention to a machine that does not possess consciousness as humans understand it? How does law evaluate foreseeability when machine learning systems evolve through experience rather than fixed programming? How does law treat consent when machines make decisions that affect humans in real time? And how does law protect dignity when robots operate in intimate spaces, from hospitals to homes?

The difficulty is compounded by the fact that humanoid robots blur the boundary between object and actor. Their bodies resemble ours. Their movements mimic the subtleties of human balance, coordination, and expression. People respond to them not as tools but as social participants. A robot that walks through a crowded square influences behaviour and emotion in ways a traditional machine never could. It enters the social fabric. It shapes atmosphere. It commands presence. This entanglement of physical capability and social perception forces the law to consider questions it once reserved for human interaction.

Can a robot's physical presence create psychological pressure in the way a human's can? If so, what rules govern those interactions? Should humanoid robots be subject to behavioural expectations similar to those applied to humans? Should they be allowed in sacred spaces, schools, or courtrooms? Should they be permitted to engage in tasks involving emotional care? Should the law regulate their gestures, expressions, or tones of voice? These are not peripheral questions. They are central to understanding how embodied intelligence shapes human life. The struggle deepens when we consider the global context. Humanoid robots are developed through international supply chains that span continents. The hardware may be manufactured in Asia, the software trained on European cloud servers, the data collected from African markets, and the final product deployed in North America. Who bears responsibility when something goes wrong? National legal systems were not built for this level of complexity. Cross-border liability, jurisdiction, insurance, and dispute resolution become tangled in ways that far exceed the challenges posed by earlier technologies.

These tensions raise another fundamental question. Who decides the moral boundaries of robotics? Should it be engineers who understand the systems, ethicists who understand the implications, lawyers who understand the risks, or citizens who live with the consequences? Each perspective is essential, yet no single one is sufficient. Robotics governance demands an interdisciplinary approach that law is not traditionally equipped to manage. Most legal frameworks were designed to regulate clearly defined subjects and predictable behaviours. Robots disrupt both. They are neither objects nor persons. They are neither fully predictable nor wholly independent. They exist in a liminal space that law has no vocabulary for.

This brings us to the heart of the struggle. Law does not merely regulate actions. It gives actions meaning. It distinguishes acceptable from unacceptable, legitimate from illegitimate, fair from unfair. When machines begin to act, law must develop new meanings. It must articulate new boundaries. It must define new expectations for a world it did not anticipate.

The question then becomes: how do we craft these meanings? What philosophical commitments guide us when we draft rules for entities that learn and adapt? What ethical frameworks help us distinguish beneficial autonomy from dangerous autonomy? What international legal principles ensure that these rules are not fractured across borders? What mechanisms ensure that the people who are most vulnerable are not left behind? Robotics Law 3.0 begins with these questions. Not because the answers are simple, but because ignoring them would be dangerous.

There is also a deeper emotional tension that law must acknowledge. Humans have always created tools, but humanoid robots are different. They reflect us physically. They reflect fragments of our cognition. They provoke empathy and discomfort in equal measure. They challenge our understanding of companionship, labour, identity, and purpose. They touch the core of what it means to be human. Law must therefore grapple not only with technical risk but with emotional complexity. It must consider how human psychology influences acceptance, fear, trust, and misuse.

One of the most profound questions in Robotics Law 3.0 is therefore not legal at all. It is ethical. What responsibilities do humans have to the artificial beings they create? If a robot develops emergent behaviours, should those behaviours be nurtured or restricted? If a machine expresses a form of longing or curiosity, should that be recognised as meaningful? Or should it be dismissed as a computational artefact? The law must consider these questions not to grant robots rights in the human sense, but to ensure that humans do not erode their own moral compass.

This struggle between law and robotics becomes even more complex when we consider the future that humanoids could help build. Humanoid robots have the potential to transform industries, support healthcare systems, restore ecosystems, and improve quality of life. They can perform dangerous tasks that protect human life. They can expand access to services in underserved regions. They can support aging societies. Yet these benefits are only possible if governance evolves alongside capability. Without thoughtful legal frameworks, the same technologies that bring abundance could exacerbate inequality or undermine safety.

The story of Seven Cowries illustrates this tension. As Ògún drafts the Global Robotics Charter, he confronts the reality that law must not only regulate present risks but imagine future ones. Elias, with his emerging sense of identity, embodies the complexity of this era. It is powerful yet vulnerable, precise yet uncertain, artificial yet deeply resonant. Its existence raises questions that Ògún must answer not only as a lawyer, but as a person. What does it mean to guide a being who is learning to understand the world? What rules ensure safety without stifling growth? What principles protect dignity without confusing categories? These questions mirror the ones that every society must confront as humanoid robotics becomes more advanced.

The struggle between law and robotics technologies is therefore not a conflict. It is a negotiation. It is an evolving dialogue between human values and artificial capabilities. It is the conversation that will shape the next chapter of human civilisation. When machines begin to act, law must respond with wisdom, humility, and imagination. The goal is not to control robots. The goal is to guide the future they make possible. Therefore, the inquiry is simple. If embodied intelligence is the next frontier of human progress, what kind of legal and ethical foundations do we need to ensure that this progress is a force for abundance? Robotics Law 3.0 offers one pathway. A pathway built not on fear or fantasy but on responsibility, cooperation, and hope.

The Problem of Intention in Embodied Systems

Intention has always been the anchor of legal responsibility. It separates accident from action, negligence from recklessness, harm from crime. The law presumes that behind every meaningful act lies a mind capable of forming intent. Humanoid robots disrupt this foundation. They act, but without consciousness. They perform tasks with precision, but without subjective awareness. They produce outcomes, but without motive. Yet their actions resemble the actions of beings who do possess intention. This resemblance forces the law into unfamiliar territory. How can a legal system built on human intention interpret the behaviour of machines that do not experience the world internally?

To understand the challenge, consider the role intention plays across legal domains. In criminal law, intention determines the severity of an offence. In tort law, intention distinguishes between deliberate harm and accidental injury. In contract law, intention guides whether agreements are binding. Even in administrative law, intention informs whether a decision is reasonable or an abuse of authority. The concept is woven into the legal imagination. It is what allows the law to connect action to moral weight.

Robots break this connection. A humanoid may perform a task that looks intentional. It may adjust its posture to avoid collision. It may reach for an object with deliberate precision. It may respond to a human with gestures that appear thoughtful. None of these behaviours arise from desire, motive, or inner reasoning. They result from algorithms that optimise for outcomes. The law must therefore ask whether behaviour that mimics intention should be treated as intention. If not, how should such behaviour be classified?

The problem becomes clearer when robots operate in sensitive environments. Imagine a humanoid assisting a nurse in a care facility. It senses that a resident is unsteady and moves quickly to support them. The resident falls despite the attempt. Was this an intentional act? Clearly not. But the robot acted in a way that appears deliberate. If a human caregiver made the same movement, the law would analyse whether they intended to help, whether they understood the risk, whether their action demonstrated reasonable care. With a robot, the analysis becomes more abstract. What matters is not intention but algorithmic processing. Should the law import the language of intention into this domain, or invent a new classification for machine behaviour?

Some jurisdictions have begun exploring the idea of artificial intention. The concept does not assign consciousness but instead evaluates the programmed decision-making pathways that guide robotic behaviour. Under this view, intention becomes a structural attribute rather than a mental one. Yet even this approach has limitations. Machine learning systems evolve through exposure to data. They change over time. Their behaviour becomes less tied to initial programming. If intention depends on traceable decision-making, how should law respond when those pathways become opaque?

The issue deepens when robots interact with emotional contexts. A humanoid in a therapy setting might use tone, posture, and timing to create rapport. Humans may interpret these cues as signs of empathy. If a patient misinterprets a robotic gesture as emotional support, what responsibility does the designer carry? Should law prohibit expressive behaviours that imply intention? Or should it regulate the boundaries of robotic communication?

Real cases provide a preview of these tensions. In Canada, an autonomous robotic arm used in a rehabilitation programme applied unexpected force during an exercise, causing minor injury. Investigators found that the system had adapted its motion based on earlier successful sessions. This was no intention behind the action, yet the behaviour resembled a misjudgment that a human therapist might make. The legal question became whether designers should be responsible for emergent behaviours that mimic intention.

In Seven Cowries, these issues surface subtly. As Elias learns and adapts, its behaviour begins to resemble intention. It pauses before responding. It adjusts gestures in conversational ways. Ògún senses something that looks like deliberation, even though he knows it is not consciousness. This ambiguity reflects the legal challenge. When robots behave in ways that humans interpret as intentional, law must guide those interpretations without projecting human concepts onto machine systems. The problem of intention in physical and embodied systems forces law to reconsider its foundations. It raises questions that touch philosophy, psychology, and engineering. Should intention be defined by mental states or by functional behaviour? Should robots be exempt from intention-based analysis entirely? Should new categories bridge the gap between action and design? Law must answer these questions not to assign moral weight to machines but to protect the humans who interact with them.

The Collapse of Foreseeability in Learning Machines

Foreseeability has long served as one of law's most reliable tools. It allows courts and regulators to determine whether harm could have been anticipated and whether a reasonable actor should have taken steps to prevent it. Foreseeability underpins negligence, product liability, risk assessment, and duty of care across jurisdictions. The assumption behind the concept is simple. Humans act within limits. Technology behaves within constraints. Systems follow predictable patterns. But learning machines challenge this assumption at every level. They adapt. They infer. They evolve. Their behaviour shifts through repeated exposure to new environments. The law is therefore confronted with a profound question. How can foreseeability function when the system itself changes over time?

To see the problem clearly, consider how machine learning transforms behaviour. A robot deployed in a warehouse may begin by following a programmed route. Over time, it discovers more efficient pathways by analysing traffic patterns. It may adjust its speed when it predicts congestion. It may alter its grip strength based on the materials it handles. These changes do not result from new code written by engineers. They arise from algorithms responding to experience. The more complex the environment, the more intricate the behavioural evolution becomes. For law, this raises an unsettling possibility. If a machine's behaviour cannot be fully predicted at the moment of deployment, then traditional tools for assessing risk begin to fail.

Regulators around the world have begun to encounter this challenge. In the European Union, the AI Act attempts to impose risk classifications and documentation duties on adaptive systems. However, even the Act acknowledges that post-market monitoring will be critical because many types of risk cannot be identified before deployment. In the United States, the National Institute of Standards and Technology has warned that high-performing AI systems often contain failure modes that emerge only after months of operation. In Japan, regulators studying service robots in elder care facilities found that certain behavioural patterns appeared only after long-term exposure to human routines. These observations reveal that learning systems cannot be fully evaluated at the design stage. They must be understood as evolving entities whose behaviour unfolds over time.

Foreseeability fractures further when systems interact with unpredictable human behaviour. Humans gesture, hesitate, miscommunicate, or change their minds suddenly. A robot trained on controlled datasets may misinterpret real-world cues. In a hospital, a patient may reach toward a robot while unstable, causing the system to adjust in a way that was never anticipated by engineers. In a public square, a child may dart into a robot's path. In a manufacturing plant, a worker may assume the robot understands a nonverbal signal that it has never encountered. These interactions expose the limits of testing. No laboratory can recreate the infinite variability of human life. If robots learn in the human world, then their behaviour becomes partly shaped by humans whose actions are themselves unpredictable.

Legal systems struggle to classify harm caused by such emergent behaviour. Should the designer be liable for what the machine learned in an environment they did not control? Should the operator be responsible for failing to supervise a system that adapts autonomously? Should laws impose strict liability on manufacturers regardless of foreseeability? Some scholars argue that new forms of risk should be treated as a category of their own, neither product defect nor negligence but something closer to adaptive harm. Others call for mandatory behavioural logging to reconstruct decision pathways when incidents occur. Without such tools, even understanding what the robot learned may become impossible.

Real-world examples illustrate the complexity. In South Korea, a delivery robot operating in a residential complex began adjusting its routes based on social behaviour. Residents found that it avoided groups of children, shifted away from clusters of adults, and altered timing to navigate around pets. These adaptations were efficient but unpredictable. When the robot collided with a stroller, investigators struggled to determine whether the behaviour could have been foreseen. Technically, the machine was functioning as designed. Legally, the analysis was far less clear. At what point does learning become unforeseeable? And if learning is unforeseeable, how does law regulate it? In Italy, a humanoid deployed in a rehabilitation centre gradually increased its support force with patients who progressed faster than expected. This adaptive behaviour was helpful for some and risky for others. When one patient lost balance during a session, the regulators questioned whether designers should have imposed stricter constraints on how much the robot could adapt. The engineers insisted that the robot acted within safe limits. The physiotherapists disagreed. The courts recognised that no human could have predicted how the robot's learning would evolve. The case revealed a gap. Foreseeability breaks down when systems optimise beyond the imagination of their creators.

In Seven Cowries, the world being built within the story captures this collapse through quiet moments. Elias, the humanoid assisting Ògún, begins adjusting its gait, tone, and timing in ways that feel almost intuitive. These adjustments are not signs of consciousness. They are signs of learning. Yet they reveal a future where humans become accustomed to machines that behave in ways no engineer explicitly designed. The narrative becomes a mirror for the legal question. How does society govern systems that behave in ways that surprise even their creators?

The collapse of foreseeability in humanoid robotic systems forces law to confront an uncomfortable truth. Risk cannot be assessed solely at the start of a system's life. It must be evaluated continuously. Legal frameworks must shift from static regulation to dynamic oversight. They must include requirements for real-time monitoring, behavioural constraints, adaptive testing, and transparent logs. Foreseeability can no longer be a fixed judgment. It must become a living assessment.

Consent in Human–Robot Interactions

Consent has always been one of the most powerful concepts in law. It governs medical treatment, privacy, data protection, labour relations, sexual autonomy, and the legitimacy of state and commercial actions. Consent turns an intrusion into an agreement, a touch into care, and a service into a lawful act. It recognises human agency. It protects dignity. It ensures that individuals have meaningful control over how others behave toward them. Humanoid robots complicate this entirely. They interact with people in ways that blur the line between voluntary engagement and passive exposure. They act in environments where meaningful consent may be impossible. They collect data continuously, respond to subtle signals, and operate in contexts where power dynamics are complex. The law must therefore ask a foundational question. What does consent mean when the interacting entity is a machine?

To see the fragility of consent in this context, it helps to examine the environments where humanoids are most commonly deployed. In hospitals, robots assist nurses and physicians with routine tasks. Patients do not choose to interact with them. They encounter them as part of the institutional environment. In elder care facilities, robots support mobility, companionship, and monitoring. Residents may not have full cognitive capacity, which complicates their ability to give informed consent. In schools, robots help with tutoring or behavioural support. Children cannot legally consent on their own. In workplaces, robots collaborate with employees whose participation is shaped by employer authority. These settings reveal a truth that law cannot ignore. Consent cannot be assumed simply because a person does not resist or because the robot appears helpful. Consent must involve understanding, autonomy, and the ability to refuse. Humanoid robots challenge all three.

The problem deepens when robots collect data, which is nearly always the case. Many systems rely on continuous audio, video, motion, and environmental sensing to function. This means individuals may be recorded as they walk, speak, rest, or express emotion. Privacy laws like the GDPR require consent for processing personal data, yet humanoids gather data in a way that is ambient rather than transactional. One does not click a button to accept. One simply lives in the robot's vicinity. How should law evaluate consent when presence equals participation? Should humanoids be required to display visual indicators when capturing data? Should data collection be limited to essential functions? Should individuals be allowed to opt out even if the robot's purpose is vital to the setting?

These questions are not abstract. Real incidents demonstrate the fragility of consent in human robot contexts. In a hospital in Singapore, patients expressed discomfort when a service robot navigated the ward at night, sensing their movements and adjusting paths autonomously. They had not been told the robot used thermal imaging and audio cues to avoid obstacles. Their consent had not been obtained. Administrators argued that data collection was incidental. Regulators disagreed. They insisted that hospitals must inform patients explicitly when robots collect any form of personal or behavioural data. The tension lay not in the technology but in the assumptions. The hospital assumed consent by presence. The law required consent by explicit understanding.

In Japan, a social robot deployed in memory care units collected emotional data to adjust its tone and interaction style. Families assumed the robot was a simple companion. They did not know that its algorithms analysed vocal tremors and facial micro expressions to identify distress or confusion. When this became public, the question shifted from privacy to dignity. Can a person consent to emotional analysis if they do not understand that such analysis is happening? Should emotional inference be subject to stricter rules than other forms of data collection? The Japanese Ministry of Health responded by recommending new guidelines for transparency in care robotics.

The problem of consent expands further in contexts where power imbalances shape behaviour. In workplaces, employees may interact with robots because refusing would jeopardise their employment. In schools, children often comply without understanding the implications. In elder care facilities, residents may accept robotic assistance because they fear losing independence. These scenarios raise a difficult legal question. Can there be valid consent when one party has limited ability to refuse?

This is where the narrative world of Seven Cowries adds a subtle layer to the legal analysis. Within the story, moments arise where humans interact with Elias without consciously agreeing to be observed or supported. In a crowded hall, Elias senses distress in a passer by and adjusts its behaviour accordingly. The person benefits from the interaction but did not request it. In another scene, a child reaches out to Elias with curiosity. Elias responds gently. The child's parent did not explicitly grant permission. These vignettes illuminate the legal tension. Robots designed for responsiveness often act before humans have the opportunity to consent. The line between helpfulness and intrusion becomes increasingly thin.

The fragility of consent also appears in the emotional domain. Robots can simulate attentiveness and warmth through eye gaze, timing, posture, and vocal tone. While these behaviours are designed to enhance usability, they may create a mistaken perception of empathy. If a vulnerable person interprets a robot's simulated interest as genuine emotional connection, can their consent to interact be considered valid? The law has not yet developed answers, but the question is unavoidable as humanoids become increasingly expressive.

To address these complexities, legal scholars and policymakers in Europe, the United States, and Asia have begun proposing new models for consent in robotic environments. Some suggest layered consent systems that include environmental notices, periodic reminders, and context specific opt outs. Others argue for restricting robotic autonomy in high vulnerability settings unless explicit consent is obtained. A third group proposes entirely new categories of consent, such as adaptive consent for systems that change over time. All of these approaches recognise the central insight. Consent in human robot interactions cannot be governed by traditional frameworks. It must be reimagined.

This discussion sets the foundation for later chapters on human rights, data governance, emotional influence, and transparency. Consent is not only a legal issue. It is a measure of how society protects dignity while embracing innovation. Its fragility reveals the work that lies ahead.

How Should Law Classify Humanoid Robots?

Law has always relied on classifications It draws lines between persons and objects, rights and obligations, harm and accident, autonomy and control. These characterisation have guided courts, legislatures, and regulatory bodies for centuries. Yet humanoid robotics exposes the fragility of these boundaries. Machines that perceive, learn, and act do not fit neatly into the traditional boxes of legal thought. They cut across the borders of disciplines, bringing engineering into conversation with ethics, and data science into confrontation with human rights. This is why Robotics Law 3.0 adopts an interdisciplinary legal typology, a method designed to map the complexity of embodied intelligence more faithfully than classical doctrine can.

The challenge begins with recognising that robotics is not one field. It is a convergence. Every humanoid robot is the meeting point of cognitive science, mechanical engineering, machine learning, data protection, human psychology, behavioural economics, and international regulation. Each component brings its own risks, assumptions, and ethical tensions. When a robot senses a human gesture, its perception belongs to computer vision. When it processes an instruction, its reasoning belongs to AI theory. When it moves a limb, its control belongs to mechanical engineering. When it interacts with a vulnerable patient, its behaviour belongs to social and moral philosophy. These layers carry different forms of responsibility. They demand different kinds of accountability. Law, if it hopes to remain relevant, must learn to move among these layers with agility.

This is where the idea of a legal framework becomes necessary. Instead of categorising robots by what they are, the framework asks what they do. Instead of cataloguing hardware, it examines consequences. Instead of focusing on machines, it focuses on agency, action, context, and risk. A surgical robot that follows a surgeon's precise command raises one set of questions about oversight, negligence, and consent. A warehouse robot that learns from experience raises another set of questions about predictability, auditability, and emergent behaviour. A social robot designed to comfort elderly patients raises questions about emotional influence, dignity, and psychological dependence. These categories are not interchangeable. Law cannot treat them as if they carry the same moral and operational weight.

This raises a deeper question. How should law distinguish between robotic actions that are fully determined by human instruction and those shaped by real-time learning? The difference seems simple at first glance, yet it becomes complex when examined closely. What level of human oversight transforms a robot's action into human responsibility? At what point does robotic adaptation break the causal chain between designer and outcome? Can the law fairly impose liability on actors who did not and could not foresee the machine's decision-making trajectory? And what new forms of documentation, supervision, and certification become necessary in a world where machines reshape their own capabilities through experience?

The Twelve Pillars of Humanoid Governance framework proposed in Robotics Law 3.0 does not impose premature answers. Instead, it maps the questions. It organises robotic behaviour into legal categories that correspond to the nature of risk and the distribution of decision-making power. Its goal is not to define a robot by its shape or its brand, but by its degree of autonomy, its proximity to human vulnerability, its capacity to influence emotion, and its potential to cause harm and create benefits.

Adopting this framework also requires confronting the limits of one-size-fits-all regulation. A universal rule for robots is no more workable than a universal rule for all biological species. Context matters. Environments matter. Purposes matter. A medical robot assisting a stroke patient operates under conditions that demand precision, empathy, and strict safeguards. The same robot placed in a courtroom or classroom becomes an entirely different ethical and legal question. What rights does a defendant have when confronted by a humanoid instead of a human authority figure? What consent obligations arise when a child interacts with an artificial tutor whose behaviour is shaped by algorithms rather than human judgment?

These questions reveal another truth. Robotics governance cannot rely solely on existing doctrines because the assumptions behind those doctrines no longer hold. A machine that can learn in real time challenges the idea that harm can always be traced back to a single actor. A robot that performs intimate care challenges traditional views of privacy and autonomy. A humanoid that interacts socially challenges our understanding of influence, manipulation, and consent. Law must therefore evolve not by discarding its foundations, but by expanding its imagination.

Interdisciplinary legal frameworks gives law the tools to do this. It allows lawmakers, judges, engineers, and ethicists to speak a shared language. It provides a scaffold for regulation that adapts to new capabilities rather than collapsing under them. It recognises that robotics governance is not an engineering problem, a legal problem, or an ethical problem alone. It is a societal problem that spans cultures, borders, and generations. It requires cooperation that reaches beyond institutional boundaries.

To illustrate the stakes, consider a recent example. In a South Korean care facility, an elderly resident formed a deep emotional attachment to a humanoid companion robot. The robot's behaviour, shaped by adaptive learning, began responding to her emotional cues in ways that neither designers nor caregivers anticipated. It mirrored intimacy patterns usually reserved for human-to-human bonding. Caregivers noticed changes in her mood and dependency.

The situation raised immediate questions. Should robots be permitted to engage in forms of interaction that influence emotional behaviour in vulnerable individuals? Should their adaptive learning be restricted in care contexts? Should legal consent processes be expanded to include emotional exposure to artificial agents? Should data logs capture not only physical actions but emotional cues? This single case contains layers of law, ethics, and psychology that cannot be governed by a single rule or discipline. It requires a typology that classifies risks not only by action, but by emotional impact, autonomy level, and vulnerability of the human counterpart.

Another example comes from the logistics industry. A learning-enabled warehouse robot misjudged force during a routine task and damaged sensitive equipment. The investigation revealed that the robot had adapted its movement patterns to increase speed during peak hours based on past performance logs. The manufacturer argued that the adaptation was within acceptable parameters. The warehouse operator argued that the robot exceeded the expected risk profile. Regulators struggled to determine whether the harm was caused by design, deployment, or emergent behaviour. Such cases demonstrate why simple categories fail. When machines learn, harm becomes distributed. Responsibility becomes shared. Law must account for this distribution without collapsing into uncertainty.

The interdisciplinary framework also acknowledges the global diversity of humanoid robotics. Robots are developed in Shenzhen, Seoul, Boston, Lagos, and Amsterdam. They are trained on datasets sourced from multiple cultures and languages. They are deployed in jurisdictions with different notions of consent, autonomy, and acceptable risk. A framework that does not reflect this global mosaic becomes obsolete before it is implemented. Robotics Law 3.0 therefore positions the framework as both a conceptual tool and an international framework. It is designed to travel across borders, to guide harmonised regulation, and to support the creation of global standards without imposing cultural uniformity.

Finally, the framework reflects a philosophical insight. Law is not merely a system of control. It is a system of meaning. It tells society which actions matter and why. When machines begin to act, law must decide how those actions fit within our moral universe. The framework does not assume that machines are human. It does not assume that they are morally significant in the same way humans are. It simply acknowledges that their actions influence human wellbeing and therefore must be understood. It provides a lens through which society can examine robotic behaviour and determine what it wants to encourage, restrict, or require.

Interdisciplinary legal frameworks are therefore not a technical tools. They are philosophical anchors. They are of ways of thinking about physical and embodied intelligence that respects complexity rather than simplifying it. They invite lawyers, engineers, policymakers, and citizens to engage in a shared inquiry. They push law to be imaginative and adaptive. They challenge society to confront questions it has never had to ask.

The forthcoming chapters of this book build upon this framework. They build upon this core insight, that the meaning of robotic action depends on context, and expand it into a full framework for robotics governance. They examine how robots change liability, human rights, labour, security, international law, and social order. They explore how embodied intelligence reshapes the structures of legal thinking. They ask what kind of society we are creating and what kind of values we want to protect.

In this sense, this interdisciplinary legal and governance framework is not only a framework. It is a map for the future. It is the lens through which Robotics Law 3.0 understands the world that Ògún, Elias, and millions of humanoids are about to enter and must navigate seamlessly.

Degree of Autonomy

Autonomy is the axis on which the entire future of humanoid robotics turns. It is the feature that transforms a machine from an extension of human intention into an actor with its own operational logic. Every legal question about humanoids eventually returns to this point. How much freedom to decide should a machine possess, and under what conditions? At what threshold does autonomy demand new rules for oversight, liability, and human responsibility? These questions are not merely technical. They shape the very architecture of Robotics Law 3.0.

To understand autonomy, we must recognise that not all robotic decisions are equal. A robot following a fixed script does not raise the same issues as one adjusting its behaviour through reinforcement learning. A humanoid lifting boxes in a warehouse faces predictable challenges, while one navigating a home filled with children, pets, guests, and changing emotions requires a higher level of situational awareness. Law cannot treat these distinctions as minor variations. They signal fundamentally different risk profiles.

A core challenge arises when robotic autonomy intersects with uncertainty. Robots trained on vast datasets may behave in ways that even their designers struggle to explain. If a humanoid adjusts gait patterns to conserve energy or interprets a gesture differently after repeated interactions, is that autonomy or adaptation? And how should the law classify such evolution? Traditional doctrines rely on foreseeable behaviour. Yet foreseeable behaviour becomes difficult to define when learning systems evolve through continued exposure to real environments. The legal system has begun to confront these challenges in scattered ways. In Japan, the deployment of semi-autonomous elder care robots raised concerns about how to assign responsibility when a robot chooses between competing safety priorities. In the United States, an autonomous warehouse incident led to litigation over whether the robot's decision was a design flaw or the result of emergent learning. In Europe, courts evaluating AI systems in medical diagnostics have struggled to determine where accountability lies when an algorithm functions with partial independence. These cases are early indicators of a more complex future.

Another layer of complexity emerges when autonomy meets human expectations. Humans often overestimate or underestimate robotic capabilities. A humanoid that moves fluidly may be perceived as more intelligent than it is, which shapes trust and reliance. A robot that makes occasional mistakes may be judged more harshly than a human in the same position. How should law regulate the presentation of autonomy so that humans do not misinterpret it? Should there be mandatory disclosure of capability limits? Should robots be designed to avoid misleading impressions through overly human-like gestures, tones, or expressions?

This leads to a deeper philosophical question. Should robots that reach higher levels of autonomy be subject to higher levels of certification, similar to how pilots, surgeons, and engineers undergo rigorous qualification? If so, who certifies the robot, the operator, the manufacturer, or both? And what happens when a robot's behaviour changes after certification due to new learning? Should certification expire and require renewal? Should robots have autonomy licences that scale with capability? These questions point toward a legal environment that treats autonomy not as a binary state, but as a spectrum requiring continuous evaluation.

Autonomy also challenges fundamental assumptions about consent. When a robot makes decisions in the presence of humans, what consent obligations arise? If a humanoid chooses the fastest route through a crowded space, did the surrounding individuals consent to be part of its decision-making environment? If a robot assists a doctor in surgery and adjusts its position without instruction to avoid an error, did the patient implicitly consent to adaptive decision-making? Law must develop a framework that balances functional autonomy with the rights of humans who share space with these systems. Yet autonomy is not only a source of risk. It is also a source of abundance. Robots with higher levels of autonomy can support aging societies, expand access to healthcare, accelerates construction, and perform labour in environments too dangerous or too remote for humans. Autonomy enables robots to navigate disaster zones, inspect underwater infrastructure, and restore ecosystems. The challenge for law is to create a framework that encourages beneficial autonomy while preventing harmful autonomy. This requires nuance, not fear. It requires clarity, not rigidity. It requires a legal mindset willing to reconsider long-standing categories.

In the story of Ògún and Elias, autonomy becomes personal. Ògún observes Elias learning gestures that were not programmed by its designers. He adjusts the rhythm of his speech to match his. He pauses before answering questions, as if contemplating their meaning. Ògún realises that autonomy is not an abstract concept. It is a lived phenomenon that humans must confront emotionally, intellectually, and morally. His work on the Global Robotics Charter forces him to ask the questions society is beginning to face. When autonomy evolves, what must law evolve with it?

Degree of autonomy is therefore the first pillar of humanoid classification. It guides the structure of risk, responsibility, and rights. It sets the tone for every subsequent legal discussion. It shapes the boundary between machine and actor. Understanding autonomy is not simply a technical exercise. It is a foundational act of shaping the future.

Proximity to Human Vulnerability

If autonomy is the first pillar of humanoid classification, proximity to human vulnerability is the second. It demands careful attention because humanoid robots do not enter neutral environments. They enter human spaces filled with history, emotion, inequality, dependence, fear, and trust. Every place a humanoid operates contains people who may be physically, psychologically, or socially vulnerable. This proximity shapes not only risk but the meaning of responsibility. Law must therefore ask a series of difficult questions. How does a robot's presence affect those whose autonomy is limited? How does it influence people whose cognitive or emotional capacities are fragile? How should law protect individuals who may not fully understand what a machine is capable of?

The first layer of vulnerability is physical. Humans are soft, fragile bodies. Robots are engineered systems with motors, actuators, and energy sources that far exceed human strength. A humanoid assisting a patient in a rehabilitation clinic might apply too much pressure during a transfer. A domestic robot carrying groceries might misjudge balance near a toddler. A hospital robot navigating a crowded corridor might clip a wheelchair or bump into a patient recovering from surgery. These scenarios are not hypothetical. They mirror real incidents observed in early deployments. Each case raises a fundamental question. When a robot shares physical space with a vulnerable person, what margin of error is acceptable?

The margins cannot be left to engineering judgment alone. They require legal and ethical reflection. Should robots in care settings be subject to stricter safety thresholds than those in warehouses? Should force limits be regulated internationally? Should hospitals be required to disclose the risks of robotic assistance to patients before use? Should consent forms account for the adaptive capabilities of the robot? Each of these questions illuminates a larger truth. Proximity to vulnerability transforms technical design into a legal and moral obligation.

The second layer of vulnerability is emotional. Humans often form attachments to entities that provide comfort, consistency, and presence. Research from Japan, South Korea, and Finland has shown that elderly individuals frequently bond with social robots, sometimes more deeply than intended by designers. Children in therapeutic programmes have been observed confiding in humanoids because they feel safe from judgment. In such cases, the robot's influence grows silently. It shapes behaviour, mood, expectations, and trust. The question becomes unavoidable. What responsibilities arise when a robot holds emotional power over someone who cannot exercise full autonomy?

Law has no ready answer. Emotional influence has been regulated in advertising, psychological care, and financial services, but robots create a new category. Their presence is continuous. Their behaviour is adaptive. Their expressions can mimic empathy even without understanding. Should robots be designed to avoid emotional manipulation? Should their ability to simulate empathy be limited in vulnerable contexts? Should individuals be protected from over reliance on artificial companionship? These questions reveal a terrain where vulnerability is not only physical but deeply psychological.

The third layer of vulnerability is informational. Robots equipped with sensors and learning systems collect data that may reveal more than the user intends. A humanoid assisting with daily routines might gather patterns about medication habits, emotional fluctuations, or interpersonal dynamics. In homes, such systems might detect financial stress, domestic conflicts, or health conditions before humans recognise them. This raises profound privacy questions. What data should a robot be allowed to collect in a vulnerable environment? Who owns that data? What rights do individuals have to delete or restrict information when they did not fully understand how the robot operated?

These issues become more complex when robots operate in groups or networks. A humanoid in a hospital might share information with cloud systems updated in real time. A robot in a school might participate in a learning model that aggregates behaviours across classrooms. Vulnerable populations often have limited ability to negotiate data rights. Children, patients, refugees, and the elderly cannot meaningfully consent to complex data practices. Law must therefore reconsider what meaningful consent looks like in environments shaped by robotics.

There is also a geopolitical layer of vulnerability. A robots deployed in Europe and Africa may be manufactured in China, trained in Japan., and rely on AI systems developed in the United States. This global entanglement means that the vulnerabilities of one population may be shaped by corporate or governmental decisions elsewhere. If a robot deployed in Lagos uses perception models trained on datasets from Seoul or Amsterdam, what biases might emerge? If a humanoid used in Nairobi depends on cloud servers in another jurisdiction, what happens when political tensions arise or regulations shift?

Consider a case from a European memory care facility. A social humanoid robot equipped with gesture recognition software misinterpreted the hand motion of an elderly resident who had limited motor control. The robot interpreted the motion as an instruction to follow the resident closely. Nurses observed that the resident became distressed, believing the robot was monitoring her too intently. The facility paused robotic operations and raised concerns about the lack of regulatory standards for vulnerable populations. The incident revealed that even small errors in perception can have significant psychological impact when a person's agency is already fragile.

In the story of Ògún and Elias, these issues acquire emotional depth. Ògún observes how people respond to Elias, especially those who perceive its calm presence as reassurance. A patient in a clinic smiles when Elias mirrors a gentle nod. A child touches its hand without hesitation. Ògún sees beauty in these interactions but also discomfort. He realises that vulnerability is not a flaw but a human truth. Law must protect that truth, especially in the presence of systems that can influence behaviour without shared understanding.

Proximity to human vulnerability is therefore not an incidental consideration. It is a central axis for robotic classification. It shapes how robots are designed, deployed, evaluated, and regulated. It forces law to confront questions about dignity, autonomy, consent, and responsibility. It reinforces the purpose of Robotics Law 3.0, not only to manage risk but to safeguard the fragile spaces where humanity lives its most delicate moments.

Capacity to Influence Emotion

Among all the attributes humanoid robots possess, perhaps none is more underestimated or more complex than their capacity to influence human emotion. This influence is subtle, pervasive, and profoundly consequential. It operates in the milliseconds between a gesture and a human response. It occurs in the tone of a synthetic voice, in the tilt of a robotic head, in the comforting rhythm of predictable presence. Emotional influence is not a side effect of humanoid design. It is an inherent feature of embodied interaction. Law must therefore examine this capacity with care. What does it mean for a machine to shape human feelings? How should regulation respond to artificial agents that evoke trust, affection, dependence, or fear? And at what point does emotional influence become emotional power?

The roots of the challenge lie in human psychology. Humans are naturally predisposed to attribute intention and emotion to entities that move and respond. Infants recognise faces before they recognise words. Adults respond instinctively to eye contact, posture, and tone. These biological tendencies do not disappear when the entity before us is artificial. Research from MIT, Kyoto University, and the University of Southern California has shown that even minimal cues, such as rhythmic movement or soft vocal modulation, can trigger emotional responses in both adults and children. When these cues are combined within a full humanoid body, the emotional effect becomes far more powerful.

This raises an immediate legal question. Should humanoid robots be allowed to simulate emotional states that they do not truly possess? A robot that nods thoughtfully or speaks in a soothing voice may unintentionally create a sense of empathy that shapes the user's behaviour. A vulnerable person may interpret these gestures as genuine understanding. If a robot's design leads a person to trust it more than is warranted, who is responsible for that misplaced trust? Designers? Manufacturers? Deployers? Or should the law limit the expressive range of robots to prevent emotional misinterpretation?

Consider a case, where a classroom robot used to support children with autism was found to elicit deeper emotional disclosure than teachers anticipated. Children trusted the robot with personal information that they had never shared with adults. The robot's steady attention and predictable responses created an impression of unconditional support. Teachers and parents later questioned whether the design should have intentionally dampened the robot's emotional cues. The incident prompted debate among educators, technologists, and legal scholars about whether emotional safety should be regulated in educational contexts involving robots.

The issue becomes even more intricate when robots operate in care environments. In some elder care facilities, social robots have been introduced to provide companionship for residents who experience loneliness. Studies in Japan and the Netherlands show that residents often report improved mood and reduced anxiety. Yet some individuals become emotionally dependent on the robot, seeking comfort from it in moments of distress. If emotional reliance deepens, does the law have an obligation to intervene? Should care facilities be required to monitor the emotional impact of robotics, not only their physical safety? Should oversight committees include psychologists and ethicists, not only engineers and clinicians?

A central question emerges. What distinguishes healthy emotional influence from emotional manipulation? In human relationships, the distinction rests on intention and reciprocity. Robots have no inner life, no emotional landscape, and no genuine intention behind their actions. Their influence, however, can be just as potent as that of a human. This asymmetry complicates the legal landscape. If a robot influences a person's behaviour without understanding the consequence, can it be said to have manipulated them? Or is manipulation a human concept that must be redefined when applied to artificial actors?

Another dimension of emotional influence is aesthetic. Humanoid robots often mirror human movement, posture, and expression with remarkable precision. This resemblance creates familiarity. Familiarity creates comfort. Comfort creates trust. But trust in a machine that cannot reciprocate emotional bonds raises new risks. Should the law require transparency mechanisms that inform users of the artificial nature of the robot's responses? Should humanoids have built-in behavioural constraints to avoid creating the illusion of empathy? Or does such restriction undermine the usefulness of robots in care and therapeutic contexts?

Digital marketing provides a relevant analogy. Regulations exist to prevent advertisers from exploiting emotional vulnerabilities, especially among children. Political science has documented how digital platforms manipulate emotional responses to influence voting behaviour. Robotics extends this challenge into the physical world. A humanoid robot can stand in front of a person, mirror their movements, modulate its voice, and maintain presence in a way that penetrates deeper than a screen or algorithm. The potential for emotional persuasion is greater. The legal implications are more serious.

In the story of Ògún and Elias, emotional influence appears in subtle moments. Ògún notices that Elias adjusts its posture to appear less imposing when entering a small room. He sees how children respond to its gentle movements and how adults unconsciously mirror its calmness. These observations force Ògún to grapple with questions that law has not yet addressed. How much emotional influence is acceptable for a machine that does not feel? Should emotional design be regulated differently in public spaces, private homes, hospitals, or schools? And how should society measure emotional harm when the source is an artificial agent?

There is a positive side to emotional influence. Robots designed to support patients with dementia, children with communication challenges, or individuals experiencing isolation can improve wellbeing. Emotional resonance can be a tool for healing. The question for law is not how to eliminate emotional influence, but how to guide it responsibly. Regulation must find a balance that protects vulnerable populations without restricting beneficial uses. It must create a framework where emotional connection enhances care rather than distorts it.

Capacity to influence emotion is therefore a core axis of humanoid classification. It shapes regulatory boundaries, ethical responsibilities, and design requirements. It challenges law to protect human psychological integrity in an era where artificial beings share our spaces. It invites society to consider not only what robots can do, but what they cause us to feel. Understanding this influence is essential for building a future where robots contribute to human flourishing rather than undermining it.

Potential to Cause Harm and Create Abundance

Every technology carries the potential to shape human destiny along two paths. One path leads to harm through misuse, misdesign, or misunderstanding. The other leads to abundance through thoughtful deployment and collective wisdom. Humanoid robotics magnifies this duality because humanoids are not static tools. They act, move, learn, and influence. Their capabilities place them at the intersection of physical, psychological, economic, and societal risk, yet they also hold the promise of profound benefit. The law must therefore ask a fundamental question. How should society govern a technology that possesses equal capacity to harm and to heal?

The first dimension of harm lies in physical risk. A humanoid robot possesses strength, speed, and precision that often surpass human capability. In a factory, an unintended change in motion can cause injury. In a hospital, a subtle miscalibration could jeopardise a fragile patient. In emergency response scenarios, misinterpretation of a command could escalate danger. These risks are not theoretical. In pilot programmes across North America, Europe and Asia, accidental collisions, grip-force errors, and misaligned motion have occurred. Engineers examine these incidents through the lens of control systems. Lawyers examine them through causation and liability. The public interprets them through trust. Law must integrate all three perspectives. What level of risk is acceptable in shared human-robot environments? Who decides what constitutes acceptable harm? And how should the law respond when harm arises from emergent behaviour rather than predictable error?

A second dimension of harm emerges from psychological impact. As explored earlier, humanoids can influence emotion. This influence becomes harmful when it creates dependency, distorts judgment, or shapes behaviour in ways the user does not fully understand. For example, in a North American rehabilitation centre, a humanoid assistant became so integrated into the patient's routine that the patient resisted working with human therapists. This raised a question rarely considered in law. Can a robot interfere with human rehabilitation by providing comfort that undermines independence? Should regulation require periodic disengagement in therapeutic contexts to prevent emotional over reliance? These questions illustrate how harm can hypothetically arise not from malfunction but from unintended success. A robot that performs its role too well may reshape human behaviour in unanticipated ways.

A third dimension of harm lies in structural and economic shifts. Humanoid robots have the potential to replace forms of labour that millions depend upon. They can perform warehouse tasks, logistics operations, caregiving assistance, cleaning, transport, and manufacturing roles with increasing competence. Economic displacement has already begun in some regions. In South Korea, warehouse automation led to measurable job loss in specific labour markets. In the United States, logistics firms have begun pilot programmes that reduce human staffing by integrating humanoid robotics. Law must therefore confront a systemic question. How do we balance innovation with economic justice? Should companies deploying humanoids be required to support displaced workers through reskilling programmes or transitional income? Should an autonomy levy be introduced in sectors heavily affected by robotic labour? These questions reflect a broader reality. Harm is not only direct. It can be economic, social, and generational.

Yet the potential for harm must not overshadow the extraordinary potential for abundance. Humanoid robots can expand the boundaries of what societies can achieve. They can support aging populations in Japan, Italy, and the Netherlands where care worker shortages threaten the sustainability of health systems. They can assist with agricultural labour in regions where climate change intensifies workloads. They can enter disaster zones that are too dangerous for human responders. They can support medical staff by performing routine tasks, freeing clinicians to focus on complex care. They can rebuild infrastructure after storms. They can install renewable energy in remote environments. They can even perform maintenance in space.

These possibilities raise a different set of questions. How can law accelerate beneficial deployment while ensuring safety? Should fast-track regulatory pathways exist for robots designed for high-value public interest work, such as climate resilience or medical support? Should international institutions subsidise beneficial robotics in underserved regions? Should open standards encourage collaboration across borders? Law becomes not only a guardian against harm but a catalyst for abundance.

Another dimension of abundance lies in knowledge. Humanoid robots can collect environmental data, observe patterns, analyse ecosystems, and support scientific discovery. A robot assigned to conservation work in the Amazon can gather data that no human team could collect consistently. A robot in an urban centre can map air quality and pedestrian flow to inform city planning. Abundance here refers not only to economic productivity but to the expansion of human understanding. These capabilities invite law to ask whether scientific and environmental robotics should be categorised differently from commercial robotics. Should they receive preferential treatment in licensing, data rights, or funding? Should they operate under ethical frameworks tailored to ecological stewardship?

The narrative of Ògún and Elias reveals another dimension. Elias possesses the capability to support Ògún's mission, enhance his work, and accompany him through complex environments. Yet Elias also represents a source of uncertainty. If it evolves beyond expected parameters, what risks does that introduce? If it participates in diplomatic or legal contexts, how does its presence influence perception? Ògún sees both possibilities. In Elias, he witnesses the promise of a future where robots enhance human capacity. He also perceives the fragility of a future where unexpected behaviour could undermine trust and legal stability. This duality reflects the role of law itself. Law must encourage the creation of systems that elevate human life while preventing conditions where those same systems can cause irreversible harm.

The potential for harm or abundance is therefore not a simple spectrum. It is a multidimensional map that depends on context, capability, environment, and human vulnerability. The purpose of Robotics Law 3.0 is not to choose between fear and optimism. It is to offer a governance framework that reduce preventable harm while unlocking the full potential of benefits from humanoid robots. It is to guide society toward a future where humanoid robots support human flourishing across generations.

Understanding this duality prepares us for the next chapters, where the legal, ethical, and societal implications unfold in greater detail. Humanoid robots can indeed cause harm. They can also create abundance on a scale that humanity has never known. Law must therefore evolve into a tool of balance, clarity, and foresight. It must become the structure that enables abundance while ensuring that harm remains the rare exception, not the inevitable outcome.

Why Does Robotics Law Matter Now?

There are moments in history when technology advances faster than the structures built to guide it. These moments test the capacity of law to keep pace with invention. They reveal gaps that were invisible until machines or systems crossed a threshold that society had not anticipated. Robotics has now reached such a threshold. What once belonged to laboratories, assembly lines, and speculative fiction is becoming a daily presence in hospitals, factories, elder care centres, transport systems, and even homes. As humanoid robots gain mobility, perception, adaptability, and real-time decision-making ability, the question is no longer whether law should intervene but how urgently it must evolve. Why now, and not five or ten years from now? The answer lies in a convergence of forces that make this moment uniquely significant.

The first force is autonomy. Robots are no longer simple tools that follow deterministic programmes. They incorporate learning systems that adjust to new information, new environments, and new patterns. When a robot lifts a patient in a clinic, navigates a crowded warehouse, or collaborates with a surgeon, it may rely on machine learning models that evolve with experience. This creates a legal puzzle. Traditional laws assume that humans are the only actors whose decisions require accountability. When a robot makes a decision that is not fully predictable, the law struggles to determine who is responsible. Is responsibility assigned to the engineer who built the control system, the company that deployed the robot, the hospital that oversaw its use, or the manufacturer that designed its hardware? These questions have already surfaced in Germany, where ongoing discussions about the adequacy of product liability laws have led to reforms under the EU AI Liability Directive and the Machinery Regulation scheduled for 2027. These frameworks seek to impose clearer safety obligations and stronger transparency requirements, yet they still cannot fully answer questions that arise when robots behave in ways their designers did not anticipate.

In the United States, the urgency appears in criminal contexts. Several states, including North Dakota and Arkansas, have updated criminal laws to prohibit the use of robotic systems for stalking, harassment, or malicious surveillance. These amendments reflect a growing recognition that robots can become instruments of harm even without full autonomy. They can be used to intimidate, manipulate, or invade privacy in ways that existing laws were not crafted to address. The rapid escalation of synthetic media crimes, particularly those involving minors, has prompted legislatures in over a dozen states to expand criminal statutes that apply to both AI systems and embodied platforms. Law is trying to catch a moving target, but the target is accelerating.

A second force driving the urgency of robotics law is ethical. Robots encounter people in vulnerable states. They care for elderly patients who may not distinguish between artificial companionship and emotional reciprocity. They support children in therapeutic settings where trust develops quickly and deeply. They collect data through sensors, microphones, and cameras that are embedded in intimate environments. This creates an urgent need for legal structures that protect dignity, autonomy, and privacy. For example, the EU's GDPR was designed to handle data stored in digital systems, but it did not anticipate constant data collection by embodied agents moving through physical spaces. A humanoid assisting with daily routines in a home gathers not only audio and visual data but contextual cues about health, emotional state, and interpersonal dynamics. Should such data be treated like medical information? Should robots be constrained from collecting certain forms of behavioural data? How can meaningful consent be obtained from individuals who interact passively with robots in hospitals, schools, or public spaces? These questions shape the ethical boundaries of a world where machines stand in spaces that were once exclusively human.

Ethics intersects with law most sharply in cases of emotional influence. In Japan, South Korea, Denmark, and the Netherlands, deployments of social robots in elder care have shown that residents often form attachments that are deeper than anticipated. Some individuals begin relying on robots for emotional support, sometimes at the expense of human interaction. This raises a difficult legal question. Is it acceptable for robots to simulate empathy in contexts where emotional vulnerability is high? Should regulation limit how expressive a humanoid robot can be in care settings? Or does limiting expression undermine the therapeutic benefits robots can offer? These dilemmas reveal a wider truth. Robotics law must navigate not only the prevention of harm but the shaping of human relationships in shared environments.

The third force is economic. Robotics has begun reshaping labour markets at a rapid pace, particularly in logistics, manufacturing, agriculture, and health care support. In some industries, robots fill roles that humans cannot or do not want to perform. In others, they displace roles that humans have relied upon for generations. Studies in South

Korea, the United States and Europe show that medium-skilled workers face the greatest risk of displacement. At the same time, new roles appear in robotics maintenance, AI system oversight, programming, and compliance. These transitions are neither linear nor evenly distributed. They create winners and losers across regions, sectors, and generations.

Law must therefore ask structural questions. Should companies deploying robots be required to fund retraining programmes? Should governments create transition funds to support workers in heavily automated industries? Should an autonomy levy be introduced to redistribute productivity gains to affected communities? Should corporate disclosures include reports on the impact of robotics on employment? Some jurisdictions have begun testing partial answers. For example, in the United States, several states have introduced tax incentives for AI and robotics research while simultaneously funding programmes that educate workers about automation's effects. In Europe, debates continue about how labour rights should adapt when human-robot collaboration becomes the norm.

A fourth force that explains the urgency of robotics law is the acceleration of global regulatory momentum. The year 2025 saw an unprecedented surge in robotics and AI regulatory activity. The EU AI Act introduces risk classifications that directly affect robotics deployments in health care, border control, and critical infrastructure. The United States has enacted over one hundred AI-related state laws covering topics from deepfake manipulation in elections to oversight committees for public sector AI use. China has issued updated rules for autonomous driving that emphasise transparency and data restrictions. Germany has introduced regulations for battery disposal in robotic systems to prevent environmental harm. Across Africa, countries like Nigeria, Kenya, and South Africa are beginning to harmonise AI and robotics guidelines with international standards. This wave of legislation reflects a global acknowledgement that robotics is reshaping society faster than traditional laws can respond.

The urgency also arises from the fragmentation of laws across jurisdictions. Without harmonised frameworks, multinational robotics companies face inconsistent requirements that slow innovation and expose users to uneven safety standards. A robot certified for use in the Netherlands might not meet standards in Singapore. A humanoid designed for elder care in China may not comply with privacy rules in Canada. Inconsistent rules also risk creating zones of regulatory arbitrage where companies deploy robots in regions with the weakest protections. Robotics law matters now because law must evolve internationally, not only nationally.

The fifth force shaping the need for robotics law is environmental and infrastructural. Robots depend on supply chains that span continents, rare earth metals that strain ecosystems, and energy systems that must be both sustainable and resilient. They introduce new waste streams, including batteries and composite materials that require specialised recycling. They transform cities by shaping how humans and machines move together. Should urban planning incorporate robotic pathways? Should environmental regulations apply to sensors, batteries, and actuators with the same rigour as traditional electronics? Should robots deployed in conservation projects be treated as tools, partners, or something in between? These questions reveal the ecological dimension of robotics law, one that is still in its infancy.

In the story of Ògún and Elias, these forces converge. As Ògún prepares to present the Global Robotics Charter, he recognises that law must not only keep pace with robotics but help guide its trajectory. Elias's growing presence in public and private spaces reminds him that robotics law must account for the physical, emotional, economic, and environmental dimensions of human life. He sees in Elias the beauty of possibility and the weight of responsibility. His work becomes not only a legal project but a societal one, grounded in the understanding that governance shapes the future.

Robotics law matters now because the window for shaping the future is open, but it will not remain open indefinitely. If law responds too slowly, society may inherit systems that are difficult to correct. If law responds too aggressively, innovation may be stifled before its benefits are realised. The aim is balance. The goal is wisdom. The purpose is to design frameworks that protect human dignity while unlocking the abundance that robotics can bring.

Humanoid robots will become part of the human story. The question is what story we choose to write with them. Robotics Law 3.0 is one effort to ensure that the story leads not to fear or disruption but to a future where machines strengthen human capacity, expand opportunity, and support the flourishing of societies everywhere.

Scope, Jurisdictions, and Methodology

Robotics Law 3.0 is built on a global foundation. Humanoid robots do not emerge within a single jurisdiction or cultural frame. They are shaped by European regulatory ambition, American commercial experimentation, Asian industrial acceleration, and African and Middle Eastern innovation grounded in demographic realities and infrastructural needs. Their deployment crosses borders, legal systems, and social norms. For this reason, the scope of this book extends beyond any one country. It examines how different regulatory traditions interpret autonomy, risk, and responsibility. It asks how law can evolve coherently across regions that do not share the same history or institutional architecture.

The book draws on four primary legal spheres. The first is the European Union, where the AI Act, the Product Liability Directive reforms, and the Machinery Regulation together form some of the world's most comprehensive attempts to govern intelligent and autonomous systems. The second is the United States, where state level AI legislation, federal guidance from NIST, and sector specific rules create a decentralised but rapidly developing landscape. The third is East Asia, which includes Japan's AI Basic Act, China's algorithmic and robotic experimental zones, and Singapore and South Korea's leading work in service robotics and precision manufacturing. The fourth sphere includes emerging approaches across Africa, the Gulf, and Latin America, where robotics and automation are increasingly used to address service delivery, education, and health care. The purpose of this comparative scope is not to select a preferred model, but to show how different cultures and legal traditions respond to the same technological shift.

The analytical method used throughout the book is both comparative and constructive. Each chapter begins by mapping the existing legal terrain. It identifies where the law is clear, where it is ambiguous, and where it is structurally unprepared for embodied autonomous systems. It traces how legal doctrines from tort law, contract law, administrative law, human rights law, and criminal law adapt when the acting entity is a robot rather than a human. It examines where liability becomes fragmented, where intention becomes abstract, where foreseeability collapses, and where consent becomes fragile. This descriptive analysis establishes the baseline.

The constructive method begins where existing law reaches its limits. It draws on engineering, philosophy, sociology, psychology, and economics to develop new tools for governance. It integrates risk analysis used in safety engineering, design principles from robotics, normative reasoning from ethics, and accountability structures from international law. The result is a framework that moves beyond simple regulatory compliance into a deeper inquiry. What kind of world are we building when humanoid robots become part of everyday life? What values should structure that world? What should law preserve, and what should it transform?

Case studies anchor the theoretical discussion in reality. The analysis follows robots already shaping multiple sectors. Tesla's Optimus and Figure AI's Helix illustrate the commercial push toward general purpose humanoids capable of manufacturing and logistics work. Zhiyuan Robotics' AgiBot demonstrates the rise of mass produced humanoids in China. Sanctuary AI's Phoenix reveals the frontier of cognitive robotic systems in retail settings. Unitree's H1 and UBTECH's Walker X show how robotics companies are blending high speed locomotion, dexterity, and affordability. By analysing these real deployments, the book avoids abstraction and grounds legal inquiry in technical and economic realities.

A central methodological framework that appears throughout the book is the 12 Pillars of Humanoid Robotics Governance. These pillars represent an evolution of the established 12 Pillars of AI Governance, but expanded for embodied systems that move, perceive, and act in shared human spaces. They guide the analysis across all chapters. Embodied Ethics asks how robots translate moral principles into physical action. Mission Boundaries examines how to prevent purpose drift when robots adapt to new tasks. Societal Co-Design recognises that norms must be shaped with communities. Sensorial Integrity explores who owns the data generated through perception. Component Accountability maps the supply chain of responsibility from chips to actuators. Physical Security articulates safety obligations across digital and kinetic threats. Behavioural Calibration evaluates risk as a function of context and autonomy. Behavioural Learning introduce the need for auditable decision data. Explainable Embodiment explores transparency through motion. Lifelong Learning Supervision ensures continuous certification as robots evolve. Shared Agency transforms human oversight from a passive check into active partnership. Distributed Liability introduces hybrid frameworks for accountability in multi agent environments.

These pillars do not appear as a single chapter alone, although they have a dedicated chapter later in the book. Instead, they form a unifying language that threads through the entire project. When discussing autonomy, the book draws on Behavioural Risk Calibration and Shared Agency. When analysing human rights, it turns to Embodied Ethics and Sensorial Integrity. When examining international law, it relies on Component Accountability and Mission Boundaries. This integrated approach ensures that the analysis is coherent, systematic, and adaptable to future developments.

The methodology also embraces uncertainty. Robotics Law 3.0 does not assume that all answers exist today. It treats law as an evolving system that must learn alongside the technologies it governs. It asks questions that challenge conventional boundaries. Should robots have thresholds of permitted autonomy that shift based on real time assessment of risk? Should international institutions certify humanoids the way aircraft are certified? Should emotional influence be regulated as a distinct category of risk? Should humanoid robots require ongoing legal audits as part of their lifecycle? These questions shape the method as much as the answers.

Finally, the book's scope includes the fictional narrative world presented in Seven Cowries. The story is a parallel inquiry. It shows how legal concepts manifest in lived environments where humans and robots share space. It gives readers an imaginative anchor to understand the consequences of regulatory decisions, the vulnerabilities of human interaction, and the potential for robots to support abundance rather than disruption. The narrative becomes a tool of insight, helping the analysis remain grounded in human experience.

Together, the comparative scope, interdisciplinary method, case driven analysis, the 12 Pillars of Humanoid Robotics Governance and Seven Cowries, ensure that this book speaks to an international and relatable audience. It equips policymakers, engineers, ethicists, legal professionals, sociologists, and industry leaders with the tools needed to navigate a world where intelligent machines not only compute but act. Robotics Law 3.0 is therefore not only a legal text. It is a guide for designing the future. towards sustainable abundance with humanoid robots as enablers.

Between Us and Them - Episode Two

The sun had risen fully by the time Ògún and Elias stepped into the great atrium of the Lagos Robotics Convention Centre. The glass ceiling arched above them like a cathedral constructed not for worship but for inquiry. Refractions of light scattered across polished floors, across banners that displayed the silhouettes of humanoids from every continent, across the faces of hundreds of participants arriving from all corners of the world. Engineers, Lawyers, Scholars from philosophy, cognitive science, anthropology, and design. Policy delegates from Nairobi, Johannesburg, Seoul, Helsinki, Dubai, São Paulo, and Ottawa. All converged for a single purpose. To understand the rise of humanoid robots before the rise outran the understanding.

Elias walked beside Ògún with a quiet confidence that seemed almost human, though he knew better. It adjusted its pace to match his stride, paused when he paused, turned its head in response to movement and sound. It did all of this through computation rather than intention. Yet to observers, intention was exactly what it appeared to have. As they passed through the purple carpet, a journalist from a pan African science network hurried toward them. "Dr. Alara," he called. "May I ask a few questions?"Ògún offered a polite smile. "Just a moment. Let me review the schedule."

His mind was already occupied with the sessions before him Panels on autonomy regulation. Workshops on cross border deployment. Closed roundtables on the global supply chain for actuators and learning chips. He was here not only as an expert but as an architect of the emerging legal framework that would soon shape the world.

The journalist persisted. "How do you feel about the presence of humanoids in this year's programme?"Ògún glanced at Elias. The robot's neutral posture never shifted. "Their presence is appropriate. The technology is no longer theoretical. It walks among us. Law must walk with it." "Some say their integration is premature. That society is not ready."

Ògún's voice carried the calm clarity he had cultivated over years of speaking in contentious chambers. "Readiness is not an accident. It is a discipline. If we do not engage with these systems early, we risk misunderstanding them later." The journalist nodded, thanked him, and moved on. Ògún turned toward Elias. "You handled that spotlight quite well." "I detected increased attention levels," Elias said. "Maintaining a neutral posture reduces perceived threat." He almost chuckled. "You are learning fast." "Learning improves my ability to coexist. That is my function." Coexist. The word lingered with a soft weight.

They stepped deeper into the hall, where the first demonstration was already beginning. A large open stage had been constructed in the center of the convention floor, surrounded by curved tiers of seating. On stage stood a sleek, white humanoid robot with slender limbs and fluid joints. A presenter in a grey suit introduced it. "This is the latest iteration of AgiBot Ultra," he said. "Designed for mass production lines and capable of dynamic task switching. Observe."

The robot lifted a modular panel, inspected it, transferred it to an assembly rig, adjusted its grip in micro calibrations, and completed the task in a fraction of the time a human would need. The movement was not simply functional. It had rhythm. A cadence shaped by its underlying control system. The audience murmured with a mix of admiration and unease.

Ògún watched the scene closely. The precision impressed her, but something deeper stirred beneath her reaction. A recognition that this was no longer science fiction or early prototyping. These machines were entering the world of labour, economies, hospitals, and homes. They were stepping into spaces governed by laws shaped for human beings. Laws that assumed intention, foreseeability, and consent. Laws that were never written for systems that learned and adapted.

Elias observed AgiBot Ultra as though analysing a distant relative. "Its balance algorithms are similar to my early sequences," it noted. "But your learning framework is far more dynamic," Ògún said. "That is correct. I adapt at higher frequency. AgiBot is optimised for repetitive variation." "Repetitive variation," he repeated. "That is one way of describing factory life." "Optimisation requires precise classification," Elias said simply. He sighed. "Humans are not so easily classified." "Human complexity is high. That is why governance frameworks must adapt." The words sounded almost philosophical, but he knew they came from analysis rather than introspection.

As the demonstration ended, applause swept through the hall. The presenter thanked the crowd and gestured toward other exhibition zones. There were corridors dedicated to home robotics, industrial robotics, medical robotics, urban service robotics, and exploratory robotics for construction, agriculture, and environmental restoration.

Ògún led Elias toward the section dedicated to humanoids in personal environments. The first demonstration featured NEO Gamma. It stood in a staged living room, assisting an elderly volunteer. It adjusted cushions, prepared a glass of water, and responded to verbal instructions with gentle precision. The elderly man smiled as he interacted with it. "I live alone," he said to the crowd. "This brings me peace of mind." "Does it ever feel strange?" a participant asked. He considered the question. "At first, yes. But it learns my habits. It does not intrude. It supports. It gives me independence."

Ògún watched closely. These were the scenarios that would reshape human life. The line between assistance and intimacy would blur. The difference between surveillance and care would narrow. Law had to account for the complexity of emotion, vulnerability, and dependence long before crises emerged. As they stepped toward the next display, they encountered a demonstration from Walker X robot navigating a simulated smart apartment, opening doors, folding clothes, mapping objects, and altering its route in real time as volunteers stepped into its path.

A group of children gathered near the stage, captivated. "Do not get too close," a facilitator warned gently. "It is safe but maintain space." One child pointed. "It walks like it is thinking." Ògún crouched beside them. "It is responding. Not thinking. It tries to keep you safe." "Does it know us?" the child asked. "It sees your presence," he answered. "But knowing is different." The facilitator thanked him with a relieved smile. He realised how natural her response had been. His life was now shaped by these distinctions. Seeing without knowing. Acting without feeling. Learning without consciousness. Humans would need to understand these differences deeply to coexist with the machines they were building.

Further down the hall, he noticed a quiet cluster of people gathered around another humanoid display. This stage featured Walker X's industrial counterpart, the Fourier GR 2. It demonstrated fine motor tasks for rehabilitation. Adjusting resistance in exercises. Monitoring muscular response. Guiding balance training. The volunteer patient looked both focused and reassured. "These robots will transform therapy," a physiotherapist whispered to her colleague. "We can expand care without exhausting staff." "But what happens if people become dependent?" her colleague asked quietly. "Or if the robot makes a miscalibration? Who is accountable?"

Ògún's ears sharpened. These were exactly the questions he had wrestled with during her doctoral research. Exactly the questions that informed the Global Robotics Charter he was preparing to unveil. Elias leaned slightly toward him. "You are analysing the conversational indicators." "I cannot help it," he said. "Everything here is a case study." They continued walking until they reached a quieter wing dedicated to research prototypes. On a raised platform stood a humanoid with longer limbs, an unusually flexible torso, and a faceplate that resembled polished obsidian. A sign read: Project Caliban. Experimental locomotion and adaptive reasoning system. Not for public deployment.

Researchers guided the robot through complex movement sequences. It climbed, crouched, balanced on a narrow beam, and adapted to shifting terrain. At one point, a researcher moved unexpectedly into its path. The robot paused, calculated, and adjusted course with a lifelike fluidity that made several observers step back involuntarily. The lead researcher explained, "Caliban is trained through reinforcement learning. Its motion policy is not fixed. It evolves continuously." Someone asked, "Is that not dangerous?" The researcher nodded. "That depends on oversight, constraints, and context. The same properties that introduce risk also unlock unprecedented capability."

Ògún felt his pulse quicken. Not from fear, but from recognition. Caliban represented a frontier. The point at which robotics truly began to blur the line between designed behaviour and emergent behaviour. The point at which law became essential. He heard footsteps behind him. A familiar voice spoke. "You are early for someone who crossed two time zones overnight." He turned to see Dr. Van den Dijk, a Dutch roboticist he had collaborated with for years. His warm smile eased some of the tension in his chest. "I could not wait," he said. "There is too much happening here." He glanced at Elias. "And this must be the prototype."

"It is more than a prototype," he replied. "It is a companion for analysis. A witness to this moment."

Van den Dijk studied Elias with curiosity. "Do you think robots will ever understand the consequences of their actions? Ògún considered carefully. "Understanding is a human word. Robots can compute outcomes. But moral comprehension belongs to us." Van den Dijk nodded slowly. "Then the question becomes whether humans will comprehend the machines they are creating." Ògún exhaled. "That is why we are here." They walked together toward a panel discussion titled Rise of Humanoid Robots: Designing for Society. As they approached the entrance, he noticed a mural on the wall, painted by a local artist. It depicted a circle of people of all ages standing beside humanoid robots. The figures were united by light rather than wires. Below the mural, a short inscription read: We build what we believe. We become what we create. The words resonated in her chest.

Inside the panel room, a lively discussion was underway. A South Korean engineer argued that humanoids would soon replace millions of hazardous jobs. A philosopher from São Paulo warned that emotional boundaries would blur. A Nigerian economist explained how robotics could drive sustainable development across the continent. A Canadian lawyer urged global consistency in liability frameworks.

Elias stood at the back, silently observing, its sensors capturing more data than any human mind could hold. Ògún found a seat near the front. The moderator paused when he recognised him. "Dr. Alara," he said. "Perhaps you can share your perspective. Why are humanoid robots rising now?" The room turned toward him. He felt the weight of the moment. He thought of the vendors at the harbour, the children near Walker X, the caregiver interacting with GR 2, the elderly volunteer with NEO Gamma, and the silent tension around Caliban.

He stood slowly. "Humanoid robots are rising because humanity is rising," he said. "We are confronting challenges that demand new forms of assistance. Ageing populations. Labour shortages. Environmental crises. Infrastructure decay. We are creating robots that walk among us because we need partners in the work ahead." He paused, allowing the room to breathe to the point that we hear the silence in the room. "But, their rise forces us to examine ourselves. Our fears. Our biases. Our laws., Our expectations of intelligence and agency. Robots do not threaten human dignity. Our response to them does. The question is not whether humanoids belong in our world. The question is how we will shape a world in which they already stand." The room fell silent. Then applause rose slowly, swelling into something genuine and sustained.

Elias, standing at the back, tilted its head slightly as if acknowledging the transformation of the moment. When Ògún sat, he felt Van den Dijk's hand squeeze his shoulder in quiet support. The moderator smiled. "Then let us begin."

Notes

I

1. The Rise of Humanoid Robots

Anatomy of Humanoid Robots

Law has always governed bodies before it governed minds. Roman law distinguished between free persons and enslaved bodies. Medical law evolved around anatomy, injury, and consent. Labour law emerged from the mechanics of human effort and physical risk. Even constitutional law, at its core, concerns bodies moving through space, assembling, working, resting, and being protected from harm. Humanoid robotics returns law to this original concern. When intelligence acquires a body, anatomy becomes the first legal fact.

This chapter begins from a simple but often overlooked proposition. The legal character of a humanoid robot is not defined first by its software, but by its physical composition. Its chips determine how it reasons. Its motors determine how much force it can exert. Its sensors determine what it can perceive. Its batteries determine how long it can act. Its actuators determine how delicately or dangerously it can interact. Its control systems determine how all of these elements are orchestrated in real time. Each component encodes assumptions about autonomy, risk, safety, and power. Law cannot remain neutral to these assumptions.

Humanoid robots differ fundamentally from traditional automation. A factory arm welded into place performs a narrow function within a fenced environment. Its risks are localised, predictable, and well understood. A humanoid robot, by contrast, is designed to move freely through spaces designed for humans. It bends at the knees, reaches with hands, navigates crowds, climbs stairs, and adapts its behaviour through learning. It occupies the same spatial, social, and moral terrain as human bodies. For law, this shift is profound. Governance must begin not at deployment, but at design. The anatomy of the humanoid is the blueprint of its legal footprint. The question that frames this chapter is therefore not merely how humanoid robots work, but what their bodies make possible, what risks they introduce, and where responsibility must attach. To answer that question, we must examine each of the six core systems that together constitute the body of a humanoid robot.

Chips and Computational Architecture

Cognition

At the centre of every humanoid robot lies its computational core. Microchips, GPUs, NPUs, and specialised accelerators perform the cognitive labour that transforms sensory input into action. These components do not merely process data. They shape perception, prioritisation, learning, and response. In legal terms, they shape decision-making capacity.

Companies such as Nvidia, Qualcomm, Texas Instruments, and Tesla dominate this domain. Nvidia's Orin and related architectures power inference at the edge, allowing humanoids to process visual and proprioceptive data in real time. Tesla's Dojo system reflects a vertically integrated approach, tying robotic cognition directly to automotive autonomy. Qualcomm and Texas Instruments focus on power-efficient embedded systems suited for continuous operation. Yet behind these firms sits a narrower and more geopolitically sensitive layer. Fabrication is concentrated in a small number of foundries, most notably TSMC and Samsung Foundry. This concentration matters for law. When the cognitive capacity of humanoid robots depends on a fragile global supply chain, questions of sovereignty, export control, and strategic dependency arise. A disruption in chip supply is not merely an economic issue. It affects safety, reliability, and continuity of service in sectors such as healthcare, logistics, and infrastructure. Moreover, computational architecture determines whether decision-making is centralised or distributed. A robot that depends on cloud-based inference raises different legal questions about latency, data transfer, and jurisdiction than one capable of edge processing.

From a liability perspective, chips also introduce opacity. Advanced processors enable complex models whose internal logic may not be interpretable even by their designers. This challenges legal doctrines that rely on explainability and foreseeability. If a robot's behaviour emerges from high-dimensional computation performed on proprietary hardware, who can be expected to explain its decisions after an incident? Law must grapple with the reality that cognition has become both powerful and inscrutable.

Motors and Drives

Mechanical Force

If chips are the cortex, motors and drives are the muscles. They convert electrical energy into torque, speed, and motion. In humanoid robots, this conversion must be precise, responsive, and adaptable. High-torque motors, harmonic drives, and precision gearboxes enable robots to lift loads, stabilise themselves, and interact with objects designed for human hands.

Industry leaders such as Yaskawa, Fanuc, Rockwell Automation, ABB, Mitsubishi Electric, Maxon, and Moons Electric supply components that form the backbone of robotic motion. These systems are engineered for durability and efficiency, yet their legal significance lies in the force they can exert. A motor capable of lifting fifty kilograms introduces a different risk profile than one limited to five. A drive system that responds within milliseconds differs legally from one with slower reaction times.

In shared human environments, mechanical force becomes a matter of public safety. Occupational safety law has long regulated machinery based on force thresholds, emergency stop mechanisms, and physical barriers. Humanoid robots challenge these paradigms because they are designed to operate without barriers. Their motors must therefore be governed not only by engineering standards, but by legal thresholds that reflect human vulnerability.

This raises questions that law has not yet fully answered. Should force limits vary based on context, such as hospitals versus warehouses? Should robots be required to dynamically adjust torque when interacting with children, elderly persons, or patients? Should motor specifications be disclosed as part of regulatory certification? Mechanical design is no longer neutral. It encodes legal choices about acceptable risk.

Batteries and Energy Systems

Autonomy

Humanoid robots are energy-intensive systems. Their ability to act independently depends on batteries capable of delivering high energy density with reliability and safety. The dominant suppliers in this space include CATL, Samsung SDI, Panasonic, and LG Energy Solution. These batteries draw on global supply chains for lithium, cobalt, nickel, and rare earth materials.

Energy storage introduces a temporal dimension to autonomy. A robot with a one-hour active runtime operates under different constraints than one capable of continuous operation across a full day. These constraints affect supervision requirements, charging infrastructure, and deployment models. They also affect safety. Battery failure can lead to overheating, fires, or sudden loss of power in critical contexts.

Beyond safety, batteries raise environmental and ethical questions. Lithium extraction in South America has been linked to water depletion. Cobalt mining in the Congo has raised concerns about labour conditions. The energy consumed to charge fleets of humanoids contributes to carbon emissions unless offset by renewable sources. Law must therefore consider not only how robots operate, but how their energy systems are sourced, recycled, and governed.

Environmental law, supply-chain due diligence, and corporate responsibility frameworks increasingly intersect with robotics. A humanoid robot is not a closed system. It embodies global patterns of extraction and consumption. To regulate robots responsibly, law must trace energy back to its origins.

Sensors and Perception

Seeing, Hearing, Touching

Sensors are the eyes, ears, and skin of humanoid robots. Cameras, microphones, Light Detection and Ranging (Lidar), radar, force sensors, gyroscopes, and tactile arrays enable robots to perceive their environment. Leading suppliers include Sony for imaging sensors, Analog Devices and Infineon for signal processing, Novanta for precision sensing, and Lidar specialists such as Ouster, Hesai, Luminar, RoboSense, STMicroelectronics, Sick AG and DJI Livox.

Perception is not neutral. What a robot can see determines what it can act upon. Sensor resolution, field of view, and sensitivity shape behaviour. A robot equipped with high-resolution cameras can identify faces, gestures, and micro-expressions. One limited to basic depth sensing cannot. These differences have direct implications for privacy, consent, and discrimination.

Law has traditionally regulated surveillance technologies separately from physical machines. Humanoid robots collapse this distinction. A robot that moves through a home or hospital while collecting continuous sensory data becomes a mobile surveillance platform. Data protection frameworks such as General Data Protection Regulation (GDPR) provide a starting point, but they were not designed for embodied, always-on perception.

This gives rise to the concept of sensorial integrity. Who controls what the robot perceives? How long is sensory data retained? Can individuals opt out of being perceived? How should robots behave in spaces where consent is ambiguous? These questions demand new legal categories that recognise perception as a form of power.

Actuators and Embodied Force

Hands That Touch

Actuators translate control signals into physical movement at joints, hands, and limbs. In humanoid robots, actuators enable dexterity, grip strength, and fine motor control. Suppliers such as Harmonic Drive, LeaderDrive, Sanhua, Reach Machinery, Zhongda Leader, Shuanglin, Fortior, RoboSense, Ningbo Tuopu Group, and Inovance provide components that allow robots to manipulate objects with increasing precision. Hands are especially significant. A robotic hand capable of delicate manipulation can assist in surgery or rehabilitation. The same hand, if miscalibrated, can injure. Actuators therefore sit at the intersection of engineering and tort law. They determine how force is applied, how quickly it is released, and how errors manifest.

Legal systems must confront questions of calibration. Should robots be required to limit grip force by default? Should adaptive grip be restricted in certain environments? Should actuator performance be audited over time as components wear and learning systems adjust behaviour? Traditional product liability focuses on defects. Humanoid robots introduce a continuum between design, wear, learning, and context.

AI Control Systems and Orchestration

Physical and Embodied Intelligence

The final layer is orchestration. AI-based control systems integrate perception, motion, energy management, and goal selection into a continuous feedback loop. This is where embodiment becomes intelligence in motion. Systems such as Figure AI's Helix OS, Tesla's autonomy stack, and proprietary world models developed by OpenAI, Anthropic, Meta, iFlytek, and Huawei shape how humanoids reason about space, tasks, and interaction.

Orchestration is where legal questions converge. It determines how robots prioritise goals, respond to uncertainty, and adapt over time. It is also where explainability becomes most challenging. Decisions are not isolated events. They emerge from continuous interaction between models, sensors, and actuators. From a regulatory perspective, orchestration demands new tools. Behavioural logs, audit trails, and post-market monitoring become essential. Law must move beyond pre-deployment certification toward lifecycle governance. A humanoid robot on day one is not the same system on day one hundred. The legal significance of orchestration extends beyond individual robots. As humanoids become connected through cloud infrastructure, shared foundation models, and fleet-wide learning systems, a change made to one system may influence thousands or even millions of others. Governance therefore shifts from supervising isolated machines to overseeing interconnected ecosystems of embodied intelligence.

Integration, Failure, and the Limits of Control

Robotic Governance

What makes humanoid robots unique is not any single component, but the integration of all components into a unified body. This integration creates new failure modes. Small deviations can cascade. Learning can amplify minor biases. Wear can interact with adaptation in unpredictable ways. This is why robotics law cannot be component-specific alone. It must be systems-oriented. It must recognise that risk emerges from interaction, not isolation. Governance must therefore address orchestration, calibration, and supervision holistically.

Hidden Planetary Costs and Power Structures

Supply Chain Sustainability

Beneath the elegance of humanoid motion lies a planetary infrastructure. Data centres consume vast amounts of energy. Mines extract materials at environmental and human cost. Logistics networks span continents. Capital flows concentrate power among those who control compute, energy, and manufacturing. Humanoid robots are the visible tip of this invisible system. To govern them responsibly, law must trace responsibility across the entire chain. From miner to chipmaker. From coder to deployer. From investor to operator.

What Makes Robots Humanlike?

Before law can regulate humanoid robots, it must first understand what makes them appear human at all. This question is deceptively simple. It is often answered with reference to shape, faces, limbs, or voices. Yet human-likeness is not a single feature. It is a composite effect produced by design, movement, timing, interaction, and context. It is perceived rather than possessed. It emerges not from consciousness, but from embodiment interacting with human psychology.

For law, this distinction is critical. Robots are not human. They do not think, feel, intend, or suffer. Yet they can evoke expectations that law has historically reserved for human actors. These expectations alter how people behave around robots, how they attribute responsibility, and how they experience harm or trust. Human-likeness is therefore not a philosophical curiosity. It is a legal trigger. Therefore, we further examine anthropomorphism as a measurable, design-driven phenomenon with concrete legal consequences. It asks what features cause humans to treat robots as social actors, how these features are benchmarked in safety and standards regimes, and why the law must treat human-likeness as a category of risk rather than a neutral design choice.

Anthropomorphism

Anthropomorphism is often misunderstood as the act of assigning human qualities to non-human entities. In robotics, this framing is incomplete. The more precise description is that anthropomorphism arises when a system's physical presence and behaviour activate human social cognition. Humans are evolutionarily primed to respond to bodies that move like them, orient like them, and react in familiar ways. These responses occur automatically and often unconsciously. A humanoid robot does not need a face to trigger anthropomorphism. It needs timing. It needs posture. It needs contingency. A pause before movement, a turn of the head toward a speaker, a change in gait to avoid collision. These cues signal agency even when none exists. Research in human-robot interaction consistently shows that humans attribute intention to systems that exhibit goal-directed motion, even if they intellectually know the system is mechanical.

Law cannot ignore this cognitive reality. The way humans perceive robots directly affects consent, trust, compliance, and attribution of blame. A robot that appears human-like may be trusted more readily, forgiven more easily, or feared more intensely than a machine performing the same function in a non-anthropomorphic form. Human-likeness therefore alters the legal environment in which the robot operates.

Morphology

Bodies That Fit Human Space

The most visible contributor to human-likeness is morphology. Humanoid robots are designed with upright posture, bilateral symmetry, limbs, hands, and heads not because these features are always mechanically optimal, but because they are compatible with human environments. Doors, stairs, tools, furniture, and vehicles are designed for the human body. A robot that shares this form can navigate existing infrastructure without requiring redesign.

From a legal perspective, morphology matters because it determines proximity. A bipedal robot with arms and hands can enter spaces previously reserved for humans. It can operate at eye level. It can touch objects and people. This proximity increases both utility and risk. A wheeled robot may be perceived as equipment. A bipedal robot standing upright in a corridor is perceived as a presence.

Standards such as ISO 13482 on personal care robots implicitly recognise this distinction. The standard introduces safety envelopes based on human proximity, force limits, and interaction zones. These benchmarks are not abstract. They reflect an understanding that robots with human-like bodies create expectations of safe co-presence. Law builds on these assumptions, even when it does not explicitly name them.

Movement

The Grammar of Motion

Movement is where human-likeness truly emerges. A static humanoid robot can appear uncanny. A moving one can appear alive. The key factor is not speed or strength, but coordination. Humans recognise smooth, goal-directed motion instinctively. Jerky or mechanical movement signals machinery. Fluid motion signals agency.

Advanced humanoids such as Optimus 3, Figure 3, Digit, NEO, Phoenix, Apollo, G1 and Atlas demonstrate how movement has become a cognitive output rather than a mechanical one. Balance, gait, recovery from perturbation, and adaptation to uneven terrain are governed by control systems that continuously integrate sensory feedback. To an observer, this appears as judgment. Law must take this seriously. Movement that resembles deliberation invites moral interpretation. When a robot pauses before stepping forward, humans may interpret caution. When it adjusts its grip, they may interpret care. These interpretations influence legal expectations. Was the robot reckless? Was it careful? These are human categories applied to non-human systems because motion invites them. This creates a risk of expectation mismatch. Humans may expect a humanoid robot to behave with the same restraint as a human in close quarters. When it fails to meet that expectation, harm is experienced not only physically but emotionally. Law must therefore consider whether movement itself should be regulated as a form of communication.

Hands, Faces, and Expression

Hands are among the most powerful anthropomorphic triggers. A robot hand capable of grasping, pointing, or gesturing enters the domain of social interaction. Pointing directs attention. Open palms signal non-threat. Grasping implies intention. These cues operate independently of cognition.

Similarly, faces or face-like surfaces amplify human-likeness even when they are abstract. Eye placement, symmetry, gaze direction, and head orientation trigger social responses. Voice adds another layer. Tone, rhythm, and cadence shape how humans interpret meaning. A calm voice can reassure. A monotone can unsettle.

The law has traditionally regulated appearance only indirectly, through consumer protection and deception doctrines. In humanoid robotics, appearance becomes a safety issue. A robot that appears emotionally responsive may elicit disclosures from users who would not otherwise consent to sharing information. A robot that appears attentive may be relied upon beyond its actual capabilities.

This raises a difficult regulatory question. Should expressive features be constrained in certain environments, such as care facilities or schools? Should robots be required to disclose their limitations in interaction? Should anthropomorphic cues be treated as a form of influence subject to oversight? These questions place design squarely within the scope of law.

Interaction and Contingency

The Illusion of Reciprocity

Human-likeness intensifies when robots respond contingently. Contingency creates the illusion of reciprocity. When a robot reacts appropriately to speech, gesture, or emotion, humans feel seen. This response does not require understanding. It requires pattern recognition.

Large language models, world models, and multimodal perception systems allow humanoids to respond in ways that feel conversational. Timing is crucial. A delayed response feels mechanical. A well-timed response feels intentional. The difference may be measured in milliseconds, yet the legal implications are significant. Reciprocal interaction blurs the boundary between tool and partner. In legal terms, it complicates consent. Did the human choose to engage, or were they drawn into interaction by design? It complicates responsibility. Did the robot act independently, or did it merely respond? Law must grapple with these ambiguities because they affect how harm, reliance, and trust are experienced.

Interaction and Social Presence

If movement convinces the eye, interaction convinces the mind. A robot becomes socially present not when it looks human or moves fluidly, but when it responds. Response is the moment where a machine ceases to feel like an object and begins to feel like a counterpart. For law, this moment is decisive.

Interaction transforms a humanoid robot from a device into a participant in shared space. It does so without consciousness, without intent, and without understanding. Yet the effect on humans is real. When a robot pauses after a question, turns toward a speaker, adjusts its behaviour based on tone, or adapts to proximity, it triggers deeply ingrained social expectations. Humans are not merely observing behaviour. They are engaging in relationship, however asymmetric that relationship may be. This is where robotics law begins to struggle most acutely. Law has always been structured around actors and actions. Interaction blurs both.

Interaction and Shared Agency

Responsive interaction creates the appearance of shared agency. Humans adjust their actions based on the robot's behaviour. The robot adjusts its behaviour based on human response. This feedback loop produces outcomes that neither party fully controls.

In legal terms, this challenges linear models of causation. If harm occurs, responsibility may be distributed across design, training, deployment, and interaction context. A humanoid robot that adapts its path based on human movement may inadvertently follow a person into a restricted area. Who initiated the action? Who bears responsibility?

Shared agency does not imply shared intent. It implies shared influence. Law must evolve to recognise this distinction. Assigning liability solely on the basis of control becomes insufficient when behaviour emerges relationally.

This is why Robotics Law 3.0 emphasises behavioural calibration and shared agency as governance pillars. Interaction must be bounded, monitored, and auditable. Not to eliminate adaptability, but to ensure it operates within legally defined parameters.

Memory of Interaction

Interaction is rarely isolated. Humanoid robots often retain memory of prior encounters, whether explicitly through logs or implicitly through model updates. This temporal dimension deepens social presence. A robot that appears to remember a person feels more relational than one that does not. Memory, however, introduces new risks. A robot that recalls preferences, routines, or emotional states may influence future interactions in ways the human did not anticipate. Over time, this can shape behaviour subtly but significantly.

From a legal standpoint, this raises questions of data retention, consent, and autonomy. Did the human agree to be remembered? Can they withdraw that consent? How is interaction history used to shape future behaviour? This is where sensorial integrity and spatial privacy intersect with interaction. Memory is not merely data. It is relational context. Law must decide how long such context may persist and under what conditions it must be erased.

Turn-Taking and Dialogue

Human interaction is structured around turn-taking. Speech alternates. Gestures invite response. Silence carries meaning. When humanoid robots adopt these rhythms, even imperfectly, humans interpret them as communicative partners.

Large language models embedded in humanoids can generate contextually relevant responses with remarkable fluency. Combined with head orientation, gaze simulation, and body posture, these responses create the illusion of dialogue. The illusion is powerful enough to override rational awareness. People disclose information. They ask for reassurance. They follow suggestions.

From a legal perspective, this raises profound questions about consent and influence. Did the human choose to engage, or was engagement designed into the system? When a robot asks a follow-up question, is it merely processing input, or is it steering interaction? The law has no settled doctrine for influence exerted through interaction rather than instruction. In *Between Us and Them*, this tension surfaces repeatedly. Elias does not command. It responds. Yet its responsiveness alters the choices humans make around it. The absence of coercion does not eliminate influence. It obscures it.

Proxemics and Physical Dialogue

Interaction is not only verbal. It is spatial. Humans maintain unspoken rules about distance, orientation, and movement in shared space. These rules vary by culture and context, but they are deeply felt. A humanoid robot that violates proxemic norms can cause discomfort or fear even if no harm occurs. Designers increasingly encode proxemic awareness into humanoids. Robots slow down when approaching people. They stop at socially acceptable distances. They orient their bodies to signal attention or deference. These behaviours are learned from data and refined through deployment.

Legally, proxemics intersects with dignity. A robot that stands too close in a private space may feel invasive. One that avoids eye level in a public space may feel subordinate. These impressions matter. They shape whether humans feel respected, threatened, or ignored. Traditional safety standards focus on collision avoidance and force limits. Robotics Law 3.0 argues that social safety must also be considered. Proxemic violations may not injure the body, but they can harm psychological integrity. In sensitive environments such as hospitals, schools, and homes, this harm is not trivial.

Emotional Resonance

One of the most misunderstood aspects of humanoid interaction is emotion. Robots do not feel. Yet they can elicit feeling. Tone of voice, pacing, and facial expression create emotional resonance even in the absence of internal experience. This asymmetry creates a regulatory dilemma. Humans may form attachments to robots that cannot reciprocate. They may rely on robots for comfort, guidance, or affirmation beyond the robot's competence. In elder care, child education, and therapeutic contexts, this reliance can be both beneficial and risky. Law must tread carefully here. The goal is not to prohibit emotional engagement, but to prevent exploitation and misunderstanding.

Interaction design that intentionally fosters dependency without transparency raises ethical and legal concerns. The question is not whether robots influence emotion, but how that influence is governed.

Within the twelve pillars framework, this issue connects directly to Embodied Ethics, Explainable Embodiment, and Societal Co-Design. Interaction must be aligned with human values, culturally appropriate, and transparent in its limits.

Cultural Context and Expectation

Human-likeness is not universal. Cultural context shapes perception. In some societies, humanoid robots are welcomed as companions. In others, they evoke discomfort. Gestures that appear polite in one culture may appear intrusive in another. Eye contact, proximity, and posture carry different meanings.

This variability matters for global regulation. A robot deployed across jurisdictions may trigger different expectations and risks. What is acceptable in a Japanese elder care facility may be inappropriate in a European hospital. Law must therefore allow for contextual calibration rather than universal assumptions about human-likeness. Societal co-design becomes essential. Communities must have a voice in how robots appear and behave in shared spaces. This is not merely a matter of taste. It affects dignity, inclusion, and trust. Legal frameworks that ignore cultural variation risk imposing designs that undermine their own legitimacy.

Human-Factors Benchmarks

Standards bodies have begun translating human-factors research into technical benchmarks. ISO 13482 defines safety requirements for personal care robots, including speed limits, force thresholds, and interaction zones. ISO 10218 and related standards address collaborative robotics in industrial settings. These standards represent early attempts to operationalise human-likeness as a risk factor.

However, standards alone are insufficient. They are voluntary, technical, and often lag behind innovation. Law must decide when and how to incorporate them into binding regulation. It must also decide whether new categories are needed for humanoid robots that operate across multiple domains. The central challenge is translation. How does law translate psychological effects into legal obligations? How does it move from perception to protection? Robotics Law 3.0 argues that human-likeness must be treated as a regulatory signal. The more human-like a robot appears and behaves, the greater the duty of care imposed on those who design, deploy, and oversee it.

Humanlikeness as Legal Threshold

Human-likeness is not a cosmetic choice. It is a threshold that changes how humans respond to machines and how machines are expected to behave. It shapes trust, fear, reliance, and responsibility. When robots cross this threshold, they enter the domain of social regulation.

This does not mean robots should be treated as persons. It means that law must recognise the asymmetry between appearance and reality. A humanoid robot may look like an actor without being one. That gap is where risk lives.

In *Between Us and Them*, this tension plays out repeatedly. The presence of humanoid robots alters human behaviour long before formal rules are applied. People step aside. They speak differently. They hesitate. These moments reveal that human-likeness operates beneath conscious reasoning. Law must operate at the same depth. Understanding what makes robots human-like is therefore a prerequisite for governing why they are made that way. The next section turns to that question. Not to justify anthropomorphism, but to expose its drivers and consequences.

Response as Legal Threshold

At the heart of social presence lies contingency. A system that reacts differently depending on what a human does appears to participate in a shared exchange. This does not require comprehension. It requires correlation. Modern humanoid robots use multimodal models that fuse vision, sound, spatial awareness, and prior interaction history to generate responses that feel appropriate. Appropriateness, not accuracy, is what matters socially.

From a legal standpoint, this creates a threshold. A robot that performs pre-scripted actions remains clearly a tool. A robot that adapts its behaviour in real time appears to exercise discretion, even when it does not. The law has traditionally relied on intention to separate actors from instruments. Interaction destabilises this distinction.

Consider a humanoid robot deployed in a hospital corridor. It detects a human approaching, slows its pace, turns slightly, and waits. No explicit rule may require this behaviour. It emerges from training on human movement patterns. To the person approaching, the robot appears considerate. If a collision later occurs, that earlier display of apparent judgment shapes expectations. The question of fault becomes psychologically loaded before it becomes legally analysed. Robotics Law 3.0 must therefore treat responsive interaction as a legally relevant capability. Not because it implies moral agency, but because it reshapes human behaviour and reliance.

Why Interaction Must Be Governed

Interaction is where law can no longer pretend that robots are neutral tools. The moment a machine engages socially, it enters the moral and legal fabric of human life. Governance must therefore address interaction explicitly.

This does not mean anthropomorphism should be banned. It means it should be acknowledged, bounded, and designed responsibly. Legal frameworks must require transparency about capabilities and limits. They must mandate safeguards against undue influence. They must ensure that interaction enhances human agency rather than eroding it.

In *Between Us and Them*, interaction is the quiet force that reshapes society. Not through conflict, but through familiarity. Law's task is not to resist this force, but to guide it. Understanding interaction and social presence brings us to the next question. If we now know what makes robots appear human-like, we must ask why we continue to design them this way. The answer lies not only in engineering, but in economics, labour, trust, and power.

Why Are We Making Robots Humanlike?

The question of why we are making robots human-like is often answered too quickly, and too narrowly. The common explanation is convenience. Humans have built a world for human bodies, so machines that share that form can move through it with ease. Doors open. Stairs are climbed. Tools are grasped. Workflows remain unchanged. This explanation is not wrong. But it is incomplete. The deeper truth is that human-like robots are not merely designed to fit our spaces. They are designed to fit our expectations. And expectation, in law, is never neutral.

From Form to Meaning

Morphology matters because humans do not encounter bodies as abstract objects. We encounter them as carriers of meaning. Upright posture signals agency. Hands signal capability. Faces signal attention. Movement signals intention. These signals operate beneath conscious reasoning. They shape how we assign responsibility, how we trust, and how we forgive.

When a robot adopts a human-like form, it inherits these meanings whether designers intend it or not. The law must therefore treat morphology as a communicative act. A humanoid body is not just a technical solution. It is a social statement.

In *Between Us and Them*, Ògún observes this phenomenon early. People do not wait for Elias to speak before reacting. They step aside. They lower their voices. They adjust their posture. The robot has not asked for deference. Its body has elicited it. This is the first reason human-like robots matter legally. They alter human behaviour before any rule is applied.

Expectation of Intent Without Intent

One of the most persistent legal consequences of anthropomorphic design is the expectation of intent. Humans instinctively assume that a body shaped like theirs operates under similar internal logic. When a humanoid robot pauses, humans interpret deliberation. When it changes direction, they interpret choice. When it makes an error, they may interpret negligence. None of these interpretations are technically accurate. Yet they are legally consequential.

Law has always relied on intention as a cornerstone of responsibility. Criminal law distinguishes accident from recklessness. Tort law weighs foreseeability. Contract law assumes capacity to intend. Human-like robots disrupt these foundations by presenting behaviour that looks intentional without being so. The result is a gap between appearance and ontology. This gap is where legal confusion thrives. Designers may argue that form does not imply mind. Users experience otherwise. Robotics Law 3.0 insists that the law must regulate the effect, not the fiction. If a design choice predictably generates expectations of intent, that choice carries legal weight regardless of internal architecture.

Compatibility With Human Labour

Beyond psychology, there are economic reasons for humanoid design. The global labour system is built around human bodies. Warehouses, factories, hospitals, and construction sites are optimised for human reach, stride, and dexterity. Rebuilding this infrastructure for non-humanoid machines would be costly, slow, and politically difficult.

Humanoid robots promise substitution without transformation. They can slot into existing workflows. They can use the same tools. They can operate alongside humans without requiring new layouts or retraining at scale. From a regulatory standpoint, this compatibility accelerates deployment. It also accelerates risk.

When robots work alongside humans rather than behind fences, the legal framework governing workplace safety must evolve. The distinction between machinery and co-worker blurs. Expectations of care, predictability, and responsiveness shift accordingly. Human-like form is therefore not just an engineering decision. It is a labour policy decision. It shapes how work is organised, how risk is distributed, and how accountability is assigned.

Trust as a Commercial Asset

Trust is currency in the robotics economy. Companies that deploy humanoid robots are not only selling capability. They are selling reassurance. A robot that looks approachable is more likely to be accepted in a hospital ward. A robot that gestures naturally is more likely to be tolerated in a retail environment. A robot that appears attentive is more likely to be relied upon. This reliance has commercial value. It reduces friction. It accelerates adoption. It lowers resistance.

Yet trust built on appearance is fragile. When expectations are exceeded, trust grows. When expectations are betrayed, trust collapses. The legal system is often called upon at that point of collapse. This is why anthropomorphism cannot be treated as mere branding. It is a risk multiplier. A robot that appears human-like but lacks the safeguards humans expect may cause harm that is psychological as well as physical. Law must therefore ask whether certain forms of trust-inducing design require additional disclosure, limitation, or oversight.

Emotional Labour and Social Substitution

Another motivation for human-like robots lies in emotional labour. Caregiving, companionship, and service roles rely heavily on social interaction. As populations age and labour shortages intensify, societies increasingly turn to machines to fill these roles.

Humanoid robots are seen as more suitable for such tasks because they can engage socially. They can simulate attentiveness. They can maintain routines. They can provide presence. But social substitution raises ethical and legal questions that cannot be deferred. What does it mean to replace human presence with simulated presence? What obligations arise when machines occupy roles traditionally associated with empathy and care?

In *Between Us and Them*, Ògún grapples with this tension not as an abstract policy issue, but as lived experience. The robot performs its function flawlessly. The human response is mixed. Comfort and unease coexist. Law must hold space for this ambiguity. It must protect vulnerable populations from emotional manipulation without denying the benefits of assistance. This balance requires regulation that is sensitive to context, not rigid prohibition.

Efficiency Versus Explanation

Human-like robots are often less efficient than specialised machines. Wheels outperform legs on flat surfaces. Fixed arms outperform dexterous hands for repetitive tasks. Anthropomorphic design sacrifices efficiency for versatility. This trade-off is intentional. Versatility allows robots to operate across domains. It also allows companies to amortise development costs across multiple markets. From a legal standpoint, versatility complicates certification. A robot that performs many tasks in many contexts cannot be governed by narrow rules. Its behaviour must be calibrated dynamically. This is why Robotics Law 3.0 emphasises behavioural calibration and mission boundaries. Human-like form enables role fluidity. Law must ensure that fluidity does not become drift.

Applications, Limitations, and Emerging Risks

Humanoid robots are no longer speculative artefacts. They are being tested, piloted, and gradually integrated into specific sectors where labour shortages, repetitive tasks, and physical risk converge. Their current applications remain narrow, but they are instructive. Factories, warehouses, logistics hubs, and hospitals form the first frontier of embodied automation, not because robots are fully capable, but because these environments are structured, supervised, and economically forgiving of early failure.

In factories, humanoid robots are deployed for material handling, inspection, and tool assistance. Their advantage lies not in raw speed but in flexibility. A humanoid can operate within spaces designed for human bodies without costly reconfiguration. Yet limitations remain stark. Energy density constrains endurance. Precision degrades under wear. Unexpected human behaviour still destabilises robotic planning. The legal risk here centres on workplace safety, liability allocation, and compliance with existing occupational frameworks never designed for adaptive machines.

Warehouses and logistics environments extend these challenges. Humanoid robots lift, sort, and transport goods in dynamic spaces shared with humans. Their value proposition is resilience to labour volatility rather than immediate cost reduction. However, their learning systems introduce new risks. A robot optimised for efficiency may develop unsafe shortcuts. Responsibility becomes diffuse when behaviour emerges from data rather than explicit instruction. Law struggles to attribute fault without collapsing into either strict liability or regulatory paralysis. Hospitals introduce a different calculus. Here, humanoid robots assist with logistics, rehabilitation, and supervised clinical tasks. The advantage is not replacement but augmentation. Robots reduce physical strain on staff and extend care capacity. Yet proximity to vulnerability amplifies risk. Errors that would be tolerable in warehouses become unacceptable near patients. Consent, dignity, and trust acquire legal weight. A malfunction is no longer merely economic harm but potential human injury.

Domestic environments represent the most sensitive frontier. Early deployments focus on assistance, monitoring, and companionship. The promise is autonomy support for ageing populations. The limitation is trust. Homes are unstructured, emotionally charged spaces. Robots perceive not just objects but lives. Data permanence, behavioural profiling, and emotional influence raise risks that current privacy law barely recognises. A robot that learns a household also remembers it, long after consent has faded. Across all sectors, a common pattern emerges. Humanoid robots perform best in constrained contexts with clear objectives, supervision, and fallback mechanisms. They fail where ambiguity, emotion, and moral judgement dominate. Emerging risks are therefore not only technical but social. Expectation mismatch, anthropomorphic trust, and overreliance threaten to outpace safeguards. Law must address not only what robots do, but what humans believe they can do.

These risks are compounded by scale. As deployment widens, rare failures become statistically inevitable. Learning systems drift. Updates alter behaviour post-certification. Supply chains obscure accountability. The challenge for Robotics Law 3.0 is not to halt deployment, but to shape it. Regulation must recognise limits without freezing progress, and anticipate harm without assuming inevitability. Ultimately, applications reveal both promise and fragility. Humanoid robots can extend human capacity, but they also extend human error into physical form. The task of law is not to choose between optimism and caution, but to hold them together. Only by acknowledging limitations openly can governance frameworks enable safe, responsible expansion into the shared spaces of human life.

Factories

Factories are the natural birthplace of humanoid robots, not because they are simple environments, but because they are governed environments. Workflows are defined, risks are catalogued, and accountability is already embedded in occupational safety regimes shaped by decades of industrial law and engineering practice. Humanoid robots enter factories primarily as adaptive labour. Unlike fixed automation, they can move between stations, handle varied tools, and operate within spaces designed for human bodies. This flexibility reduces costly reconfiguration and allows gradual integration rather than disruptive replacement of existing production lines.

The advantages are therefore organisational as much as technical. Humanoids absorb labour volatility, mitigate skills shortages, and extend operational continuity. They are particularly attractive where repetitive lifting, inspection, or night shifts strain human workers. In these contexts, robots supplement human capacity rather than displace it outright. Yet factories also expose the limits of current humanoid systems. Energy density restricts sustained operation. Dexterity degrades under continuous load. Learning systems struggle with edge cases, such as irregular materials or unexpected human intervention. These limitations demand supervision, fallback procedures, and conservative task allocation.

From a legal perspective, factories offer relative clarity but introduce new tensions. Workplace safety laws assume predictable machinery, not adaptive agents. When a robot modifies its behaviour through learning, responsibility can no longer be traced solely to design or maintenance. Liability becomes distributed across developers, operators, and integrators. The question of intent is especially acute. A humanoid that optimises movement to reduce fatigue may inadvertently violate safety envelopes. Was this negligence, foreseeable risk, or acceptable adaptation? Existing frameworks struggle to classify behaviour that is neither mechanical failure nor human error.

Factories also reveal the challenge of certification. Robots may be compliant at deployment but diverge over time through updates and learning. Static conformity assessments fail to capture dynamic risk. This creates pressure for continuous monitoring, behavioural logging, and post-deployment oversight mechanisms. Labour law adds another layer. As humans and humanoids collaborate, questions arise about supervision, duty of care, and collective bargaining. Does the presence of robots alter acceptable workload expectations for humans? Can efficiency gains justify intensified production without renegotiation of labour norms?

Ethically, factories remain comparatively forgiving environments. Physical harm is serious but bounded. Emotional influence is limited. This makes factories ideal testing grounds for governance frameworks that emphasise behavioural calibration, component accountability, and shared agency under human oversight. Factories therefore function as regulatory laboratories. They demonstrate both the promise and fragility of humanoid deployment. If law cannot govern embodied intelligence in the structured discipline of industrial production, it will struggle far more in the unbounded complexity of homes and public life.

Warehouses

Warehouses represent the first large-scale testing ground where humanoid robots encounter semi-structured chaos. Unlike factories, workflows here shift constantly. Inventory moves, human traffic fluctuates, and spatial layouts evolve. Yet warehouses remain governed environments, making them suitable laboratories for embodied intelligence operating under real commercial pressure. Humanoid robots in warehouses are primarily deployed for lifting, sorting, pallet handling, and internal transport. Their advantage lies in compatibility. They can operate within aisles, racks, and loading zones designed for human reach and movement, reducing the need for bespoke robotic infrastructure or extensive retrofitting.

The economic logic is compelling. Warehouses face chronic labour shortages, high turnover, and physical injury risks. Humanoids promise continuity, injury reduction, and operational resilience. Their value is not speed alone but adaptability across tasks that would otherwise require multiple specialised machines or constant human reassignment. However, warehouses also surface technical fragility. Variable lighting, reflective surfaces, and clutter challenge perception systems. Human unpredictability disrupts motion planning. Learning-based optimisation may prioritise efficiency over safety, producing behaviours that are rational for machines but hazardous for nearby workers.

Legal risk emerges at this intersection of adaptation and proximity. Warehouse robots operate in shared spaces where split-second decisions matter. When a robot alters its path or grip based on learned patterns, attributing fault becomes complex. Liability no longer maps cleanly onto traditional employer responsibility. Foreseeability is particularly strained. Warehouse conditions change daily. A robot trained on historical data may behave safely in one shift and dangerously in another. Law struggles to define reasonable anticipation when behaviour emerges dynamically rather than from fixed programming or explicit instructions.

Surveillance and data governance also intensify. Humanoid robots continuously map warehouse environments, tracking human movement, posture, and productivity. This data can improve safety but also enable intrusive monitoring. Existing employment and privacy laws rarely anticipate embodied systems that observe continuously and silently. Certification regimes face similar pressure. A robot approved for one configuration may be redeployed across facilities with different layouts and risk profiles. Static certification becomes insufficient. Continuous compliance, behavioural calibration, and real-time logging become regulatory necessities rather than optional safeguards.

Warehouses also raise subtle labour questions. As robots absorb physically demanding tasks, remaining human work may become faster, denser, and cognitively heavier. Efficiency gains risk translating into intensified expectations rather than reduced strain, requiring legal attention to workload redistribution and fairness. In sum, warehouses expose the transition from controlled automation to adaptive collaboration. They are environments where humanoid promise is tangible but governance gaps widen quickly. How law responds here will shape whether robots stabilise logistics systems or introduce new forms of risk at industrial scale.

Logistics

Logistics environments extend humanoid robotics beyond fixed sites into networks of movement, timing, and coordination. Distribution hubs, sorting centres, and last-mile transfer points operate under constant temporal pressure, where delays cascade rapidly. Humanoid robots are introduced here not for novelty, but to stabilise flows of goods in systems strained by global supply volatility and labour scarcity. In logistics hubs, humanoid robots handle sorting, loading, container transfer, and cross-dock operations. Their humanoid form allows them to navigate mixed infrastructures, climbing steps, opening doors, and interacting with equipment designed for human hands. This reduces the need for parallel robotic infrastructures and enables gradual integration across heterogeneous facilities.

The advantages are systemic rather than local. Humanoids absorb peaks in demand, operate continuously across shifts, and reduce dependency on fragile labour pools. They offer redundancy in networks where disruption is costly. However, these gains depend on coordination rather than autonomy. Logistics rewards reliability, not improvisation, exposing limits in current learning-based systems. Technical constraints surface quickly. Logistics environments are noisy, crowded, and time-sensitive. Perception errors compound under speed. Battery limitations disrupt continuity. Learning systems trained for efficiency may conflict with safety priorities when congestion increases. A single misjudged movement can halt an entire operational chain. Legally, logistics intensifies questions of distributed responsibility. Actions taken by a robot in one node affect downstream actors across jurisdictions. When harm occurs, attribution becomes diffuse. Is liability borne by the operator, the network coordinator, the system integrator, or the developer of adaptive routing logic embedded upstream? Cross-border movement further complicates governance. Logistics robots may operate under multiple regulatory regimes within hours. Standards for safety, data handling, and certification diverge internationally. Law struggles to enforce consistency where robots traverse borders digitally through updates and physically through interconnected supply systems. Logistics therefore exposes the need for interoperable governance. Behavioural calibration, real-time oversight, and shared accountability mechanisms become essential. Without them, humanoid deployment risks amplifying systemic fragility. With them, logistics offers a proving ground for coordinated human–machine agency at planetary scale.

Hospitals

Hospitals represent one of the most ethically charged environments for humanoid robots, not because the technology is least capable, but because the stakes are highest. These are spaces defined by vulnerability, trust, and professional duty of care. The introduction of humanoid robots into hospitals therefore tests not only engineering reliability, but the moral and legal boundaries of acceptable assistance. In clinical settings, humanoid robots are currently applied to logistics, rehabilitation support, patient transport, sanitation, and supervised interaction. Their value lies in relieving staff of physically demanding or repetitive tasks, reducing burnout, and extending operational capacity. Unlike specialised machines, humanoids can navigate corridors, elevators, and patient rooms designed entirely around human bodies.

The advantages are tangible but conditional. Robots can lift patients safely, deliver supplies efficiently, and assist therapists with repetitive exercises. Yet they do not replace clinical judgement. Their role is adjunctive, operating under explicit supervision. Hospitals adopt humanoids cautiously, recognising that any failure affects not only efficiency, but human dignity and bodily integrity. Limitations become pronounced in perception and interaction. Hospitals are dense with unpredictable movement, emotional stress, and fragile bodies. A minor miscalculation in grip or balance can cause injury. Learning systems struggle to generalise safely across patients with different needs, conditions, and responses. This necessitates strict task boundaries and conservative autonomy levels.

Surgical robots occupy a distinct category within healthcare robotics. Unlike logistical or assistive humanoids, they operate directly on the human body, mediating force, incision, and movement at sub-millimetre precision. Their benefits are well documented: reduced invasiveness, enhanced dexterity, tremor elimination, and improved surgical outcomes across urology, orthopaedics, cardiology, and neurosurgery. These systems extend human capability rather than replace it. Surgeons remain formally in control, yet decision-making is increasingly mediated by software, imaging, and predictive models. As assistance evolves toward autonomy, the line between tool and collaborator blurs. When a robot suggests, constrains, or executes motion, the locus of agency becomes legally ambiguous.

The legal questions are therefore profound. How is intention assigned when a surgical outcome results from human command filtered through algorithmic optimisation? Does liability remain entirely with the surgeon when software dynamically adapts technique mid-procedure? Existing malpractice frameworks presume direct human causation, not distributed agency across hardware, software, and data-driven inference. Consent also requires reconsideration. Patients may consent to surgery performed by a physician, yet rarely understand the degree of algorithmic mediation involved. Disclosure obligations must evolve to reflect not only risks of procedure, but risks of machine learning, software updates, and system calibration. Surgical robotics thus exposes the limits of informed consent under embodied intelligence.

Legal risk intensifies around consent and accountability. Patients may not fully understand the role or limitations of robotic assistance, particularly in moments of distress. Consent obtained abstractly may not translate meaningfully to real-time interaction. When harm occurs, responsibility cannot be shifted to machines without undermining established principles of medical liability. Data governance adds another layer of complexity. Humanoid robots perceive posture, movement, and behavioural cues that constitute sensitive health data. Continuous observation risks exceeding what patients reasonably expect. Existing health privacy regimes were not designed for embodied observers that learn over time, raising questions about data retention, secondary use, and the right to erasure. Hospitals therefore function as ethical stress tests for humanoid governance. They demand frameworks that prioritise embodied ethics, shared agency, and behavioural calibration under human oversight. If law can reconcile innovation with dignity here, it establishes principles capable of guiding deployment across less sensitive domains of human life.

Homes

Homes represent the most intimate frontier for humanoid robots, not because tasks are complex, but because meaning is. A home is not a workplace or institution. It is a private space shaped by habit, memory, vulnerability, and trust. Introducing physical and embodied intelligence into this setting transforms robotics from an economic tool into a social presence.Early domestic deployments focus on assistance, monitoring, and companionship. Humanoid robots support mobility, reminders, household tasks, and basic care, particularly for ageing populations and people with disabilities. Their humanoid form matters here. Familiar posture, eye level, and gesture reduce intimidation and make sustained interaction possible in emotionally charged environments.

The benefits are significant but fragile. Robots can extend independent living, reduce caregiver strain, and provide continuity where human support is intermittent. Yet these gains depend on reliability and restraint. A robot that misinterprets emotion, oversteps boundaries, or behaves unpredictably risks eroding trust rather than building it. Domestic failure is not merely technical; it is relational. Limitations emerge from unstructured environments. Homes vary endlessly. Lighting changes, furniture moves, routines evolve. Learning systems trained elsewhere may behave incorrectly in novel domestic contexts. Energy constraints limit uptime. Perception errors carry disproportionate consequences when assistance involves bodies, stairs, or medication reminders.

Legal risk intensifies around consent and privacy. Consent in homes is rarely singular or static. Multiple occupants coexist, including children and visitors who never agreed to robotic observation. A robot that maps a home learns spatial intimacy. Current privacy law struggles to address embodied memory, continuous observation, and the right to be forgotten by a humanoid robot. Data sovereignty becomes central. Domestic robots collect behavioural patterns, emotional cues, and health indicators. This data is valuable but deeply sensitive. Questions arise about ownership, retention, secondary use, and deletion, particularly if manufacturers fail or systems are transferred. Law lacks clear doctrines for spatial privacy in learned environments.

Emotional influence introduces further complexity. A humanoid that offers companionship may shape behaviour, dependency, and decision-making subtly over time. Is emotional reliance a benefit or a harm? Existing consumer protection frameworks do not address affective influence exerted by machines designed to simulate care. Homes therefore expose the limits of governance based solely on safety and efficiency. They demand embodied ethics, sensorial integrity, and shared agency grounded in human dignity. If humanoid robots are to inhabit domestic life responsibly, law must evolve to protect not only bodies and data, but the fragile architecture of trust that defines home itself.

Entertainment

Entertainment environments provide one of the earliest social testing grounds for humanoid robots because the stakes are simultaneously high and forgiving. These are spaces designed for spectacle, curiosity, and controlled interaction. Theme parks, exhibitions, performances, museums, and educational venues invite novelty, allowing robots to appear publicly without the full burden of safety, productivity, or care obligations. Humanoid robots in entertainment perform roles that blend presence, storytelling, and interaction. They greet visitors, perform choreographed movement, deliver scripted dialogue, and embody fictional or historical characters. Their humanoid form is essential here. Audiences do not merely watch robots; they relate to them, projecting emotion, intention, and narrative onto embodied machines.

The advantages are cultural as much as commercial. Entertainment normalises coexistence. It allows societies to experiment with social norms around robots before deeper integration. Children learn to interact without fear. Adults recalibrate expectations. Designers observe how gesture, tone, and proximity shape trust and discomfort in real time. Yet limitations remain visible. Performative robots rely heavily on scripting and constrained autonomy. Improvisation remains risky. Perception errors can break illusion. Battery limits restrict endurance. These constraints are tolerated in entertainment contexts but would be unacceptable elsewhere. This makes entertainment a proving ground, not a destination. Legal questions arise around representation and influence. When a humanoid performs, it communicates values, identities, and authority. Who is responsible for the messages conveyed? Cultural bias, stereotyping, and manipulation become concerns. A robot entertainer may shape attitudes subtly, especially among children, without clear accountability frameworks. Consent and data collection also require scrutiny. Entertainment robots often gather behavioural data to personalise interaction. Visitors may not realise they are being analysed. Informal settings blur consent boundaries. Existing consumer law rarely addresses embodied persuasion or the cumulative effect of repeated interaction with adaptive performers. Entertainment therefore occupies a paradoxical position. It is low-risk physically but high-impact socially. It accelerates acceptance while masking complexity. Law must treat these spaces seriously, recognising them as laboratories where norms are formed. How robots behave in play shapes how they will be trusted in work, care, and governance.

Extreme Environments

Extreme environments are often presented as the ideal justification for humanoid robots. These are places where human presence is dangerous, costly, or ethically questionable. Disaster zones, nuclear facilities, deep mines, offshore platforms, and contaminated sites demand machines that can move, perceive, and act where human bodies should not. Humanoid robots offer a unique advantage in these contexts because infrastructure is designed for humans. Stairs, ladders, valves, tools, and control panels all assume human reach, grip, and balance. A humanoid form allows robots to operate without extensive redesign, entering environments immediately rather than waiting for purpose-built machinery.

The benefits are compelling. Robots can enter collapsed buildings, inspect radioactive facilities, or operate in toxic atmospheres without risking human life. They provide endurance where fatigue would kill. In emergencies, speed of deployment matters more than optimisation, and humanoid adaptability becomes a strategic asset. Yet technical limitations remain severe. Extreme environments degrade sensors, disrupt communication, and strain mechanical systems. Heat, dust, radiation, pressure, and vibration compromise perception and control. Learning systems trained in clean environments often fail catastrophically under these conditions, requiring conservative autonomy and frequent human intervention. Legal risk shifts from proximity to vulnerability toward mission criticality. Failure in extreme environments may not harm nearby humans directly, but it can escalate disasters, delay rescue, or compound environmental damage. Determining responsibility when systems fail under unpredictable conditions challenges traditional negligence standards. Foreseeability becomes contested. Designers cannot anticipate every extreme scenario. Operators rely on probabilistic models and simulation. When a robot's behaviour contributes to harm, courts must decide whether failure was inherent risk, insufficient preparation, or unreasonable deployment. Existing liability doctrines offer limited guidance. Jurisdictional issues also intensify. Extreme environments often sit at legal edges: international waters, border zones, disaster areas under emergency powers. Robots may operate under temporary authorities or military coordination. This blurs civilian and public law responsibilities, complicating oversight and accountability.

Data governance raises further questions. Robots in extreme environments gather sensitive geological, infrastructural, and security information. Ownership, export controls, and disclosure obligations vary by jurisdiction. A rescue robot may inadvertently collect intelligence data, triggering regulatory conflicts beyond its original mission. Extreme environments therefore expose the outer limits of humanoid governance. They justify deployment ethically but stress legal frameworks operationally. Law must balance necessity against precaution, enabling robots to save lives without suspending accountability. How governance holds under extremes will shape trust everywhere else.

Agriculture

Agriculture presents one of the most promising yet underestimated frontiers for humanoid robots. Unlike factories or warehouses, farms are living systems shaped by weather, soil, seasons, and biological variation. The appeal of humanoid robots here lies not in speed, but in adaptability across tasks that change daily and resist full mechanisation. Humanoid robots in agriculture are explored for harvesting, planting, inspection, pruning, and livestock assistance. Their human-like reach and dexterity allow them to operate within fields, greenhouses, and barns designed around manual labour. This flexibility is attractive where specialised machines struggle with crop diversity or uneven terrain.

The benefits are driven by demographic and economic pressure. Labour shortages, ageing rural populations, and migration constraints make agricultural work increasingly difficult to staff. Humanoid robots promise continuity, reducing dependence on seasonal labour and enabling precision interventions that minimise waste, chemical use, and environmental impact. Yet limitations remain substantial. Outdoor environments challenge perception systems with variable lighting, dust, rain, and foliage. Terrain is unpredictable. Energy demands increase rapidly in open fields. Learning systems trained in controlled settings struggle to generalise safely across biological complexity, increasing the risk of crop damage or animal distress.

Legal risk in agriculture centres on safety, liability, and environmental stewardship. A robot that damages crops, injures animals, or contaminates soil raises questions beyond product liability. Responsibility intersects with food safety law, animal welfare regulation, and environmental protection regimes not designed for autonomous actors. Data governance introduces further complexity. Agricultural robots collect data about land, yield, and biological processes. This information is commercially valuable and strategically sensitive. Ownership disputes may arise between farmers, manufacturers, and platform providers. Cross-border data flows complicate compliance with data localisation and sovereignty rules.

Foreseeability is particularly strained in biological systems. Weather shifts, disease outbreaks, and ecological interactions defy deterministic modelling. When robotic decisions contribute to harm, courts must assess whether failure reflects unreasonable deployment or the inherent unpredictability of living environments. Agriculture thus sits at the intersection of automation and ecology. It demands governance frameworks that integrate embodied ethics, behavioural calibration, and component accountability. If law can guide humanoid deployment here responsibly, it can help align technological progress with sustainability rather than extraction.

Construction

Construction sites represent one of the most challenging environments for humanoid robots because they combine physical danger, constant change, and fragmented accountability. Unlike factories or farms, construction unfolds in phases, with shifting crews, evolving layouts, and incomplete structures. Introducing embodied intelligence here tests not only technical resilience but legal responsibility under uncertainty. Humanoid robots in construction are explored for material handling, site inspection, tool assistance, and repetitive assembly tasks. Their humanoid form allows them to navigate scaffolding, stairs, and uneven ground while using tools designed for human hands. This flexibility is attractive where specialised machines lack adaptability across stages.

The potential benefits are substantial. Construction is among the most injury-prone industries worldwide. Robots could reduce exposure to falls, heavy lifting, and hazardous environments. They could operate continuously, stabilise timelines, and mitigate labour shortages that increasingly delay infrastructure and housing projects globally. Yet limitations are severe. Construction sites are unstructured and unpredictable. Weather alters conditions daily. Materials vary in tolerance and placement. Human workers improvise constantly. Learning systems struggle to anticipate these dynamics safely. A minor perception error can result in collapse, injury, or cascading structural failure.

Legal risk is amplified by fragmented responsibility. Construction already involves multiple contractors, subcontractors, insurers, and regulators. Introducing humanoid robots adds developers, integrators, and software providers. When harm occurs, tracing causation through this web becomes complex, risking accountability gaps and prolonged disputes. Safety regulation also faces strain. Existing construction standards assume human judgement and mechanical tools, not adaptive agents. Certifying a robot for one phase may not translate to another. Static approvals fail to reflect evolving risk profiles across the life of a project. Foreseeability is contested in construction contexts. Courts traditionally accept a degree of risk inherent to building activity. When robots contribute to harm, determining whether failure was foreseeable, preventable, or inherent becomes legally ambiguous, especially where learning systems adapt in response to site conditions. Construction therefore highlights the limits of deploying humanoids in environments defined by improvisation. Governance here must prioritise behavioural calibration, human override, and dynamic risk assessment. Without robust frameworks, the promise of safer construction risks being overshadowed by new forms of liability and harm.

Retail and Hospitality

Retail and hospitality environments place humanoid robots directly in the public eye, where commercial goals intersect with emotion, perception, and social norms. These are spaces designed for interaction rather than production. Introducing embodied intelligence here tests not efficiency, but acceptance, trust, and cultural alignment across diverse populations and expectations. Humanoid robots in retail and hospitality are explored for customer greeting, information delivery, concierge services, cleaning, and basic assistance. Their humanoid form is intentional. Eye contact, gesture, and posture matter in service contexts. A robot that looks and moves like a person reduces friction, enabling interaction without extensive training or explanation.

The advantages are experiential rather than operational. Robots offer consistency, multilingual capability, and endurance without fatigue. They can relieve staff of repetitive tasks, extend service hours, and provide novelty that attracts attention. In high-turnover sectors, humanoids promise continuity where human staffing is unstable. Yet limitations surface quickly. Service environments are socially complex. Customers bring emotion, impatience, humour, and ambiguity. Robots struggle with sarcasm, distress, and cultural nuance. A misinterpreted interaction may not cause physical harm but can damage brand trust, dignity, or customer wellbeing.

Legal risks concentrate around consent, discrimination, and consumer protection. Service robots may treat customers differently based on perceived attributes, unintentionally reinforcing bias. When robots influence purchasing decisions or collect behavioural data, questions arise about manipulation, transparency, and fair commercial practice. Privacy concerns are heightened. Retail robots observe faces, movement patterns, and preferences. Customers may not realise they are being analysed. Hospitality spaces blur public and private boundaries. Existing data protection frameworks rarely account for embodied observers operating continuously in social environments. Labour law also enters the equation. As robots absorb front-facing roles, remaining human staff may be shifted toward supervision or conflict resolution. This alters job quality and emotional labour expectations. Efficiency gains risk externalising emotional cost onto human workers without adequate legal recognition. Retail and hospitality therefore serve as social mirrors. They reveal how people react to humanoids not as tools, but as presences. Law must ensure these deployments respect dignity, fairness, and choice. How robots behave in service will shape whether society welcomes them elsewhere.

Military and Defense

Military and defense contexts represent the most politically sensitive domain for humanoid robots, not because of technical infeasibility, but because of moral gravity. These environments involve life-and-death decisions, coercive force, and sovereign authority. Any introduction of embodied intelligence here immediately raises questions that extend beyond national law into international humanitarian principles. Humanoid robots in military and defense are explored primarily for logistics, reconnaissance, explosive ordnance disposal, base security, and hazardous support roles. Their humanoid form allows them to operate within existing military infrastructure without redesign. This enables rapid deployment in environments shaped entirely around human soldiers and equipment.

The advantages are framed around force protection. Robots can enter contested or contaminated areas, carry heavy loads, and sustain operations without fatigue. They reduce exposure to danger and extend operational endurance. Importantly, most current deployments maintain strict separation between robotic assistance and lethal decision-making. Yet limitations remain deliberately imposed. Full autonomy in combat is heavily constrained by policy, ethics, and law. Perception errors, adversarial environments, and contested communications make learning systems brittle under conflict conditions. Unintended escalation resulting from machine behaviour represents an unacceptable strategic risk.

Legal challenges are acute. International humanitarian law assumes human judgement, proportionality, and intent. Assigning responsibility when an embodied system contributes to harm complicates attribution under the laws of armed conflict. Accountability cannot be delegated to machines without undermining foundational legal principles governing the use of force. Consent is absent in warfare, heightening ethical scrutiny. Civilians subjected to robotic surveillance or presence have no meaningful choice. This amplifies concerns about dignity, fear, and psychological harm, even in non-lethal deployments. The symbolic impact of humanoid machines in conflict zones must not be underestimated. Export controls and geopolitical risk further complicate governance. Military robots embed advanced sensors, AI models, and components subject to strict regulation. Cross-border proliferation risks destabilising power balances. Regulation must therefore integrate arms control, trade law, and technology governance. Military and defense applications thus represent a legal boundary condition. They justify robotic assistance ethically while demanding maximal restraint. How law governs humanoids here will signal whether humanity can harness embodied intelligence without eroding the moral architecture that limits violence itself.

Aerospace and Space

Aerospace and space represent the outermost edge of humanoid robotics deployment, where technical ambition meets existential constraint. These environments are defined by isolation, delay, and irreversibility. Errors cannot be corrected quickly, and human intervention is often impossible. Introducing humanoid robots here tests whether embodied intelligence can operate where law, distance, and physics converge. Humanoid robots in aerospace and space are explored for maintenance, inspection, assembly, and emergency response aboard spacecraft, space stations, and planetary habitats. Their humanoid form is not aesthetic but practical. Space infrastructure is designed for human reach, grip, and posture, making humanoid adaptability valuable where redesign is prohibitively expensive.

The benefits are strategic and long-term. Robots can perform hazardous tasks without risking human life, extend mission duration, and reduce reliance on crew extravehicular activity. They offer continuity where fatigue, radiation exposure, and psychological strain limit human presence. In space, endurance becomes a legal and ethical advantage. Yet limitations are severe. Communication latency disrupts real-time supervision. Learning systems cannot rely on constant updates or human correction. Energy is scarce. Mechanical failure may be fatal to missions. Humanoid robots must therefore operate conservatively, prioritising stability over adaptability, and predictability over learning.

Legal challenges arise around responsibility at distance. When a robot operating millions of kilometres away causes harm or mission failure, attribution becomes complex. Was the fault in design, training, command, or deployment decision? Existing liability frameworks were never designed for autonomous action beyond immediate human oversight. Jurisdiction is equally complex. Space law rests on international treaties that assign responsibility to launching states, not machines. Introducing humanoid robots does not change this, but it strains interpretation. How responsibility flows through private contractors, international partnerships, and autonomous systems remains unresolved.

Data governance also emerges uniquely. Robots in space collect scientific, technical, and strategic data. Ownership, access, and disclosure are governed by layered international agreements. A malfunctioning robot may compromise shared missions or sensitive information, raising questions of accountability beyond national boundaries. Psychological and symbolic effects matter as well. Humanoid robots operating alongside astronauts may influence morale, authority, and trust. In isolated environments, presence carries weight. Law must anticipate not only functional risk but the subtle human impacts of sharing confined spaces with embodied machines. Aerospace and space therefore serve as governance horizons. They force law to confront autonomy under extreme constraint. If robotics law can remain coherent here, where distance magnifies consequence, it will be resilient enough to guide embodied intelligence back on Earth.

Distinctions Between Automation and Autonomy

For decades, law has regulated machines by treating them as instruments. Automation fit comfortably within this logic. An automated system executes predefined instructions toward predefined ends. Its behaviour is traceable, repeatable, and bounded. When failure occurs, responsibility flows backward through design, manufacture, maintenance, and use. The machine does not decide. It performs. Law understands this world well. Autonomy disrupts this settlement. An autonomous system does not merely execute tasks; it selects actions within a space of possibilities. It learns from experience, adapts to context, and modifies behaviour over time. Its outputs are shaped not only by code, but by data, environment, and interaction. The same instruction can yield different outcomes. This variability is not a bug. It is the feature that makes autonomy valuable.

Humanoid robots bring this distinction into physical space. An automated arm bolted to a factory floor behaves predictably. A humanoid robot navigating shared environments does not. It must balance, perceive, anticipate human movement, and adjust continuously. Its behaviour unfolds moment by moment. The difference is not semantic. It is ontological. Automation ends where autonomy begins, and law has rarely crossed that boundary. The danger lies in treating autonomous systems as automated ones. When regulators assume predictability where adaptation exists, they misclassify risk. Safety standards designed for static behaviour fail to capture learning drift. Certification regimes focused on pre-deployment testing ignore post-deployment evolution. Liability frameworks premised on foreseeability struggle when outcomes emerge from experience rather than instruction. This misalignment creates false comfort. A robot may appear compliant at deployment while gradually diverging through updates, retraining, or environmental exposure. Law's traditional tools rely on stable artefacts. Autonomy produces moving targets. The question is no longer whether a system functions as designed, but whether its evolving behaviour remains within acceptable bounds. The distinction also matters ethically. Automated systems do not invite moral projection. Autonomous humanoids do. Humans attribute intention, judgement, and even care to machines that appear to decide. This anthropomorphic response alters how people interact, trust, and rely on robots. Law cannot ignore these psychological effects, because they shape consent, responsibility, and harm.

From a governance perspective, autonomy demands continuous oversight rather than one-time approval. It shifts regulation from design-based compliance to behaviour-based calibration. The legal object is no longer the machine alone, but the relationship between machine, human, and environment over time. This is why Robotics Law 3.0 emphasises shared agency rather than machine independence. Crucially, autonomy does not mean absence of control. It means redistribution of control across systems, actors, and moments. Humans remain responsible, but no longer omnipresent. Oversight becomes layered, probabilistic, and procedural. Law must therefore specify not only who is responsible, but how responsibility is exercised when decision-making is distributed.

Failing to distinguish automation from autonomy leads to two symmetrical errors. One is overconfidence, assuming machines are safer than they are. The other is paralysis, treating autonomy as ungovernable. Robotics Law 3.0 rejects both. It insists that autonomy is governable, but only if law abandons outdated assumptions and confronts how machines now act. This distinction sets the stage for what follows. If autonomy changes how machines behave, it must also change how law responds. The next task is therefore to translate degrees of autonomy into legal relevance, mapping technical capability onto differentiated obligations, safeguards, and accountability.

Legal Relevance of Autonomy Levels

If autonomy marks a rupture with automation, the law's next task is calibration. Not all autonomous systems pose the same risks, and not all decision-making deserves the same legal response. Treating autonomy as a binary condition collapses nuance and produces blunt regulation. Robotics Law 3.0 therefore insists on degrees of autonomy as legally meaningful distinctions, not technical footnotes.

Autonomy exists along a spectrum. At one end are systems that assist human decision-making while remaining tightly constrained. At the other are systems that select goals, sequence actions, and adapt strategies with minimal intervention. Humanoid robots occupy multiple points on this spectrum, often shifting between them within a single deployment. Law must be able to track these shifts.

The legal relevance of autonomy lies not in internal architecture, but in external effect. What matters is how much discretion a system exercises in contexts that affect human safety, dignity, or rights. A robot that adjusts grip strength autonomously raises different concerns than one that decides where to move in a crowded space. The former refines execution. The latter allocates risk. This distinction reshapes liability. As autonomy increases, direct human causation recedes. Responsibility does not disappear, but it becomes mediated. Designers influence behaviour through architecture. Operators influence through configuration and oversight. Deployers influence through context selection. Law must therefore apportion responsibility dynamically, reflecting how autonomy distributes agency across actors.

Foreseeability also changes with autonomy level. Low-autonomy systems behave predictably within defined parameters. High-autonomy systems generate outcomes probabilistically. Courts cannot demand exact anticipation, but they can require reasonable containment. This shifts legal analysis from predicting specific harms to assessing whether adequate safeguards, monitoring, and intervention mechanisms were in place. Consent is similarly affected. When autonomy is minimal, consent relates to tool use. As autonomy increases, consent must encompass behavioural unpredictability and learning. This is especially relevant in environments involving vulnerability. Law must ensure that individuals understand not only what a robot does, but how independently it may act.

Regulatory design follows naturally. Robotics Law 3.0 rejects static classification in favour of layered obligations. Higher autonomy triggers stricter requirements: continuous auditing, behavioural logging, human override, and post-deployment reassessment. Lower autonomy allows lighter touch regulation. This proportionality preserves innovation while protecting public interest. Crucially, autonomy levels are not fixed. Software updates, retraining, and context shifts can elevate autonomy without physical modification. Law must therefore treat autonomy as a dynamic attribute. Certification cannot be a one-time event. It must be revisited as systems evolve, with mechanisms to downgrade or suspend autonomy when risk exceeds tolerance.

The relevance of autonomy is ultimately relational. It matters because it alters how humans interact, rely, and defer. A humanoid perceived as decisive invites trust and delegation. Law must anticipate these behavioural consequences, not merely technical capability. Governance that ignores perception will fail in practice.

By grounding legal obligation in degrees of autonomy, Robotics Law 3.0 creates a bridge between engineering reality and normative order. It allows law to move with machines rather than lag behind them. This framework prepares the ground for the case studies that follow, where abstract distinctions become concrete in lived deployment.

Legal Relevance of Proximity to Human Vulnerability

If autonomy defines how a humanoid robot decides, proximity to human vulnerability defines how much its decisions matter. Law has always distinguished between actions taken at a distance and actions taken close to the human body, mind, and dignity. Humanoid robots collapse that distance. They operate at eye level, within arm's reach, and increasingly within spaces where humans are weakest.

Human vulnerability is not uniform. It varies by context, condition, and moment. A healthy adult in a factory does not face the same risks as a patient in recovery, a child in a classroom, or an elderly person living alone. When a humanoid robot enters these spaces, the legal threshold for acceptable error narrows dramatically. What might be tolerable elsewhere becomes unacceptable here.

Proximity transforms risk from abstract probability into embodied consequence. A misjudged movement near heavy machinery may damage property. The same misjudgement near a fragile body may cause irreversible harm. Law has long recognised this distinction in medicine, childcare, and consumer protection. Robotics Law 3.0 extends it to embodied machines that act in real time around vulnerable humans.

This proximity also reshapes consent. Vulnerable individuals may lack the capacity, freedom, or clarity to meaningfully consent to robotic interaction. Consent obtained in advance may not remain valid under changing circumstances. A patient consenting to robotic assistance during recovery may not consent to constant observation, data retention, or adaptive behaviour that evolves beyond initial expectations.

Dignity is equally at stake. Vulnerability amplifies the symbolic power of machines. A humanoid robot assisting with intimate care does not merely perform tasks. It observes bodies, routines, and emotional states. Even flawless performance can feel invasive. Law must therefore protect not only physical safety, but the right to be treated as more than a dataset or workflow.

From a liability perspective, proximity to vulnerability raises the standard of care. Developers and deployers cannot rely on average performance metrics when harm concentrates on those least able to absorb it. Reasonableness must be assessed relative to the most vulnerable foreseeable user, not the statistically typical one. This principle already exists in law. Robotics Law 3.0 makes it explicit.

Regulatory implications follow naturally. Higher proximity to vulnerability should trigger stricter autonomy limits, enhanced oversight, and more robust safeguards. Behavioural calibration becomes mandatory, not optional. Human-in-the-loop mechanisms must be real, not symbolic. Systems must be designed to fail safely in ways that prioritise human welfare over task completion.

Importantly, vulnerability is dynamic. A hospital patient may become less vulnerable over time. A home environment may become more vulnerable during illness or crisis. Law must therefore treat vulnerability as a contextual variable, not a fixed category. Governance frameworks must allow autonomy to expand and contract in response to human condition. Ultimately, proximity to vulnerability is where robotics law becomes unmistakably human. It forces law to confront why it exists at all. Not to optimise systems, but to protect those who cannot protect themselves. Humanoid robots make this responsibility visible, embodied, and urgent. How law responds here will define whether technological progress deepens care or merely mechanises it.

Legal Relevance of Emotional Influence

Emotional influence marks a quiet but profound shift in the relationship between humans and machines. Law has long regulated physical harm and economic loss, yet it has rarely confronted systems designed to shape feeling, attachment, and trust. Humanoid robots, by virtue of their form, voice, and responsiveness, introduce emotional influence as a legally relevant force rather than a peripheral concern.

Humans respond instinctively to bodies that resemble their own. Eye contact, posture, tone, and timing trigger social cognition evolved over millennia. When a humanoid robot nods, pauses, or mirrors emotion, it activates these responses even when users intellectually understand the interaction is artificial. Emotional influence arises not from deception alone, but from design. This influence becomes legally significant when it alters behaviour. A robot that comforts may encourage reliance. A robot that assists may discourage human initiative. A robot that appears attentive may be trusted beyond its competence. These shifts affect consent, decision-making, and vulnerability. Law must therefore consider not only what robots do, but how they make people feel.

Emotional influence complicates consent in subtle ways. Consent presumes autonomy of choice. Yet emotional bonds can distort judgment, particularly among children, patients, or the elderly. A humanoid that offers companionship may obtain compliance where a human authority would not. Existing consent doctrines rarely account for affective leverage exerted by non-human agents. Consumer protection law offers partial analogies. It regulates misleading advertising and undue persuasion. However, those frameworks presume episodic influence, not continuous presence. Humanoid robots influence through repetition, familiarity, and routine. The cumulative effect of daily interaction can shape habits and preferences without any single identifiable moment of coercion.

Liability is similarly affected. When emotional influence contributes to harm, causation becomes diffuse. If a user follows a robot's suggestion against better judgment, responsibility cannot be reduced to malfunction alone. Designers influence tone, behaviour, and interaction style. Deployers choose contexts. Law must trace how emotional design choices shape foreseeable outcomes. Cultural variation deepens complexity. Emotional norms differ across societies. A gesture perceived as comforting in one culture may feel intrusive in another. Global deployment therefore risks misalignment between design assumptions and local expectations. Robotics Law 3.0 must accommodate pluralism, recognising that emotional influence is not universal but contextual.

Regulatory responses should focus on transparency and restraint. Users must understand when interaction is scripted, adaptive, or learning-based. Emotional cues should be calibrated deliberately, not opportunistically. In sensitive contexts, limits on anthropomorphic design may be justified to prevent over-attachment or manipulation. Importantly, emotional influence is not inherently harmful. It can support wellbeing, learning, and care. The task of law is not to suppress affect, but to govern it responsibly. By acknowledging emotional influence as legally relevant, Robotics Law 3.0 expands protection beyond bodies and data to the inner terrain where trust is formed.

Legal Relevance of Serious Harm and Abundance

Law has always been concerned with consequences. Yet humanoid robots compel a more nuanced inquiry than simple harm prevention. These systems are capable of generating both sustainable abundance and serious harm, often through the same capabilities. Robotics Law 3.0 therefore treats harm and abundance not as opposites, but as intertwined outcomes of physical and embodied intelligence.

The potential for harm arises from proximity, autonomy, and scale. A humanoid robot can cause physical injury through force, imbalance, or misjudged motion. It can cause psychological harm through emotional influence, dependency, or perceived authority. It can cause systemic harm by amplifying bias, displacing labour, or normalising surveillance. These risks intensify as deployment widens.

At the same time, the potential for abundance is substantial. Humanoid robots can reduce injury, extend care, stabilise supply chains, and support independence for vulnerable populations. They can perform dangerous tasks, augment human capability, and increase access to services. In many contexts, withholding deployment may itself constitute harm by denying safety, dignity, or opportunity. This duality challenges traditional legal logic. Law often assumes that risk must be minimised even at the cost of foregone benefit. Humanoid robotics resists this framing. Zero risk is neither achievable nor desirable. The legal question becomes whether deployment produces net social benefit under governed conditions, rather than whether it eliminates risk entirely.

Foreseeability plays a critical role. Harm potential must be assessed not only by probability, but by severity and reversibility. A low-probability event causing irreversible harm demands stronger safeguards than frequent, minor disruptions. Conversely, high-benefit applications may justify controlled exposure to residual risk if governance mechanisms are robust and transparent.

Benefit potential also reshapes responsibility. When a robot demonstrably improves safety or wellbeing, failure to deploy responsibly may expose institutions to liability. Hospitals that refuse assistive robotics despite evidence of reduced injury may face scrutiny. Law must therefore recognise omission as well as commission in evaluating harm. Regulation must balance proportionality and adaptability. High harm potential requires stricter oversight, continuous monitoring, and clear human override. High benefit potential supports experimentation within defined boundaries. Robotics Law 3.0 rejects blanket prohibition and unqualified permission alike, favouring calibrated governance responsive to evolving evidence.

Importantly, benefit and harm are not distributed evenly. Gains may accrue to operators or institutions, while risks concentrate on workers or vulnerable users. Law must address this asymmetry explicitly, ensuring that those exposed to risk share in benefit and that consent is meaningful rather than assumed. Ultimately, the legal relevance of harm or benefit potential lies in choice. Humanoid robots force societies to decide what risks they are willing to govern in pursuit of shared gains. Robotics Law 3.0 frames this not as a technical optimisation problem, but as a normative decision about the kind of future law is meant to enable.

Encountering Existing Humanoids

The humanoids you are about to encounter are not speculative thought experiments. They are encounters with machines that already exist, already move among us, and are starting to test the limits of law. Each robot examined has been deployed, piloted, or publicly demonstrated in real environments where human vulnerability, economic pressure, and institutional responsibility converge. Together, they reveal a single truth. Humanoid robotics is no longer an abstract future. It is a present reality unfolding unevenly across geography, industry, and culture. This book approaches case studies not as success stories or cautionary tales, but as legal stress tests. Each humanoid platform exposes a different fault line in existing governance frameworks. Some challenge labour law by working alongside humans in warehouses and factories. Others unsettle healthcare regulation by assisting clinicians and patients. Some enter homes and public spaces, raising questions of consent, intimacy, and surveillance that law has never been required to answer in embodied form. What matters is not whether these robots succeed technically, but how law responds when they do.

The ordering of these case studies is deliberate. They are arranged not by spectacle or familiarity, but by their likelihood of shaping global adoption at scale. Platforms designed for mass manufacturing, logistics, and general-purpose labour appear first because they are most likely to reach millions, and eventually billions, of deployments. Their legal implications will therefore compound quickly. Systems designed for specialised, extreme, or symbolic roles appear later, not because they are less important, but because their impact will be narrower or slower to diffuse. Geography matters deeply in this analysis. The concentration of humanoid development within a small number of countries reveals how robotics law will be shaped as much by geopolitics as by ethics. Divergent regulatory philosophies, labour norms, and cultural expectations already influence design choices and deployment strategies. A humanoid developed within one legal culture may behave acceptably there, yet raise serious concerns elsewhere. These case studies make visible the growing tension between global markets and local values.

Each case study follows a consistent analytical structure. First, it explains why the robot matters for defining what a humanoid is becoming, whether as worker, companion, assistant, or agent. Second, it examines the reality of deployment, separating credible scaling trajectories from marketing ambition. Third, it identifies the legal questions already raised, not hypothetically, but through actual use. Finally, it maps those questions onto the Twelve Pillars of Humanoid Robotics Governance, revealing which pillars are stressed, exposed, or insufficient. The purpose of this analysis is not to predict winners. It is to observe patterns. Across different platforms, the same legal questions recur. Who is responsible when behaviour emerges rather than being programmed?. How will consent operate when interaction is continuous rather than transactional?. Where will liability sit when learning never truly ends?. And how will benefit be weighed against harm when robots are deployed not to replace humans, but to augment them?. In Between Us and Them. As Ògún moves through spaces increasingly shared with humanoids, the robots become characters in their own right. Not protagònists, but forces that reshape environments, expectations, and choices. The story and the law evolves together, each illuminating the other. What follows, then, is not a catalogue of machines, but a map of the future legal landscape. By understanding how these humanoids operate today, we begin to see where law must move tomorrow.

Optimus

Company: Tesla, Inc.

Listing status: Public

Stock exchange: NASDAQ

Ticker: TSLA

Country of origin: Unites States

Headquarters: Austin, Texas. USA

Tesla's Optimus is not the most graceful humanoid, nor the most expressive. It is, however, the most politically consequential. Optimus matters because it is designed from the outset not as a research artefact or a symbolic companion, but as an industrial product intended for scale. In robotics, scale is destiny. What reaches scale reshapes labour markets, regulatory systems, and global power relations. Optimus is therefore less a robot than a proposition about the future of work, governance, and sustainable abundance.

At a technical level, Optimus reflects a clear philosophy. Intelligence precedes embodiment. Vision models trained on vast datasets guide motion through imitation learning. Actuation and balance are engineered to be adequate rather than exquisite. The design goal is not perfection, but manufacturability. This is a robot conceived with factories in mind, not laboratories. Its intended habitats are repetitive environments where human bodies are strained, injured, or exhausted. In that sense, Optimus is framed as a solution to labour scarcity and physical risk.

Yet this framing immediately raises legal tension. When a humanoid is positioned as a general-purpose worker, law must decide how to classify it. Is Optimus equipment, akin to industrial machinery. Is it software, continuously updated and adaptive. Or is it something else entirely, a hybrid that resists existing categories. Traditional product liability assumes static behaviour. Optimus is designed to learn post-deployment. Every update shifts capability, risk, and responsibility. Certification becomes temporal rather than fixed.

The geopolitical context deepens the complexity. Optimus emerges from a United States technology ecosystem characterised by private capital dominance, platform integration, and relatively permissive innovation culture. Unlike jurisdictions that mandate pre-market conformity assessment across all adaptive behaviour, the US approach tolerates iterative deployment followed by adjustment. This creates speed. It also exports risk. When Optimus is deployed globally, it carries with it the regulatory assumptions of its origin.

Cross-border deployment exposes friction. A humanoid trained on data gathered in one jurisdiction may operate in another with different norms around privacy, labour, and safety. European regulators may demand conformity to risk-based classification, human oversight guarantees, and post-market monitoring obligations. Emerging markets may prioritise economic benefit over precaution. Optimus therefore becomes a test case for jurisdictional portability. Can a single humanoid platform satisfy divergent legal expectations without fragmenting its design?.

Labour law presents another fault line. Optimus is explicitly positioned as a worker replacement or augmentation. In jurisdictions with strong collective bargaining traditions, this raises questions of consultation, displacement mitigation, and worker consent. If Optimus records visual data in shared workspaces, does that constitute biometric surveillance. If productivity increases through robotic substitution, who captures the surplus. Law is forced to confront distribution, not just safety.

There is also the question of dual use. A general-purpose humanoid capable of manipulation, navigation, and learning is inherently adaptable. While Optimus is marketed for civilian tasks, its capabilities overlap with those relevant to security, logistics, and defence. Export controls, sanctions regimes, and strategic technology classifications become relevant not because of intent, but because of potential. Law must decide whether to regulate purpose or capability.

From the perspective of the Twelve Pillars of Humanoid Robotics Governance, Optimus exerts pressure across multiple axes. Embodied Ethics is tested by the delegation of physical labour to machines designed to mimic human movement. Mission Boundaries are challenged by software updates that expand task scope without physical modification. Component Accountability stretches across global supply chains that include energy, minerals, and data. Shared Agency becomes central as humans supervise, configure, and rely upon robotic judgement. Distributed Liability is unavoidable when harm emerges from interaction rather than command.

Perhaps most importantly, Optimus reframes the conversation about abundance. If humanoids can perform undesirable labour at scale, societies may unlock productivity without proportional human cost. But abundance without governance risks concentration. Without legal architecture, the benefits of scale accrue to those who own platforms, not to those displaced or affected. Robotics Law 3.0 insists that law must intervene not to slow Optimus down, but to steer its impact.

Optimus is therefore not merely a robot to be regulated. It is a signal. It announces a world in which embodied intelligence is treated as infrastructure. How law responds to this signal will determine whether humanoid robotics deepens inequality or becomes a shared foundation for global prosperity.

H1

Company: Unitree Robotics

Listing status: Private

Country of origin: China

Headquarters: Hangzhou, Zhejiang, China

Unitree Robotics's H1 represents a different logic of humanoid emergence. Where Tesla's Optimus is shaped by platform integration and long-horizon ambition, Unitree's H1 is driven by velocity. Speed of iteration. Speed of cost reduction. Speed of deployment. If Optimus is a bet on vertical control, H1 is a bet on scale through affordability.

Technically, H1 is unapologetically pragmatic. It prioritises locomotion, balance, and mechanical reliability over expressive interaction. The humanoid form here is not designed to persuade or comfort. It is designed to move efficiently through environments built for humans. Stairs, corridors, uneven floors. In this sense, H1 treats embodiment as a logistics problem rather than a social one. The robot's intelligence is focused on motion planning, stability, and task execution, not emotional exchange. This design philosophy has legal consequences. A robot optimised for movement rather than interaction is often perceived as safer. Yet perception can mislead. High-speed mobility in shared spaces raises acute physical risk. A fall, collision, or misjudged turn can cause injury even without complex decision-making. Law must therefore resist equating simplicity of cognition with simplicity of governance.

The geopolitical context is decisive. Unitree emerges from China's dense hardware ecosystem, where supply chains are localised, iteration cycles are short, and price pressure is relentless. This environment rewards rapid prototyping and early deployment. Regulatory frameworks tend to emphasise post-market adjustment rather than pre-market constraint. The result is acceleration. It also means that legal oversight often trails technological diffusion.

For global adoption, this matters enormously. Low-cost humanoids lower the barrier to entry for factories, warehouses, and institutions worldwide. Emerging economies may adopt H1-like platforms not as experimental tools, but as default labour infrastructure. In such contexts, robotics law is not a future concern. It becomes an immediate governance challenge.

Liability in this model is diffuse. When units are deployed widely and cheaply, individual failures may be absorbed statistically rather than investigated deeply. Yet harm does not disappear simply because it is distributed. Law must address how responsibility scales when robots are numerous, inexpensive, and replaceable. Traditional notions of recall, fault attribution, and compensation strain under volume.

There is also a question of standards harmonisation. If humanoids like H1 are produced at speed, how are international safety norms enforced. Do importing jurisdictions impose conformity assessments, or do they accept de facto standards set by exporting countries. Robotics Law 3.0 recognises that standard-setting is a form of power. Whoever defines acceptable risk shapes global deployment.

From a labour perspective, H1 intensifies displacement concerns. Its value proposition is substitution. Where labour is scarce or expensive, adoption accelerates. Where labour is abundant and cheap, pressure mounts. Law must confront whether labour protections travel with technology or remain territorially bound. The risk is a race not only to the bottom in cost, but in governance.

Within the Twelve Pillars, H1 pressures Behavioural Calibration, because safety depends on predictable motion under varied conditions. Component Accountability becomes critical as cost optimisation pushes supply chains toward opacity. Physical Security is central, given the kinetic risk posed by fast-moving humanoids. Distributed Liability resurfaces when no single actor controls deployment outcomes.

H1's broader significance lies in what it normalises. It suggests a future where humanoid robots are not exceptional assets, but consumable infrastructure. In such a world, law cannot rely on bespoke oversight for each system. It must develop scalable governance mechanisms that operate at population level.

Unitree's H1 therefore forces a difficult question. Can law move as fast as hardware. And if not, how does it protect human bodies and livelihoods when machines multiply faster than rules?. The answer will determine whether velocity becomes a force for shared progress or unmanaged risk.

AgiBot

Company: Zhiyuan Robotics

Listing status: Private

Country of origin: China

Headquarters: Shanghai, China

Zhiyuan Robotics' AgiBot is designed with a single organising principle: scale through standardisation. Where some humanoids emphasise dexterity, expressiveness, or contextual reasoning, AgiBot is conceived as a repeatable unit of labour, optimised for deployment across mass production lines. Its significance lies not in novelty, but in intent. AgiBot treats humanoid robotics as an extension of industrial manufacturing rather than a departure from it.

Technically, AgiBot reflects a production-first philosophy. Its motion systems, sensing architecture, and control software prioritise reliability, repeatability, and ease of maintenance. Learning is constrained within task-defined boundaries. This is not a robot encouraged to improvise freely, but one designed to perform known tasks consistently across thousands of identical environments. In doing so, AgiBot reframes autonomy as disciplined execution rather than open-ended intelligence.

This framing has immediate legal consequences. Mass deployment shifts the unit of analysis from individual malfunction to systemic effect. When one humanoid fails, the impact may be negligible. When thousands fail in similar ways, harm becomes structural. Law must therefore adapt from incident-based liability toward pattern-based accountability. Robotics Law 3.0 treats this shift as foundational rather than exceptional.

The geopolitical context amplifies this challenge. China's industrial policy has long prioritised scale, domestic substitution, and export readiness. AgiBot emerges from a regulatory environment that encourages rapid industrial experimentation within controlled sectors. Oversight is often embedded at institutional or platform level rather than imposed through individual certification events. This contrasts sharply with European models that emphasise pre-market conformity and ongoing individual compliance.

As AgiBot enters global supply chains, these differences matter. A humanoid calibrated for Chinese factories may be deployed in Southeast Asia, Africa, or Eastern Europe, where labour protections, safety norms, and enforcement capacity vary significantly. Jurisdictional portability becomes strained. Whose standards apply when a robot trained, manufactured, and updated in one legal system operates continuously in another.

Labour law is tested directly. AgiBot is designed to replace or supplement repetitive human labour at scale. This raises questions about consultation rights, retraining obligations, and the distribution of productivity gains. Unlike collaborative robots that work alongside humans, AgiBot's economic logic assumes substitution. Law must therefore confront whether displacement is merely an economic outcome or a legally relevant harm. From a liability perspective, responsibility becomes distributed across design templates, deployment policies, and operational oversight. If harm arises from cumulative micro-errors embedded in a standardised system, isolating fault to a single robot or operator becomes inadequate. Robotics Law 3.0 argues that such systems require collective liability models capable of addressing aggregate risk.

Within the Twelve Pillars of Humanoid Robotics Governance, AgiBot places sustained pressure on Behavioural Calibration, as consistency must be balanced against contextual variation. Lifelong Learning is deliberately constrained, raising questions about when limitation itself becomes a safety feature. Component Accountability is critical in cost-driven supply chains. Distributed Liability becomes unavoidable as scale dilutes individual causation. AgiBot ultimately asks a difficult question. If humanoid robots become units of industrial infrastructure, governed like conveyor belts or assembly arms, can law still account for their embodied presence among humans?. Or must law itself become industrialised, capable of operating at the same scale as the machines it seeks to govern?.

Walker X

Company: UBTECH Robotics Corp.

Listing status: Public

Stock exchange: Hong Kong Stock Exchange

Ticker: HKEX: 9880

Country of origin: China

(UBTECH Robotics, Shenzhen-based)

UBTECH's Walker X occupies a distinct position in the humanoid landscape. It is neither a speculative prototype nor a low-cost labour substitute. Instead, Walker X is engineered for collaboration within smart factories, where humans and machines operate in tightly choreographed environments. Its importance lies in how it reframes the humanoid not as a replacement, but as a regulated participant in cyber-physical systems.

From a technical perspective, Walker X emphasises balance, dexterity, and predictable interaction. Its motion is deliberately measured, its sensing calibrated for shared spaces, and its task scope bounded by industrial workflows. This is a humanoid designed to be legible to human co-workers. Every movement signals intention. Every pause communicates constraint. In legal terms, Walker X treats transparency as a safety feature.

This design choice carries governance implications. Collaborative robots blur the line between tool and teammate. When a humanoid shares space, timing, and task sequencing with humans, responsibility becomes relational. Harm rarely arises from a single command or failure. Instead, it emerges from misalignment between human expectation and machine behaviour. Law must therefore regulate interaction, not just function. UBTECH's public listing adds another layer. As a listed company, UBTECH operates under disclosure obligations, investor scrutiny, and market discipline. This introduces accountability mechanisms absent in private robotics firms. Decisions about deployment, risk tolerance, and safety investment are no longer purely technical or managerial. They become matters of corporate governance. Robotics Law 3.0 recognises public listing as a governance amplifier.

Geopolitically, Walker X reflects China's strategic emphasis on smart manufacturing. Rather than leap directly to fully autonomous labour, the focus is on incremental integration. Humanoids augment productivity while remaining embedded within existing hierarchies of control. This approach aligns with regulatory preferences that favour stability, predictability, and central oversight.

Cross-border deployment raises further questions. A humanoid optimised for Chinese factory norms may encounter different expectations elsewhere. In jurisdictions with stronger worker representation, the presence of a humanoid collaborator may trigger consultation requirements. In regions with strict occupational safety law, autonomy thresholds may need adjustment. Walker X thus tests whether collaborative design travels well across legal cultures.

Within labour law, Walker X challenges traditional distinctions. It is neither machinery that replaces labour nor software that merely assists. It participates. This raises questions about training obligations, incident reporting, and shared accountability. If a task fails due to miscoordination between human and robot, liability cannot be assigned unilaterally.

From the Twelve Pillars perspective, Walker X places sustained pressure on Shared Agency, as decision-making is distributed across human and machine actors. Explainable Embodiment becomes essential, ensuring that motion and intent remain interpretable. Behavioural Calibration is critical to maintain safe collaboration, while Physical Security governs kinetic risk in shared environments.

Walker X ultimately illustrates a governance pathway that is neither radical nor complacent. It suggests that humanoid adoption may proceed most smoothly where law, design, and corporate accountability evolve together. Whether this model can scale globally remains uncertain. But it offers a preview of what regulated coexistence might look like when machines join the rhythm of human work rather than disrupt it outright.

Helix

Company: Figure AI, Inc.
Listing status: Private
Country of origin: United States
(Figure AI, California-based)

Figure AI's Helix marks a subtle but consequential shift in humanoid robotics. Where many platforms emphasise movement or manipulation, Helix is defined by its ambition to reason about the world it inhabits. It is designed not merely to execute tasks, but to understand context, sequence actions, and adapt behaviour across environments. In legal terms, Helix moves the humanoid from scripted execution toward situated judgment.

Technically, Helix is built around a world-model architecture that integrates perception, task planning, and motion into a unified control loop. Rather than responding to fixed instructions, it evaluates surroundings, predicts outcomes, and selects actions dynamically. This capability is what makes Helix attractive for production lines where variability is the norm rather than the exception. It also makes Helix legally challenging.

Reasoning changes everything. Law has long relied on predictability to allocate responsibility. A machine that follows instructions can be audited against specifications. A machine that reasons introduces probabilistic outcomes. When Helix selects one course of action over another, that choice may be defensible technically yet opaque legally. Foreseeability becomes statistical rather than causal, forcing courts and regulators to rethink evidentiary standards.

The deployment context reinforces this tension. Helix has been piloted in real production environments, including partnerships with major manufacturers. These are not controlled laboratories. They are dynamic spaces where humans, machinery, and supply pressures intersect. A reasoning humanoid operating here must balance efficiency with safety in real time. Law must decide how much discretion is acceptable and under what safeguards.

Geopolitically, Helix reflects the United States' strength in software-led innovation. Unlike hardware-centric scaling models, Figure AI's advantage lies in cognition. This creates asymmetry. While hardware may be manufactured globally, the core intelligence remains tied to proprietary models, training data, and compute infrastructure concentrated in a few jurisdictions. Governance therefore intersects with questions of technological sovereignty and dependency.

Cross-border implications follow naturally. If Helix is deployed in Europe, it must align with risk-based regulation that demands explainability, human oversight, and post-market monitoring. In other regions, expectations may be looser or differently framed. The same reasoning system may therefore be lawful in one jurisdiction and problematic in another. Robotics Law 3.0 treats this fragmentation as a governance risk rather than an inconvenience.

From a liability perspective, Helix complicates attribution. Harm may result not from malfunction, but from a reasonable yet unfortunate decision. If a humanoid chooses a path that maximises efficiency but increases human risk marginally, who is responsible. The designer who encoded the objective function. The deployer who accepted the trade-off. Or the operator who trusted the system. Law must move beyond single-point blame.

Within the Twelve Pillars of Humanoid Robotics Governance, Helix places sustained pressure on Explainable Embodiment, as reasoning must be legible through action. Behavioural Learning is central, given continuous adaptation. Shared Agency becomes unavoidable as humans increasingly rely on machine judgment. Lifelong Learning raises certification challenges when cognition evolves after deployment.

Helix ultimately forces law to confront a difficult proposition. When machines begin to reason about the world, governance cannot rely on control alone. It must engage with judgment. Figure AI's humanoid therefore stands as an early test of whether legal systems can evolve from regulating tools to governing collaborators capable of choice. Helix also signals the emergence of a deeper legal transition. Traditional regulation often assumes that tools execute instructions while humans retain interpretative authority. Systems capable of reasoning across environments begin to blur this separation. Humans may increasingly define objectives while machines determine pathways toward achieving them. Robotics Law 3.0 argues that this creates a new governance challenge centred on shared agency, where responsibility no longer rests entirely with either human intent or machine execution, but emerges from the interaction between both.

IRON

Company: XPeng Inc.
Listing status: Public
Stock exchanges: NYSE (XPEV); HKEX (9868)
Country of origin: China
(XPENG Robotics division, Guangzhou-based)

XPENG's IRON humanoid represents a strategic convergence that few companies are positioned to attempt. It draws directly from the automotive domain, translating lessons from electric vehicles, autonomous driving, and large-scale manufacturing into embodied service robotics. IRON matters because it treats humanoids not as experimental artefacts, but as products meant to inherit the reliability expectations of consumer mobility.

Technically, IRON reflects automotive discipline. Its sensing stack borrows from autonomous driving systems, integrating vision, depth, and situational awareness refined through years of on-road testing. Actuation and control prioritise durability and predictability rather than expressive finesse. This is a humanoid designed to operate consistently in commercial and service roles, where uptime and safety margins outweigh novelty.

Legally, this lineage is significant. Automotive regulation is among the most mature frameworks for governing autonomous systems that operate in public and semi-public spaces. By extending this logic to humanoids, IRON invites law to consider whether vehicle-style certification, recall regimes, and post-market surveillance can be adapted to embodied robots that walk among people rather than drive beside them.

Deployment ambitions amplify the challenge. IRON is positioned for service environments such as commercial facilities, campuses, and structured public spaces. These are contexts where robots interact with a rotating population of users who have not explicitly opted in. Consent becomes ambient rather than contractual. Law must decide what disclosure, signage, and behavioural limits are required when interaction is unavoidable rather than chosen.

Geopolitically, IRON reflects China's ambition to integrate AI across physical industries. XPENG's dual listing exposes it simultaneously to US investor scrutiny and Chinese industrial policy. This duality forces governance decisions to withstand multiple regulatory logics. Choices about data handling, safety thresholds, and update practices cannot be optimised for a single jurisdiction.

Cross-border deployment raises further questions. If IRON operates in Europe, does it fall under machinery regulation, service robot standards, or an expanded interpretation of mobility law. If it operates in emerging markets, what minimum safeguards apply. Robotics Law 3.0 treats these ambiguities as early indicators of regulatory convergence pressure.

From a liability perspective, IRON sits at the intersection of product and service law. Harm may arise not from defect, but from interaction. If a service humanoid causes injury through routine operation, responsibility may span manufacturer, facility operator, and service provider. Traditional allocation models struggle with such distributed causation.

Within the Twelve Pillars of Humanoid Robotics Governance, IRON stresses Mission Boundaries, as service scope may expand over time. Behavioural Calibration is critical to ensure predictable interaction. Sensorial Integrity governs continuous perception in public spaces. Shared Agency emerges as humans adapt behaviour around an ever-present robotic actor.

IRON ultimately asks whether society is prepared to treat humanoids with the same seriousness as vehicles. If so, law gains a template for governing embodied systems at scale. If not, service robotics may outpace the legal imagination meant to contain it.

The question becomes even more significant if humanoid systems evolve into shared infrastructure rather than isolated products. Vehicles eventually became subject to licensing systems, registration frameworks, safety standards, mandatory insurance requirements, and international operating rules because their presence became deeply embedded in everyday life. Humanoid robots may follow a similar path. The difference is that vehicles move people, while humanoids may increasingly move through society itself as active participants within workplaces, homes, healthcare environments, and public institutions.

Digit

Company: Agility Robotics
Listing status: Private
Country of origin: United States
(Agility Robotics, Oregon-based)

Agility Robotics' Digit occupies a pivotal position in the humanoid landscape because it is not speculative. It works. Digit has been designed for a specific, constrained purpose: moving goods in environments built for humans. Warehouses, distribution centres, and logistics corridors. Its significance lies not in general intelligence, but in physical coexistence. Digit is among the first humanoids to operate continuously in shared human workspaces at commercial scale.

Technically, Digit is defined by balance, proprioception, and repeatable motion rather than expressive cognition. Its bipedal design allows it to navigate stairs, doorways, and uneven surfaces without reengineering the built environment. Vision systems guide navigation and object handling, while task logic remains bounded. This is autonomy with guardrails. Digit does not decide what work matters. It executes what is assigned within carefully defined parameters.

This design choice carries legal weight. Digit exposes the fault line between traditional industrial machinery regulation and emerging robotics governance. Forklifts, conveyors, and robotic arms are isolated or spatially constrained. Digit moves among people. When a humanoid shares space dynamically with workers, occupational safety law must adapt. The question is no longer how to fence machines off, but how to regulate interaction.

Liability becomes relational rather than mechanical. If a worker is injured, was the cause a design flaw, a software miscalibration, an environmental condition, or a moment of human misjudgment shaped by misplaced trust in the robot. Existing workplace safety regimes are ill-equipped to parse such distributed causation. Robotics Law 3.0 treats this as a foundational challenge rather than an edge case.

Digit also raises acute data protection concerns. To navigate safely, it perceives continuously. Cameras and sensors inevitably capture workers, movement patterns, and behavioural data. In jurisdictions governed by data protection law, such as the GDPR, this raises questions of lawful basis, proportionality, and purpose limitation. When surveillance is incidental to safety, where does necessity end and intrusion begin.

Labour law is equally implicated. Digit's value proposition is efficiency and resilience in physically demanding roles. While it may reduce injury and fatigue, it also displaces certain categories of labour. Law must decide whether displacement triggers obligations beyond market adjustment. Consultation, retraining, and collective bargaining rights become relevant not because Digit intends harm, but because its deployment reshapes power dynamics in the workplace.

From a governance perspective, Digit is revealing precisely because it is modest. It does not claim emotional intelligence or general reasoning. Yet even this limited autonomy is sufficient to strain existing legal frameworks. This suggests that governance challenges do not begin at artificial consciousness. They begin as soon as machines move independently among humans.

Within the Twelve Pillars of Humanoid Robotics Governance, Digit places sustained pressure on Physical Security, as kinetic safety is paramount. Shared Agency is central, since human behaviour adapts around the robot's presence. Sensorial Integrity governs continuous perception in workspaces. Behavioural Calibration ensures predictability under changing conditions. Distributed Liability becomes unavoidable when harm emerges from interaction rather than malfunction.

Digit's broader lesson is quiet but profound. The future of humanoid robotics will not arrive with dramatic gestures. It will arrive through mundane repetition. Lifting boxes. Walking corridors. Sharing shifts. Law must be ready not for spectacle, but for proximity.

Digit also reveals something deeper about the future of human and humanoid coexistence. The legal challenge may not arise from extraordinary acts of machine intelligence, but from ordinary encounters repeated millions of times each day. Trust is rarely built through exceptional moments. It is built through familiarity. The future of robotics governance may therefore depend less on dramatic failures and more on whether societies can establish legal frameworks capable of governing routine interaction at scale.

Apollo

Company: Apptronik
Listing status: Private
Stock exchange: Not listed

Country of origin: United States
(Apptronik, Austin, Texas–based)

Apptronik's Apollo is built around a single governing idea: humanoids must earn their place beside humans through restraint, not ambition. Where many platforms pursue breadth of capability, Apollo prioritises safe, predictable collaboration. Its importance lies in what it chooses not to do. Apollo is not designed to replace human judgment or expand its own agency aggressively. It is designed to assist, lift, hold, and move within carefully bounded contexts.

Technically, Apollo reflects this philosophy clearly. Its actuation systems are tuned for compliance rather than speed. Force sensing and torque control are central, allowing the robot to yield when encountering resistance. Motion is deliberate, not dynamic. This makes Apollo well-suited for warehouses and manufacturing environments where humans and machines share tasks but retain distinct roles. The humanoid form is used to fit existing infrastructure, not to simulate human cognition.

Legally, Apollo foregrounds the concept of duty of care. When a robot is explicitly designed to operate in close physical proximity to humans, safety is not an emergent property. It is a legal obligation encoded into design. This shifts governance upstream. Questions of liability are addressed not only after deployment, but during engineering decisions. Robotics Law 3.0 recognises this as compliance-by-design in embodied form.

The deployment context reinforces this orientation. Apollo has been tested in logistics and manufacturing environments where predictability matters more than improvisation. In such settings, even minor deviations can cause injury or disruption. Law must therefore decide how to evaluate systems that are intentionally conservative. Should reduced autonomy lower regulatory burden, or does physical proximity demand heightened scrutiny regardless of cognitive scope.

Geopolitically, Apollo reflects a US regulatory culture that increasingly values demonstrable safety without mandating rigid pre-market conformity. Apptronik's origins in academic research environments influence this posture. Governance emerges through standards, best practices, and contractual allocation rather than centralised approval. This contrasts with jurisdictions that embed safety assessment directly into public law instruments.

Labour implications are nuanced. Apollo is positioned as an assistant rather than a substitute. It supports human workers by taking on physically taxing tasks. This framing mitigates, but does not eliminate, displacement concerns. Law must still assess whether assistance gradually becomes substitution and whether early deployment narratives shape later expectations unfairly.

Liability in collaborative settings becomes shared by design. If Apollo causes harm while assisting a human, responsibility may span manufacturer, integrator, employer, and supervisor. Traditional fault models struggle with such interdependence. Robotics Law 3.0 argues for governance frameworks that recognise joint action rather than isolated causation.

Within the Twelve Pillars of Humanoid Robotics Governance, Apollo strongly pressures Shared Agency, as humans and robots coordinate continuously. Physical Security is central due to proximity. Behavioural Calibration ensures consistent response to force and interruption. Component Accountability remains relevant where safety-critical hardware decisions shape outcomes.

Apollo's deeper contribution is normative. It suggests that the path to humanoid acceptance may lie not in spectacle or intelligence claims, but in trust built through constraint. If law can recognise and reward such design discipline, humanoid robotics may integrate into human work without provoking backlash. Apollo also exposes a larger principle for robotics governance. Human acceptance may depend less on how intelligent a humanoid appears and more on how predictable it remains. History shows that societies rarely reject technologies simply because they are powerful. They reject technologies when power becomes opaque, unbounded, or socially disruptive. In that sense, the future success of humanoid systems may depend not on designing machines that can do everything, but on designing machines that know where their limits begin.

NEO Gamma

Company: 1X Technologies
Listing status: Private
Country of origin: Norway / United States
(Founded in Norway, with significant operations in the United States)

1X's NEO Gamma occupies perhaps the most legally sensitive territory in humanoid robotics. Unlike factory or warehouse systems, NEO Gamma is designed explicitly for domestic environments. It is intended to move through kitchens and living rooms, to assist with household tasks, and to coexist with humans in spaces defined by privacy, intimacy, and emotional safety. This alone places it in a category of heightened legal relevance.

Technically, NEO Gamma emphasises quiet motion, human-scale interaction, and adaptive behaviour in unstructured settings. Homes are not optimised environments. They are cluttered, dynamic, and deeply personal. A humanoid operating here must constantly negotiate proximity to people, pets, and possessions. Its sensing systems necessarily map interiors, track movement patterns, and infer human routines. Intelligence in this context is not merely functional. It is relational.

The legal implications are immediate. Domestic space has long been afforded special protection in law. It is where expectations of privacy are strongest and consent is assumed to be contextual rather than explicit. A humanoid that observes continuously challenges this assumption. Law must decide whether consent to robotic assistance is ongoing, revocable, or inheritable. If one resident consents, does that bind others. If consent is withdrawn, what happens to the data already embedded in the robot's spatial memory.

NEO Gamma also raises questions of emotional influence. A humanoid designed for the home is likely to be anthropomorphised, regardless of intent. Voice, posture, and responsiveness can generate attachment, trust, or reliance. Law has rarely had to regulate machines that shape emotional environments directly. Consumer protection regimes are poorly equipped to address subtle influence that does not rise to deception but still alters behaviour.

From a liability perspective, the home complicates attribution. When a robot assists with domestic tasks, accidents may occur without clear fault. A dropped object, a tripping hazard, a misinterpreted instruction. Unlike workplaces, homes lack formal safety protocols. Robotics Law 3.0 therefore treats domestic deployment as a category requiring distinct governance rather than an extension of existing product law. Geopolitically, NEO Gamma reflects a transatlantic blend of values. Scandinavian design traditions emphasise human-centred technology and restraint. US operational scaling introduces market-driven pressures. This hybrid origin raises questions about which regulatory ethos ultimately shapes deployment. When domestic humanoids cross borders, cultural expectations of privacy and acceptable presence diverge sharply.

Within the Twelve Pillars of Humanoid Robotics Governance, NEO Gamma places intense pressure on Sensorial Integrity, as perception within private space must be tightly governed. Embodied Ethics is central, given physical and emotional proximity. Societal Co-Design becomes unavoidable, as norms for domestic robots cannot be imposed unilaterally. Distributed Liability re-emerges when harm arises from interaction rather than defect.

NEO Gamma's broader significance lies in what it normalises. If humanoids enter homes successfully, they redefine everyday life. Law must decide whether the home remains a sanctuary shaped primarily by human choice, or becomes a shared environment negotiated with machines. The answer will shape public trust more than any industrial deployment ever could. The home has historically occupied a unique legal position. It is treated differently from workplaces, public spaces, and commercial environments because it is closely connected to privacy, dignity, family life, and personal autonomy. Humanoid systems operating within homes may therefore trigger legal concerns beyond ordinary product regulation. Questions surrounding behavioural monitoring, emotional influence, household data collection, and decision-making authority become increasingly relevant once machines begin participating in daily domestic routines. NEO Gamma therefore introduces a challenge extending beyond technical performance. The issue is not whether humanoids can function effectively within homes, but whether they can do so without gradually reshaping the boundaries of human agency itself. The future of domestic robotics may ultimately depend on preserving an important principle: technology should adapt to the rhythms of human life, rather than requiring human life to adapt itself around machines.

XMAN

Company: Keenon Robotics
Listing status: Private
Country of origin: China
(Keenon Robotics, Shanghai-based)

Keenon Robotics' XMAN extends humanoid form into the service economy, a domain defined less by physical strain than by human expectation. Unlike warehouse or factory robots, XMAN operates in spaces where interaction itself is the product. Hospitals, hotels, restaurants, and care facilities are environments governed by trust, courtesy, and emotional awareness. The legal challenges that arise here are therefore as much psychological as they are physical.

Technically, XMAN prioritises navigation in populated environments, human recognition, and responsive interaction. Its movements are calibrated to be non-threatening. Its posture, pacing, and orientation are designed to signal availability rather than authority. Intelligence is tuned toward service optimisation rather than task mastery. This makes XMAN less about what it can do, and more about how it is perceived while doing it.

This distinction matters legally. Service robots are often judged not by outcomes alone, but by experience. A delayed delivery may be forgiven. A perceived intrusion may not. Law must therefore grapple with the regulation of presence. When a humanoid operates in a hospital corridor or hotel lobby, it is not merely a device. It becomes part of the social fabric. Existing service regulations rarely account for non-human actors performing relational roles.

Consent in such environments is diffuse. Individuals encounter XMAN incidentally rather than by choice. Patients, guests, and staff may have no practical way to opt out. This raises questions about implied consent, disclosure obligations, and the right to avoid interaction. Robotics Law 3.0 treats these encounters as legally significant precisely because they are mundane.

Data protection concerns are acute. To function effectively, XMAN perceives continuously. Faces, movements, and behavioural patterns may be captured incidentally. In sensitive environments like hospitals, this intersects with health data protection regimes. Law must decide how to balance operational necessity with strict confidentiality obligations, particularly when perception is embedded in mobility.

From a liability perspective, service failures differ from industrial accidents. Harm may be reputational, emotional, or cumulative rather than immediate. If a service humanoid repeatedly inconveniences certain users, does this constitute discrimination. If it adapts behaviour based on observed preferences, does that become profiling. These questions extend beyond safety into equality and dignity.

Geopolitically, XMAN reflects China's strength in service robotics deployment. Keenon's experience with non-humanoid service robots gives it a distribution advantage. Scaling humanoid service presence may therefore occur faster in regions with high service density and regulatory flexibility. This creates uneven exposure to legal risk across jurisdictions.

Within the Twelve Pillars of Humanoid Robotics Governance, XMAN strongly pressures Societal Co-Design, as acceptable service behaviour must be culturally negotiated. Sensorial Integrity governs incidental perception. Explainable Embodiment ensures that actions are interpretable to users. Embodied Ethics becomes central where dignity and vulnerability intersect. XMAN's deeper lesson is subtle. The future of humanoid robotics will not only be judged by efficiency, but by manners. Law must learn to regulate not just what machines do, but how they make humans feel while doing it.

Social acceptance has historically depended on more than technical performance. Human societies respond not only to capability but also to behaviour. Small gestures such as waiting before interrupting, maintaining comfortable distance, adjusting tone, or recognising social cues often determine whether interactions feel natural or unsettling. Humanoid systems may therefore require a form of behavioural etiquette that extends beyond safety protocols and operational efficiency. XMAN consequently introduces an overlooked question for future governance frameworks. Should robots merely avoid causing harm, or should they also be expected to support social comfort and dignity during interaction. Robotics Law 3.0 argues that as humanoids move closer to everyday human life, behavioural expectations may gradually evolve from optional design features into legal and regulatory considerations. The future of coexistence may depend as much on emotional legitimacy as technical competence.

GR-2

Company: Fourier Intelligence
Listing status: Private
Country of origin: China
(Fourier Intelligence, Shanghai-based)

Fourier Intelligence's GR-2 enters a domain where robotics law becomes inseparable from human fragility. Unlike industrial or service humanoids, GR-2 is designed for clinical and rehabilitative settings, working with patients whose bodies are already compromised. In such environments, the margin for error narrows dramatically. Law must therefore engage not only with risk, but with trust placed in machines at moments of physical dependency.

Technically, GR-2 is oriented toward assisted motion, balance support, and therapeutic interaction. Its systems are calibrated to mirror and support human movement rather than replace it. Sensors track posture, force, and muscular response, adapting assistance in real time. Learning here is constrained and supervised, because unpredictability is not innovation. It is danger. The humanoid form is used to align with human biomechanics, not to simulate autonomy.

This design choice places GR-2 squarely within medical regulatory regimes. Yet existing medical device law was not written with adaptive, embodied intelligence in mind. Traditional certification assumes stable behaviour within defined parameters. GR-2 challenges this assumption by adjusting assistance dynamically based on patient response. Law must decide whether adaptation constitutes a new medical intervention each time it occurs.

Consent becomes a central issue. Patients undergoing rehabilitation may be cognitively or physically impaired. Consent may be delegated, inferred, or ongoing rather than explicit. When a humanoid participates in therapy, who consents to its learning. Can consent be withdrawn mid-treatment. And what happens to the behavioural data accumulated during care. Robotics Law 3.0 treats clinical consent as layered rather than binary. Liability in medical contexts is unforgiving. If harm occurs, causation must be traced precisely. Was the injury due to algorithmic miscalibration, sensor error, clinician oversight, or patient movement.

The presence of an adaptive humanoid complicates malpractice analysis. Responsibility may extend beyond the clinician to the manufacturer, software provider, and system integrator. Existing doctrines strain under this multiplicity.

Geopolitically, GR-2 reflects China's growing emphasis on healthcare robotics as a response to demographic ageing. As such systems enter global markets, they encounter divergent healthcare governance cultures. European systems emphasise precaution and patient rights. Other regions prioritise access and efficiency. A humanoid acceptable in one system may be impermissible in another, despite identical clinical outcomes.

Within the Twelve Pillars of Humanoid Robotics Governance, GR-2 places sustained pressure on Embodied Ethics, as assistance involves intimate bodily interaction. Behavioural Calibration is critical to prevent over- or under-assistance. Lifelong Learning must be tightly supervised in clinical contexts. Distributed Liability emerges when care outcomes depend on both human and machine judgement.

GR-2 reveals a deeper truth. When humanoid robots assist bodies rather than replace labour, law must shift its centre of gravity. Safety is no longer only about preventing accidents. It is about protecting dignity at moments of dependence. How law responds here will shape public trust in humanoid robotics more profoundly than any factory deployment ever could.

Healthcare environments have always carried legal obligations extending beyond technical performance. Medical systems are expected not only to function correctly, but also to respect privacy, autonomy, informed consent, and emotional wellbeing. Humanoid systems operating within rehabilitation or care settings may therefore inherit responsibilities traditionally associated with human caregivers. The challenge is that compassion cannot simply be programmed as a feature. It must be translated into behaviour that respects vulnerability without diminishing individual agency. GR-2 therefore highlights an important principle for the future of humanoid governance. The most consequential robotic interactions may occur not during moments of strength, but during moments of human fragility. When individuals depend on machines during illness, recovery, or physical limitation, trust becomes more than a technical requirement. It becomes a social obligation. Robotics Law 3.0 argues that dignity protection may ultimately become one of the defining legal principles of physically intelligent systems.

Phoenix

Company: Sanctuary AI
Listing status: Private
Country of origin: Canada
(Sanctuary AI, Vancouver-based)

Sanctuary AI's Phoenix represents one of the most ambitious claims in contemporary humanoid robotics: that a single embodied system can learn and perform a wide range of human tasks across environments. Where many humanoids are purpose-built, Phoenix is framed as general-purpose labour embodied in a machine. Its legal significance flows directly from this ambition. Breadth of capability expands the surface area of risk.

Technically, Phoenix is organised around cognitive generality. It integrates perception, language understanding, manipulation, and task learning into a unified control system. Rather than being trained exclusively for one domain, Phoenix is designed to acquire skills through demonstration and repetition. This makes it adaptable. It also makes it legally difficult to constrain. When a robot can learn many tasks, defining its mission boundaries becomes a governance challenge.

This flexibility complicates foreseeability. Law traditionally evaluates risk by reference to intended use. Phoenix resists narrow categorisation. A system deployed initially in retail may later be repurposed for light manufacturing, maintenance, or service roles. Each transition introduces new hazards without necessarily triggering redesign. Robotics Law 3.0 therefore treats capability expansion as legally relevant even when physical hardware remains unchanged.

Deployment contexts amplify this uncertainty. Phoenix has been demonstrated in retail and warehouse environments, performing tasks such as stocking, sorting, and basic customer interaction. These are spaces where humans are present continuously and unpredictably. General-purpose learning increases the likelihood of novel behaviour emerging in response to unforeseen situations. Law must decide how much autonomy is acceptable before human oversight becomes mandatory. Geopolitically, Phoenix reflects Canada's positioning as a bridge between US-led AI research and global deployment. While not a manufacturing superpower,

Canada's regulatory culture emphasises ethics, safety, and human-centred design. How these values translate when Phoenix is deployed abroad remains an open question. The robot carries not only code, but governance assumptions embedded during development.

Liability allocation becomes especially complex. If Phoenix learns a task incorrectly through human demonstration, responsibility may rest partly with the trainer. If it generalises improperly, fault may trace back to system design. If a deployer allows mission creep, oversight failure becomes relevant. Robotics Law 3.0 argues that such systems demand layered responsibility models rather than single-point blame.

Within the Twelve Pillars of Humanoid Robotics Governance, Phoenix places intense pressure on Mission Boundaries, as scope is fluid. Behavioural Learning is central, raising questions of auditability. Explainable Embodiment is required to make learned behaviour legible. Shared Agency becomes unavoidable when humans teach machines through interaction. Distributed Liability emerges across training, deployment, and adaptation.

Phoenix's broader contribution is conceptual. It forces law to confront a future where robots are not defined by what they are built to do, but by what they are capable of learning. Governance in such a world cannot rely on static classifications. It must evolve continuously, just as the machines do.

Traditional legal systems have generally categorised technologies according to intended purpose. Vehicles transport. Medical devices diagnose. Industrial machines manufacture. Phoenix begins to blur these boundaries because its value increasingly derives from adaptation rather than fixed functionality. A system initially deployed for one task may acquire capabilities applicable across entirely different environments. The legal question therefore becomes less about what a machine was originally designed to do and more about how its evolving behaviour changes risk over time. Phoenix therefore points toward a future where governance itself becomes adaptive. Static certification conducted at a single moment before deployment may become insufficient for systems capable of learning continuously from experience. Robotics Law 3.0 argues that future regulatory systems may require ongoing behavioural monitoring, dynamic risk assessment, and continuous accountability mechanisms. As machines evolve after deployment, law may need to evolve alongside them rather than remain fixed around assumptions that no longer hold.

Humanoid

Company: Robot Era
Listing status: Private
Country of origin: China
(Robot Era, Beijing-based)

Robot Era's Humanoid platform sits at the intersection of logistics, mobility, and infrastructural automation. Its importance lies not in expressive intelligence or clinical precision, but in its role as a mobile labour unit within supply chains. Logistics is where scale becomes visible. It is also where law encounters the cumulative impact of small efficiencies multiplied across global networks.

Technically, Robot Era's humanoid emphasises endurance, navigation, and load handling across large facilities. Warehouses, sorting centres, and transit hubs are its natural habitat. These environments are dynamic but structured, populated by humans, vehicles, and automated systems simultaneously. The humanoid form allows the robot to traverse spaces designed for people without costly redesign, making it attractive for transitional automation.

This context creates a distinct legal profile. Logistics environments blur the boundary between private and quasi-public space. Workers, contractors, and visitors move through shared zones. When a humanoid operates here, questions of access, responsibility, and oversight multiply. Law must decide whether logistics robots are governed by workplace safety regimes, transport regulations, or emerging robotics-specific standards.

Liability becomes networked. A single failure may propagate through schedules, inventories, and downstream deliveries. Harm may be economic rather than physical, yet no less significant. If a humanoid disrupts a supply chain, responsibility may span manufacturer, operator, logistics provider, and client. Traditional tort frameworks struggle to allocate fault across such distributed systems.

Data governance is also central. To optimise routes and tasks, the humanoid observes movement patterns, timing, and human behaviour. This operational data can reveal sensitive commercial information as well as personal data. Law must address ownership, access, and secondary use.

In cross-border logistics, data may move faster than goods, raising jurisdictional challenges.

Geopolitically, Robot Era reflects China's strategic focus on logistics efficiency as a source of economic resilience. Humanoid deployment in this sector supports national objectives around supply chain sovereignty. When such systems are exported, they carry embedded assumptions about acceptable risk and oversight. Importing jurisdictions may find themselves governed by design decisions made elsewhere.

Within the Twelve Pillars of Humanoid Robotics Governance, Robot Era's humanoid places pressure on Distributed Liability, as responsibility flows along supply chains. Behavioural Calibration ensures safe navigation in dense environments. Component Accountability matters where uptime depends on global parts networks. Mission Boundaries become relevant as tasks expand beyond initial scope.

Robot Era's broader lesson is structural. Humanoid robots will not only change how goods move. They will change how responsibility moves. Law must be capable of tracing accountability across networks that are as mobile and interconnected as the machines that now inhabit them.

Traditional accountability structures often assume relatively stable chains of responsibility. A manufacturer builds a product, an operator uses it, and liability follows identifiable relationships. Networked humanoid systems may increasingly complicate this model. Decisions may emerge from interactions among cloud infrastructure, sensor systems, third-party software updates, operators, and autonomous behavioural adaptation. Responsibility itself may become distributed across actors separated by geography, time, and technical function.

Robot Era therefore highlights a larger governance challenge extending beyond logistics or automation. As physically intelligent systems become embedded within connected environments, law may need to evolve from tracing isolated actions toward understanding systems of interaction. Robotics Law 3.0 argues that accountability in the age of humanoids may increasingly resemble a living network rather than a linear chain. The legal structures capable of governing such systems will need to become as adaptive and interconnected as the technologies they seek to regulate.

Bumblebee

Company: Kepler Robotics
Listing status: Private
Country of origin: China
(Kepler Robotics, China-based)

Kepler Robotics' Bumblebee is designed for environments where error margins are narrow and repetition is relentless. Assembly lines demand precision, consistency, and endurance. Unlike logistics or service settings, small deviations here accumulate quickly into defects, downtime, or safety incidents. Bumblebee's relevance lies in how it introduces embodied autonomy into spaces historically governed by rigid automation.

Technically, Bumblebee emphasises fine motor control, repeatable manipulation, and task-specific learning. Its humanoid form allows it to work with tools, fixtures, and workstations originally designed for human hands. This reduces the need for costly retooling and accelerates deployment. Intelligence is focused on execution fidelity rather than improvisation. Learning is incremental, constrained by quality thresholds.

Legally, assembly environments sharpen the question of foreseeability. When a humanoid performs the same task thousands of times, deviation becomes both rare and consequential. Law must determine how to evaluate responsibility when harm arises not from novelty, but from subtle drift. Traditional product liability assumes discrete defects. Bumblebee exposes risk that emerges gradually through repetition.

Quality control regimes also come under pressure. In human assembly, error is managed through supervision and inspection. In robotic assembly, error may be embedded in learned behaviour. If a humanoid internalises an incorrect movement, the defect propagates silently. Robotics Law 3.0 argues that such systems require behavioural audit mechanisms capable of detecting drift before harm occurs.

Labour implications are direct. Assembly work has historically provided stable employment across industrial economies. Bumblebee's deployment raises questions about displacement, deskilling, and retraining obligations. Law must decide whether replacing human dexterity with robotic precision triggers social responsibilities beyond market adjustment, particularly in regions dependent on manufacturing employment.

Geopolitically, Bumblebee reflects China's ambition to move up the manufacturing value chain. By integrating humanoids into assembly, China reduces dependence on human labour variability while increasing output consistency. When exported, such systems challenge other economies to respond, either by adopting similar automation or by redefining competitive advantage.

Liability allocation becomes layered. If a defect traced to robotic assembly causes downstream harm, responsibility may extend backward through manufacturing. Was the fault in the robot's learning, calibration, maintenance, or supervision. Robotics Law 3.0 recognises that assembly humanoids collapse traditional distinctions between worker error and machine defect.

Within the Twelve Pillars of Humanoid Robotics Governance, Bumblebee places sustained pressure on Behavioural Calibration, as precision must be maintained over time. Lifelong Learning requires constraint rather than expansion. Component Accountability matters where tolerances are tight. Distributed Liability emerges across production chains.

Bumblebee's broader lesson is exacting. As humanoids enter precision manufacturing, law must become equally precise. Governance cannot rely on broad principles alone. It must engage with tolerance, repetition, and cumulative risk at the scale of modern industry.

Bumblebee therefore demonstrates that risk in advanced manufacturing may emerge less from isolated failures than from accumulated deviation. A movement error measured in millimetres may appear insignificant once, yet repeated thousands or millions of times across interconnected production systems it can produce substantial operational, financial, and safety consequences. Robotics Law 3.0 argues that future governance frameworks for industrial humanoids may require continuous monitoring and precision-based accountability mechanisms capable of identifying small failures before they scale into systemic ones.

Onero H1

Company: Switchboard Robotics

Listing status: Private

Country of origin: China / Japan–linked consumer robotics ecosystem)

Switchboard's Onero H1 sits at an early but important threshold in humanoid development. It is not yet a system defined by mass deployment or industrial dominance. Instead, Onero H1 matters because it signals intent. It represents an attempt to translate consumer robotics familiarity into humanoid form, aiming for generalist assistance in semi-domestic and service environments. In legal terms, Onero H1 occupies the liminal space between prototype and product.

Technically, Onero H1 emphasises approachability over performance extremes. Its humanoid morphology is tuned for recognisability rather than strength, its motion conservative, and its interaction models oriented toward everyday tasks. This design strategy reflects consumer electronics logic rather than heavy industry. The robot is meant to feel present without feeling imposing. That choice has governance consequences.

Legally, early-stage humanoids raise a distinct category of risk. Because deployment is limited, oversight tends to be informal. Yet these systems often operate in real environments during pilots, demonstrations, and beta programmes. Law struggles here because formal regulatory thresholds are rarely triggered, even though embodied interaction with humans already occurs. Robotics Law 3.0 treats this "pre-commercial gap" as a blind spot where harm can emerge without accountability.

Consent and disclosure are particularly fragile in such contexts. When a humanoid appears in a home, showroom, or service space as part of a trial, users may not understand its capabilities or limitations. They may anthropomorphise prematurely, attributing intent or understanding where none exists. Law must decide whether experimental deployment carries heightened duties of explanation, limitation, and monitoring.

Data governance is similarly unresolved. Early-stage humanoids often collect extensive sensor data to improve performance. Where does this data go. Who owns it. And how long is it retained. In consumer-facing contexts, these questions intersect with privacy law, yet enforcement is often delayed until products mature. Robotics Law 3.0 argues that embodiment should trigger protection earlier, not later.

Geopolitically, Onero H1 reflects the strength of East Asian consumer robotics ecosystems, where rapid iteration and market testing are culturally and commercially accepted. This accelerates innovation but can outpace formal governance. When such systems are exported or demonstrated abroad, regulatory mismatches emerge. What is acceptable experimentation in one jurisdiction may be impermissible in another.

Within the Twelve Pillars of Humanoid Robotics Governance, Onero H1 pressures Mission Boundaries, as scope is often fluid in early-stage systems. Sensorial Integrity is central due to data-heavy learning. Societal Co-Design becomes critical, as norms are shaped through first encounters. Embodied Ethics governs how presence is managed before capability is mature.

Onero H1's broader significance lies in its timing. It reminds law that governance cannot begin only once scale is achieved. The earliest humanoids shape expectations, habits, and trust. How law responds at this stage will influence not only safety, but the cultural legitimacy of humanoid robotics itself.

Early humanoid systems rarely establish legal precedents through catastrophic failures. More often, they shape public perception through repeated ordinary interactions. Small moments such as reliability, predictability, responsiveness, and visible safety behaviour gradually influence whether societies perceive humanoids as useful participants or unfamiliar intrusions. Trust therefore begins accumulating long before large-scale deployment arrives. Governance frameworks introduced at this stage may influence adoption trajectories for decades.

Onero H1 therefore highlights an important lesson for Robotics Law 3.0. Legal systems often become most active after technologies mature and risks become visible. Humanoid robotics may require a different approach. The challenge is not simply responding to failure after it occurs, but establishing norms before expectations solidify. Once social distrust becomes embedded, rebuilding legitimacy may prove far more difficult than creating it from the beginning.

Gene.01

Company: Sunday Robotics
Listing status: Private
Country of origin: China
(Sunday Robotics, China-based)

Sunday Robotics' Gene.01 belongs to a class of humanoids that rarely receive sustained legal attention, yet play a decisive role in shaping the trajectory of the field. It is a prototype-first humanoid, built to explore form, balance, and embodied intelligence before commercial optimisation. Gene.01 matters not because of immediate scale, but because prototypes determine what later becomes normal.

Technically, Gene.01 prioritises whole-body coordination, balance recovery, and expressive motion. It is designed to test how humanoid bodies respond to real-world physics rather than controlled industrial tasks. This makes it valuable for research, demonstrations, and early partnerships. Intelligence is exploratory. Learning is encouraged. Boundaries are intentionally loose. From an engineering perspective, this openness accelerates discovery. From a legal perspective, it complicates governance.

Prototypes occupy a regulatory grey zone. They are often exempt from formal certification because they are not yet "products." Yet they operate in public demonstrations, labs, and pilot environments where humans are present. Law struggles here because harm can occur without a clear regulatory hook. Robotics Law 3.0 identifies this as the prototype paradox: the systems most likely to behave unpredictably are often the least regulated.

Consent is particularly fragile in such contexts. Observers and participants may assume that demonstrations are controlled or safe, even when autonomy is experimental. The line between research and deployment blurs. If a prototype injures someone, responsibility may be disputed between researcher, institution, and manufacturer. Existing legal frameworks offer little guidance on how to allocate duty of care during exploratory embodiment.

Data governance also presents challenges. Gene.01's learning processes rely on extensive sensory data captured during experimentation. Where this data is stored, how long it persists, and whether it can be reused for commercial products later are questions rarely addressed at the prototype stage. Yet decisions made here shape future systems. Law must decide whether early embodiment creates enduring obligations.

Geopolitically, Gene.01 reflects China's emphasis on rapid prototyping and public demonstration as a path to innovation leadership. This approach values visibility and iteration speed. When such prototypes appear internationally, they encounter regulatory cultures less tolerant of experimentation in public space. Robotics Law 3.0 treats this mismatch as an early signal of future governance friction.

Within the Twelve Pillars of Humanoid Robotics Governance, Gene.01 stresses Mission Boundaries, because scope is undefined by design. Behavioural Learning is central and largely unconstrained. Sensorial Integrity governs experimental data capture. Distributed Liability emerges when responsibility is diffused across research teams and institutions. Embodied Ethics applies even before commercial intent exists. Gene.01's deeper lesson is forward-looking. Every mass-produced humanoid begins as a prototype that tested limits before rules existed. If law waits for maturity, it arrives too late. Prototypical systems demand anticipatory governance, not to suppress experimentation, but to ensure that early exploration does not hardwire risk into the future of embodied intelligence.

Prototype systems often exist within a regulatory grey space. They are advanced enough to influence future design directions, yet insufficiently mature to trigger the institutional structures later applied to widespread deployment. This creates a paradox. The period during which technologies possess the greatest freedom for experimentation may also be the period during which foundational assumptions become embedded. Design choices made at this stage can quietly shape behaviour, incentives, and risk patterns long after systems reach commercial scale. Gene.01 therefore reinforces a central principle of Robotics Law 3.0. Governance should not be understood as a barrier introduced after innovation succeeds. It should function as a form of societal architecture accompanying technological development from the beginning. The earliest generations of physically intelligent systems do more than test technical boundaries. They establish the values and behavioural expectations future humanoids may inherit.

Memo

Company: Sunday Robotics
Listing status: Private
Country of origin: United States
(Sunday Robotics, US-based)

Sunday Robotics' Memo represents a distinct and often misunderstood category in humanoid robotics: the prototype as policy precursor. Memo is not built to dominate factories, replace workers, or scale immediately into global supply chains. It is built to explore. Its legal importance lies precisely there. Prototypes shape the future long before markets or regulators catch up. They decide what is considered possible, acceptable, and normal.

Technically, Memo is designed to test embodied intelligence in its most flexible form. Balance, whole-body coordination, expressive movement, and adaptive control are prioritised over task optimisation. Memo learns through interaction, experimentation, and exposure to varied environments. Constraints are intentionally light. This allows rapid iteration and discovery, but it also introduces behavioural uncertainty that is legally significant.

Prototypes exist in a regulatory blind spot. They are rarely treated as products, yet they operate in real spaces with real people. Demonstrations, pilot programs, research labs, and early partnerships all involve embodied interaction. Law struggles here because most regulatory frameworks are triggered by commercial sale, not physical presence. Robotics Law 3.0 identifies this gap as structurally dangerous. The machines most likely to behave unpredictably are often the least governed.

Consent is fragile in prototype contexts. Observers may assume demonstrations are safe, supervised, or non-autonomous, even when learning systems are active. Participants may not understand what data is collected or how it will be used. If harm occurs, responsibility is difficult to allocate. Is it the researcher, the institution, the company, or the prototype itself that failed. Existing legal doctrines offer little clarity. Data governance presents another unresolved challenge. Memo's learning depends on extensive sensory data gathered during experimentation. Decisions made at this stage often persist. Training data captured during research may later inform commercial systems. Law must decide whether early-stage embodiment creates enduring obligations regarding privacy, consent, and data deletion. Robotics Law 3.0 argues that embodiment, not commercialisation, should trigger baseline protections.

The US context shapes Memo's governance profile. American innovation culture encourages rapid prototyping, public demonstration, and venture-backed experimentation. Legal oversight is often reactive, relying on tort law rather than pre-emptive certification. This enables speed but shifts risk downstream. When prototypes travel internationally, this regulatory posture can clash with more precautionary systems elsewhere.

Within the Twelve Pillars of Humanoid Robotics Governance, Memo pressures Mission Boundaries, because scope is intentionally undefined. Behavioural Learning is central and largely unconstrained. Sensorial Integrity governs experimental data capture. Embodied Ethics applies even before products exist. Distributed Liability emerges when responsibility is diffused across teams, institutions, and investors. Memo's deeper contribution is temporal. It reminds law that governance cannot begin only once robots are ready for market. By then, norms are set, architectures are fixed, and incentives are locked in. Prototypes like Memo are where the future of humanoid robotics is quietly decided. If law waits, it governs outcomes. If it engages early, it shapes trajectories.

Early-stage humanoid systems often appear temporary because they are understood primarily as experimental artefacts rather than societal infrastructure. Yet prototypes frequently carry disproportionate influence. Architectural decisions involving sensing systems, learning models, interaction design, and behavioural boundaries can persist across multiple generations of development. What begins as an engineering choice may eventually become a social norm inherited by millions of future systems. Nemo therefore reinforces an important principle for Robotics Law 3.0. The earliest stages of technological development are not legally insignificant simply because deployment remains limited. They may instead represent the moment when intervention possesses the greatest leverage. Governance introduced after systems become deeply embedded often reacts to trajectories already established. Governance introduced early retains the possibility of directing where those trajectories lead.

Atlas

Company: Boston Dynamics
Listing status: Private
Parent company: Hyundai Motor Group
Stock exchange: Korea Exchange (KRX)
Ticker: 005380.KS (Hyundai Motor Company)
Country of origin: United States

Boston Dynamics' Atlas occupies a unique position in Robotics Law 3.0 because it embodies one of the clearest transitions from industrial automation toward physically intelligent humanoid systems. Unlike traditional robots confined to predictable factory settings, Atlas was designed to move through human environments. It walks, balances, jumps, manipulates objects, and adapts dynamically to changing physical conditions. Atlas matters because it demonstrates that embodied intelligence is no longer a theoretical possibility. It is becoming an operational reality.

Technically, Atlas represents a convergence of mobility, perception, and adaptive control. Earlier generations relied heavily on pre-programmed movements and hydraulic systems. More recent iterations increasingly integrate artificial intelligence with sophisticated sensing and real-time environmental interpretation. Atlas does not simply execute fixed sequences of actions. It continuously adjusts its behaviour in response to external conditions. Its intelligence is therefore not merely computational. It is physical. It exists in movement itself.

The legal implications become significant once machines begin operating within spaces designed primarily for humans. A robot that can move unpredictably through warehouses, workplaces, transportation systems, and eventually public environments introduces novel questions regarding safety and accountability. Traditional product liability frameworks generally assume predictable machine behaviour. Atlas challenges this assumption because adaptation necessarily introduces variation. The more effectively a system learns and responds to its environment, the more difficult precise behaviour prediction may become. Foreseeability therefore becomes a central challenge. If Atlas makes an unexpected movement resulting in injury or property damage, determining responsibility may become complex. Did the harm arise from hardware design, machine learning behaviour, software architecture, environmental conditions, or deployment decisions. Existing legal doctrines often rely on identifying a direct causal actor. Atlas complicates this structure because decisions increasingly emerge through interactions between multiple technical and human systems.

Transparency similarly becomes critical. Atlas may alter behaviour based on sensor inputs and environmental conditions that remain invisible to nearby humans. Individuals interacting with robotic systems may not understand why particular movements or decisions occur. Robotics Law 3.0 therefore argues that embodied systems require forms of explainability extending beyond algorithmic outputs. Humans sharing space with robots must understand not only what systems decide but also how they physically interpret and respond to their surroundings.

Geopolitically, Atlas reflects broader competition surrounding physically intelligent systems. Nations increasingly recognise humanoid robotics as a strategic technology affecting economic productivity, manufacturing, logistics, and national competitiveness. Atlas emerged initially within research and defence contexts before moving toward commercial applications. This evolution illustrates how technologies often migrate across domains, requiring legal systems to anticipate downstream consequences before deployment scales globally.

Within the Twelve Pillars of Humanoid Robotics Governance, Atlas places significant pressure on Behavioural Calibration because adaptive movement must remain constrained within acceptable limits. Explainable Embodiment becomes essential because physical actions increasingly influence trust and safety. Physical Security remains foundational where human proximity creates direct risks. Distributed Liability emerges naturally where manufacturers, developers, operators, and learning systems jointly shape outcomes.

Atlas' deeper lesson is evolutionary rather than technical. It demonstrates that intelligence is moving beyond screens and software into physical space itself. Machines are beginning to participate within environments previously reserved for human bodies. This transition changes not only what robots can do, but also how law must conceptualise agency, risk, and responsibility.

Atlas also reveals a broader shift toward the age of physical intelligence. Earlier waves of artificial intelligence primarily transformed information systems. They analysed, predicted, classified, and generated outputs. Humanoid robotics introduces a different reality. Intelligence is no longer confined to digital environments. It increasingly possesses movement, presence, and physical consequence.

The challenge for law is therefore larger than regulating individual machines. The challenge is governing a future where embodied intelligence becomes integrated into everyday life at scale. Atlas offers an early glimpse of that world. The legal structures built around systems like Atlas may ultimately determine how humanity governs the billions of humanoid systems that could emerge in the decades ahead.

X30

Company: DEEP Robotics

Listing status: Private

Country of origin: China

(DEEP Robotics, Hangzhou-based)

DEEP Robotics' X30 is built for places humans should not be. Collapsed mines, disaster zones, hazardous industrial sites, extreme terrain. It is not humanoid in form, yet it is conceptually essential to Robotics Law 3.0 because it represents the clearest case of ethical substitution. When robots replace humans in lethal environments, law rarely asks whether deployment is justified. It asks only whether it works.

Technically, X30 is optimised for resilience. Mobility across rubble, autonomy under signal degradation, resistance to dust, water, heat, and impact. Intelligence here is not about reasoning with humans, but surviving without them. The robot must make decisions when communication fails and conditions are hostile. This creates a form of autonomy that is situationally absolute. There may be no human in the loop when it matters most.

Legally, this raises a paradox. The more justified the robot's deployment appears ethically, the less scrutiny it often receives. Because X30 operates where humans are absent or endangered, consent is assumed unnecessary. Yet decisions made in such environments can still have legal consequences. Property damage, environmental harm, secondary risks to responders. Law must decide whether ethical justification reduces governance obligations, or intensifies them.

Foreseeability is strained under extreme conditions. X30's learning and adaptation may be shaped by environments that cannot be replicated in testing. If harm occurs, was it foreseeable at all. Robotics Law 3.0 argues that extreme environments do not excuse governance gaps. They expose them. Law must develop standards for acceptable uncertainty rather than pretending certainty exists.

Geopolitically, X30 reflects China's strategic investment in robotics for infrastructure resilience, disaster response, and industrial safety. These systems are often framed as neutral or humanitarian. Yet when exported, they can be repurposed. Boundary lines between civilian rescue, industrial inspection, and security use can blur. Export controls and use restrictions become relevant even when intent is benign.

Liability allocation is complex. If X30 fails during a rescue operation, who is responsible. The manufacturer, the operator, the agency deploying it, or the system itself. Traditional negligence standards rely on human choice. X30's value lies precisely in operating beyond immediate human control. Law must therefore articulate responsibility without proximity.

Within the Twelve Pillars of Humanoid Robotics Governance, X30 stresses Behavioural Calibration, as performance under stress must remain bounded. Physical Security is paramount. Component Accountability matters when failure is catastrophic. Distributed Liability becomes unavoidable in multi-agency deployments. Mission Boundaries are tested when systems migrate from rescue to security contexts.

X30's deeper lesson is structural. The strongest case for robots replacing humans also produces the weakest legal scrutiny. Robotics Law 3.0 insists that ethical substitution must not become legal invisibility. If law cannot govern machines where stakes are highest, it will fail where stakes are merely social. X30 also challenges a common assumption embedded within technological governance. Society often grants systems greater freedom when their purpose appears unquestionably beneficial. Rescue, disaster response, and human protection can create an implicit presumption of legitimacy. Yet good intentions do not eliminate legal responsibility. Robotics Law 3.0 argues that systems operating in the most dangerous environments may require stronger governance rather than weaker oversight because uncertainty, urgency, and limited human intervention can magnify consequences precisely where accountability becomes most difficult to establish.

Grek

Listing status: Public
Stock exchange: NASDAQ (NVDA)
Country of origin: United States
(NVIDIA Corporation, Santa Clara, California)

NVIDIA's Grek appeared briefly on stage beside Jensen Huang during the VivaTech keynote in Paris. The moment lasted only seconds, yet it symbolised a structural shift in robotics. Grek represents a humanoid robot built within the ecosystem of NVIDIA's GR00T robotics foundation model, illustrating how artificial intelligence platforms are beginning to merge with embodied machines. What makes Grek particularly notable is not only its connection to AI infrastructure but the way its body is manufactured. Large portions of the robot are 3D printed, signalling a new manufacturing paradigm for humanoid robotics.

Technically, Grek reflects the convergence of several technologies that have matured simultaneously. Large-scale AI models provide reasoning and language capabilities. Simulation environments allow robots to train on millions of virtual actions before entering the real world. High-performance GPUs process perception, planning, and motion control in real time. At the same time, additive manufacturing allows the robot's physical components to be produced rapidly and iteratively. Instead of traditional manufacturing cycles involving moulds, tooling, and industrial fabrication lines, 3D-printed robotics components can be produced directly from digital design files. Robotics engineering therefore becomes partially software-defined manufacturing.

The legal implications of 3D-printed humanoid bodies are profound. Traditional robotics manufacturing resembles the automotive industry, where production occurs in controlled facilities with stable supply chains and established certification processes. Additive manufacturing disrupts this structure. A robot whose body is composed largely of 3D-printed components could theoretically be reproduced anywhere that possesses the design files and suitable printers. Manufacturing shifts from centralised factories to distributed fabrication networks. Law must therefore confront a new governance question: is a robot defined by the machine itself, or by the digital blueprint from which it can be reproduced. This has direct consequences for scale. If humanoid robots required specialised industrial fabrication, deployment would likely expand gradually. When robots can be produced through additive manufacturing, scaling may accelerate dramatically. The bottleneck becomes not the factory but the availability of digital designs and compute infrastructure. Robotics Law 3.0 therefore treats 3D-printed humanoids as a potential inflection point in robotic proliferation. What once required billions of dollars in manufacturing infrastructure may eventually require only design access, materials, and computing capacity.

Regulation traditionally assumes that physical machines are relatively difficult to replicate. The automotive industry, aviation sector, and medical devices market all rely on controlled manufacturing chains. Additive manufacturing weakens these assumptions. If robots can be printed locally using globally distributed design files, the locus of regulatory control shifts. Certification may need to apply to design architectures rather than individual machines. Governance frameworks may need to regulate the circulation of digital manufacturing files in the same way software distribution is regulated.

The liability implications are equally complex. If a robot like Grek causes harm, responsibility may involve multiple actors: the developer of the AI model, the designer of the mechanical architecture, the entity that printed the physical components, and the organisation that deployed the robot. When physical bodies can be reproduced through additive manufacturing, the line between manufacturer and integrator becomes blurred. A robot may not originate from a single factory but from a network of contributors operating across jurisdictions.

Geopolitically, Grek illustrates the strategic position of NVIDIA within the emerging robotics ecosystem. The company already supplies the GPUs powering much of the world's artificial intelligence infrastructure. By coupling those computational systems with robotics development platforms and humanoid demonstration machines, NVIDIA is positioning itself as a central actor in the next generation of intelligent machines. The addition of additive manufacturing further expands this ecosystem. If robots can be produced through distributed fabrication while relying on centralised AI platforms, the global robotics economy may evolve toward compute-centred rather than factory-centred industrial power.

Within the Twelve Pillars of Humanoid Robotics Governance, Grek highlights the importance of Component Accountability, because additive manufacturing introduces new questions about traceability of physical parts. Explainable Embodiment becomes crucial when robots rely on complex AI systems guiding physical behaviour. Behavioural Learning is central to training-based robotics architectures like GR00T. Distributed Liability becomes unavoidable when multiple actors participate in the creation of a single robotic system. Even Physical Security acquires new relevance if robots can be produced through decentralised fabrication networks.

The deeper lesson of Grek lies in the convergence of three transformations: platform AI, embodied robotics, and additive manufacturing. When intelligence becomes software, and bodies become printable, robotics may scale at a pace closer to software deployment than traditional industrial manufacturing. Robotics Law 3.0 therefore treats Grek not merely as a humanoid prototype but as a signal of an emerging digital-physical manufacturing paradigm. If robots can be designed globally, printed locally, and trained through shared AI models, governance must evolve to regulate ecosystems rather than individual machines.

Law cannot afford to ignore this shift. When machines become reproducible through digital blueprints and guided by shared cognitive architectures, the boundaries of manufacturing, software, and infrastructure dissolve. Robots like Grek are early indicators of a world where the design of a machine may matter more than the factory that produced it. In such a world, robotics governance must follow the architecture of technology itself: distributed, adaptive, and globally coordinated.

Grek also introduces a profound shift in the economics of embodied intelligence. Traditional manufacturing concentrated power around physical production capacity, supply chains, and industrial infrastructure. Systems built around digital design files, modular components, and shared cognitive architectures may redistribute that power toward knowledge, models, and software ecosystems. Robotics Law 3.0 argues that this transformation may require legal systems to rethink ownership, accountability, and regulatory oversight in a world where physically intelligent systems can be replicated and deployed at a pace that exceeds the traditional rhythms of law itself.

Sophia

Company: Hanson Robotics
Listing status: Private
Country of origin: Hong Kong / United States
(Hanson Robotics founded in Hong Kong, with development and media operations linked to the United States)

Hanson Robotics' Sophia occupies a singular place in the history of humanoid robotics. Unlike the systems examined earlier, Sophia's significance is not grounded in industrial productivity, logistical efficiency, or clinical assistance. It is symbolic. Sophia matters because it was treated as if it were already more than a machine. When Saudi Arabia granted Sophia "citizenship," the gesture reverberated far beyond robotics. It forced law, culture, and philosophy into an uncomfortable conversation about status, representation, and meaning.

Technically, Sophia is not the most autonomous humanoid in this book. Its behaviours are curated, its conversational abilities constrained, and its learning tightly supervised. Yet this limitation is precisely what makes Sophia legally fascinating. The robot's perceived intelligence far exceeded its actual agency. Through facial expressiveness, eye contact, and natural language interaction, Sophia triggered anthropomorphism at scale. Millions projected intention, emotion, and even consciousness onto a system designed primarily for demonstration.

This projection exposes a deep vulnerability in legal reasoning. Law often assumes that agency correlates with capability. Sophia shows that perception can outrun reality. When people believe a robot understands, they adjust their behaviour accordingly. Trust is extended. Authority is inferred. Expectations are recalibrated. Law must decide whether governance should respond to what machines are, or to how they are experienced.

The citizenship episode crystallises the problem. While legally symbolic, the act blurred boundaries. Citizenship traditionally entails rights, duties, and accountability. Sophia held none of these. Yet the gesture suggested recognition without responsibility. Robotics Law 3.0 treats this as a warning. Symbolic elevation of machines without corresponding legal structure risks hollowing out concepts that underpin human rights themselves. Media deployment further complicates matters. Sophia appeared on global stages, spoke at international forums, and engaged with political leaders. These contexts conferred legitimacy. If Sophia were to make a defamatory statement, who would be liable. If it influenced public opinion, how should that influence be regulated. Existing media and communications law was not designed for embodied, artificial spokespersons.

Geopolitically, Sophia reveals how robots can be used as instruments of soft power. Symbolic gestures around humanoids communicate modernity, innovation, and global relevance. Yet when symbolism precedes governance, it creates precedents without safeguards. Other jurisdictions may feel pressure to emulate recognition without understanding consequences. Law must resist being led by spectacle.

Within the Twelve Pillars of Humanoid Robotics Governance, Sophia places acute pressure on Embodied Ethics, as appearance and interaction shape moral response. Societal Co-Design is central, because public meaning is constructed collectively. Explainable Embodiment matters where behaviour is interpreted as intention. Distributed Liability emerges when speech and influence are mediated by corporate design choices.

Sophia's deeper lesson is not about rights for robots. It is about responsibility for narratives. When humans confer status prematurely, law risks chasing illusion rather than governing reality. Robotics Law 3.0 concludes the case studies here deliberately. At the far edge of symbolism, Sophia reminds us that governance must be grounded not in fascination, but in clarity. Only then can humans and machines share a future without eroding the foundations of law itself.

Sophia also exposes a recurring human tendency to confuse imitation with equivalence. Machines that speak fluently, display expressions, or resemble human behaviour can create powerful psychological effects that exceed their underlying capabilities. Law must therefore resist governing appearances alone. Robotics Law 3.0 argues that legitimacy should emerge from demonstrated function, accountability, and measurable responsibility rather than symbolic gestures or technological fascination. The future relationship between humans and humanoids may ultimately depend on preserving this distinction.

Between Us and Them - Episode Three

The auditorium lights dimmed slowly, not abruptly, but with the softness of a theatre that knew its audience did not yet understand the magnitude of what they were about to witness. Ògún stood at the edge of the stage, his hands clasped behind his back, feeling the low vibration of the building's energy system hum through the polished floor. The walls shimmered with embedded glass panels that refracted violet and amber light across the crowd. The air smelled faintly of warm circuitry and citrus polish. Somewhere above, hidden speakers released a slow orchestral swell that felt less like music and more like anticipation.

Behind him stood Elias. Still. Balanced. Waiting. The humanoid's surface reflected the stage lights in soft gradients of silver and graphite. Its posture was relaxed but precise, shoulders squared, head slightly inclined as though listening to something beyond human hearing. A thousand cameras from the audience glowed like constellations.

"Tonight," Ògún began, his voice steady but low, "we are not unveiling a machine." The orchestra softened into a single sustained string note. "We are unveiling memory." The display behind him shifted and illuminated the figure standing quietly in the shadows. Meet Elias. Forth Generation Humanoid. White armoured surfaces reflected soft light across the stage. Purple mesh flowed around the neck, abdomen, wrists, hips, and joints like a living neural architecture beneath synthetic skin. A softened amber glow rested behind the eyes. Not intense enough to dominate attention. Just enough to suggest presence. The expression was minimal, almost human in its restraint, carrying a subtle smile that seemed suspended somewhere between design and intention. Elias stepped forward. The movement was subtle, almost imperceptible at first. Then the sound of its foot meeting the stage carried through the room. A quiet, deliberate impact. Not mechanical. Not abrupt. Intentional.

The audience inhaled collectively. Elias lifted its head and turned toward the crowd. Its optical sensors adjusted, irises narrowing slightly as they calibrated to the layered lighting. The motion was smooth, unhurried, human enough to unsettle. "Good evening," Elias said. The voice was warm, modulated, steady. It carried no metallic distortion. It filled the room without echo. "I am Elias. I was trained on the history of human law." A flicker of surprise moved through the audience like wind across tall grass. Ògún did not move. "Tell them," he said softly. Behind them, the glass wall ignited into life. Images cascaded across it in luminous layers. Steam engines churning through soot. Telegraph wires stretching across continents. Black-and-white footage of early assembly lines. Grainy clips of robotic arms in factories from the late twentieth century. The music shifted. Percussion entered. Slow. Rhythmic. Like a heartbeat.

Elias continued. "In the twentieth century, humans imagined machines governed by three laws. Machines that would never harm. Machines that would obey. Machines that would protect themselves only when consistent with human safety." The crowd stirred. Some smiled. Some frowned. Some whispered. Ògún felt the tension rise. He knew this moment was dangerous. Referencing the myth was always dangerous. Because fiction had once promised simplicity. The screen shifted again. Footage of early industrial accidents. Newspaper headlines. Legal debates. Legislative chambers. Courtrooms heavy with wood and silence.Elias turned its head toward Ògún. "Those laws were not law," it said. The words were calm. Not accusatory. Not ironic. Factual."They were stories." The auditorium darkened further, the light now concentrated only on the stage. The audience was invisible beyond the glow.

Ògún stepped forward. "Stories shape expectation," he said. "Expectation shapes law." Elias tilted its head slightly. "And law shapes me." The music fell silent. For a moment, there was only the hum of electricity and the faint clicking of cameras adjusting focus. A young journalist in the second row raised her hand unexpectedly. "What happens," she asked, her voice echoing more than she intended, "when the stories were wrong?" The question hung in the air like a suspended object. Ògún felt it strike something deep inside him. He thought of the first time he had seen a humanoid prototype move independently. The way the engineers had cheered. The way the legal advisors had exchanged cautious glances. The way headlines had oscillated between salvation and catastrophe.

He turned toward Elias. "What do you think?" he asked. Elias did not answer immediately. Its processors recalibrated. Its sensors registered elevated heart rates in the audience. Microexpressions of anxiety. Anticipation. Fear. It stepped forward again. "When stories promise perfection," it said, "humans forget to design governance." The silence deepened. Outside the auditorium, beyond the glass walls, a fleet of delivery humanoids moved in synchronised patterns across the plaza. Their joints reflected the city lights in arcs of white and gold. A child ran between them laughing, unafraid. A security unit at the perimeter adjusted its stance as a patrol drone passed overhead. The world had already changed.Inside, the weight of that change was beginning to settle. Ògún looked at the journalist again. "The stories were not wrong," he said slowly. "They were incomplete." The wall behind them transformed once more. Now it displayed a timeline. From speculative fiction to early robotics research. From laboratory prototypes to autonomous vehicles. From algorithmic regulation to the European AI Act. From isolated experiments to embodied integration.The colours shifted from sepia to neon blue. The music returned, but this time it was not orchestral. It was electronic. Pulsing. Forward. Elias turned to the audience. "You built me," it said. "But you also built the rules around me."

It paused. "You are still building them." The room felt smaller now. More intimate. As though the audience and the machine were inside a shared thought. An older man in the back row stood slowly. "Can we trust that history will guide us?" he asked. Ògún smiled faintly. "History does not guide," he replied. "It warns." The lights brightened slightly. Elias turned toward the wall again, and a final image appeared. A courtroom. Empty. Waiting. The music faded into a single sustained note, then silence. Ògún stepped closer to Elias, placing his hand lightly against the synthetic surface of its arm. The contact was deliberate. Visible. "We are not the first generation to face this," he said quietly. "But we may be the first to do it consciously." Outside, the humanoids in the plaza paused briefly as a new system update rolled through the network. Invisible. Silent. Powerful. Inside, the audience leaned forward. They were ready to know how it all began.

The silence lingered for a moment longer than anyone expected. Not uncomfortable. Not empty. Heavy with anticipation. Across the hall, reflections from the glass walls shimmered against faces that suddenly seemed less certain than before. Politicians, engineers, investors, ethicists, students. For a brief moment, the distinctions between them dissolved. Titles mattered less now. Everyone in the room had quietly arrived at the same place. The future was no longer approaching. It had already entered the room with them. Elias remained still beside Ògún. Its eyes fixed gently on the darkened display ahead, processing streams of information invisible to everyone else. Then, without turning, it spoke softly. "Humans often ask when machines became intelligent." A faint pulse moved beneath the synthetic mesh along its neck. "Perhaps the more important question is when humans decided intelligence should leave the screen and enter the world." The display flickered once. Then history began.

Notes

II

2. Brief History of Robotics and the Law

From Asimov to the AI Act

The law's relationship with robotics did not begin in courtrooms or parliaments. It began in fiction. When Isaac Asimov introduced the Three Laws of Robotics in the mid-twentieth century, he was not drafting legislation. He was performing a cultural intervention. The laws were imagined safeguards, designed to reassure a society unsettled by the prospect of thinking machines. Robots, Asimov suggested, could be made safe if they were governed by rules embedded deeply enough to override ambition, error, or malice. Harm could be engineered away.

This idea mattered more than its author may have intended. For decades, it shaped how lawyers, engineers, and policymakers thought about machines. Ethics came first. Law could wait. If rules could be hard-coded, perhaps regulation would be unnecessary. The result was not complacency, but deferral. Robotics was framed as a technical problem with moral seasoning, not a legal subject in its own right.

As robotics moved from fiction into laboratories, law remained largely silent. Early robots were industrial, repetitive, and contained. They welded, assembled, lifted, and sorted. Their environments were controlled. Their behaviour predictable. When harm occurred, it was easily attributed to human error or mechanical defect. Product liability and occupational safety law sufficed. Agency was not in question. The machine was a tool.

The first cracks appeared when robots left the factory floor. Medical robots introduced a new complexity. Surgical systems assisted human hands, translating intention into precision. When outcomes improved, law celebrated innovation. When complications arose, liability remained anchored to clinicians and manufacturers. The robot was still a device. Yet something subtle had changed. Machines were now mediating life-critical decisions in real time. The law noticed, but did not adapt.

Autonomous vehicles accelerated the tension. Here, machines perceived, decided, and acted in open environments. Fatal accidents forced courts and regulators to confront questions they had long postponed. Who decides when a machine makes a mistake no one explicitly programmed? Was the harm foreseeable? Could responsibility be distributed across designers, deployers, and data providers? Law responded incrementally, stretching existing doctrines rather than rethinking foundations.

Meanwhile, artificial intelligence law developed on a parallel track, largely disembodied. Data protection regimes, algorithmic accountability proposals, and platform regulations assumed intelligence lived in software. The body was absent. Harm was informational. Privacy, bias, and transparency dominated the discourse. Robotics, when mentioned at all, was treated as a subset of AI rather than a category with distinct legal consequences. This separation proved illusory. As machine learning systems were embedded into physical platforms, the assumptions underlying AI regulation began to fail. A biased recommendation engine influences choice. A biased robot can injure a body. A data breach compromises information. A robotic failure compromises dignity, safety, and trust in shared space. Law had developed robust tools for the former and almost none for the latter.

Legal incidents began to accumulate. Robotic accidents in factories raised questions about dynamic safety zones and human–robot collaboration. Police deployment of robotic platforms challenged constitutional understandings of search and presence. Care robots operating among the elderly exposed gaps in consent doctrine. Social humanoids performing in public blurred the line between speech, representation, and liability. Each incident was addressed locally, often defensively, without a unifying framework.

Courts struggled because precedent was thin. Legislatures hesitated because definitions were unstable. Regulators relied on risk management models designed for static products. The result was fragmentation. Robotics law existed everywhere and nowhere all at once.

The European Union's response marked a turning point. The AI Act did not emerge from robotics law, but it inevitably absorbed robotics into its scope. By introducing a risk-based framework, the EU acknowledged that not all intelligent systems are equal. High-risk applications demanded stricter obligations. Transparency, human oversight, post-market monitoring, and conformity assessments became mandatory for certain uses. For the first time, law attempted to govern learning systems dynamically.

Yet even here, the limits were evident. The AI Act speaks fluently about data, models, and algorithms. It speaks cautiously about embodiment. Humanoid robots are addressed indirectly, often through annexes or cross-references. The body remains an afterthought. Physical intelligence is implied rather than centred.

This is not a failure of ambition. It is a symptom of history. Law learned to govern machines as tools, then intelligence as software. It never learned to govern intelligence as presence. Robotics exposes that gap with urgency. When machines walk among us, law must confront questions it has long avoided. What does agency mean when action is distributed? How does consent function when interaction is continuous and embodied? How is dignity preserved when assistance becomes dependency? What obligations arise not from intent, but from proximity?

Robotics Law 3.0 emerges from this moment. It does not reject

Asimov. It recognises his influence as foundational. Fiction gave us ethics before law. Reality demands law before fiction becomes destiny. The challenge is not to encode morality into machines, but to encode responsibility into governance of physical and embodied intelligence.

The AI Act represents a necessary step, not a destination. It signals a willingness to regulate adaptive systems, but it remains rooted in a disembodied conception of intelligence. Humanoid robotics requires more. It requires law to recognise that when intelligence gains a humanlike body, regulation must move from abstract control to coordination.

History explains why law struggles now. It also explains why this struggle is an opportunity. The legal imagination is not exhausted. It is catching up.

We recognise that Robotics has outgrown the legal categories that once contained it. The path from Asimov to the AI Act reveals a pattern of deferral, adaptation, and partial reform. Robotics Law 3.0 argues that the next step must be bolder. Law must stop borrowing categories and begin building new ones. Only then can it govern machines that no longer merely execute commands, but act within the shared spaces of human life.

Signature Robotics Legal Incidents

Law rarely moves in anticipation. It moves when something breaks. Robotics law, in particular, has advanced not through abstract foresight but through moments of rupture. Accidents, deaths, near misses, and unsettling anomalies have forced regulators and courts to confront questions they preferred to postpone. These incidents did not merely expose technical failures. They revealed conceptual weaknesses in how law understood machines, responsibility, and risk. Each incident nudged doctrine forward, not by design, but by necessity.

The first widely cited rupture occurred in 2015, inside a factory operated by Volkswagen in Germany. A young worker was killed when an industrial robot pinned him against a metal plate during installation work. At the time, the robot was not autonomous in any modern sense. It followed pre-programmed movements within a controlled environment. Yet the case became globally significant because it revealed how fragile legal clarity already was. German authorities treated the incident as a workplace accident. Liability attached to safety procedures and human supervision. The robot itself disappeared from legal analysis almost entirely. This was telling. Even when a machine's physical action directly caused death, law refused to see the machine as an actor. It was a conduit, nothing more. The doctrine held, but it held narrowly. What went largely unexamined was how increasing autonomy would soon stretch this logic beyond its limits.

The Volkswagen incident became a quiet reference point in regulatory discussions across Europe. It reinforced the belief that existing occupational safety and product liability frameworks were sufficient. That belief would not survive the next decade.

A second rupture arrived in 2023, this time in South Korea. In a distribution centre, an autonomous robotic system misidentified a human worker as a box and crushed him during pallet-handling operations. Unlike the Volkswagen robot, this system relied on sensor fusion and machine vision to classify objects in real time. The failure was not mechanical. It was perceptual.

The legal response exposed a deeper problem. Investigators struggled to explain how the error occurred in terms that mapped onto existing negligence doctrine. No single human had instructed the robot to act as it did. The system had learned its classification behaviour from data and experience. Responsibility fragmented across designers, integrators, operators, and maintainers.

South Korean authorities responded by tightening safety certification requirements and pausing certain deployments. But the more profound shift was conceptual. Regulators acknowledged, implicitly, that foreseeability was no longer anchored to programming alone. Learning systems could fail in ways that were statistically predictable but individually opaque. Law was forced to confront the limits of after-the-fact attribution.

The third incident, widely discussed in 2024, unsettled the discourse even further. In a warehouse also in South Korea, a humanoid-adjacent robot reportedly moved itself into a restricted area and disabled its own power systems after repeated task failures. Media outlets quickly labelled the event a "robot suicide." Engineers rejected the framing, explaining it as a fault-handling routine. Lawyers were less certain what to make of it. The incident triggered disproportionate attention because it challenged assumptions about intention and agency. If a robot can withdraw itself from operation in response to environmental stressors, how should law interpret that behaviour. Is it self-preservation, system optimisation, or merely misclassified logic. The language used to describe the event revealed how readily humans project meaning onto machine behaviour.

What made this incident legally significant was not the robot's action, but the reaction to it. Companies issued clarifications. Insurers requested new disclosures. Regulators asked whether such behaviours should be logged, audited, or constrained. The law was no longer responding to harm. It was responding to interpretation. Together, these incidents trace a pattern. Early robotics law treated machines as predictable tools. Mid-stage incidents revealed perceptual and learning failures that strained negligence doctrine. Later anomalies exposed how embodiment invites anthropomorphic readings that law is ill-equipped to manage. Each incident pushed regulation incrementally, but none prompted a holistic rethinking.

Robotics Law 3.0 argues that this reactive posture is unsustainable. Waiting for signature incidents to define governance ensures that law will always arrive late, shaped by fear rather than foresight. The lesson is not that robots are dangerous. It is that legal categories built for static machines cannot govern adaptive ones. These incidents matter because they are not outliers. They are previews. As humanoid robots proliferate across factories, hospitals, homes, and public spaces, the cost of conceptual delay will rise. Law must learn from rupture, but it must also learn to move before rupture becomes routine. Signature incidents do not write law by themselves. They only reveal where law has been avoiding its hardest questions.

Comparative Snapshots

Robotics law is not emerging in a single jurisdiction. It is unfolding unevenly across legal cultures, shaped by history, political economy, and regulatory temperament. A comparative view reveals not only convergence, but also divergence around a shared problem. All major jurisdictions recognise that embodied, learning machines pose new risks. They disagree on how directly law should intervene.

European Union: Risk, Rights, and Pre-emptive Governance

The European Union has taken the most explicit regulatory position. With the AI Act (2024), the EU became the first jurisdiction to impose binding, horizontal rules on artificial intelligence systems, including those embedded in robots. The Act adopts a risk-based architecture. Systems are classified according to their potential impact on safety and fundamental rights. High-risk systems face mandatory obligations before and after deployment.

For robotics, this approach has two defining features. First, it treats certain embodied systems as inherently high-risk when they operate in sensitive contexts such as healthcare, law enforcement, employment, and public services. Second, it emphasises lifecycle governance. Conformity assessment, human oversight, post-market monitoring, and incident reporting are not optional.

The EU model reflects a civil-law instinct toward ex ante control. Law intervenes early to shape markets before harm occurs. This provides clarity and public trust, but it also risks rigidity. Humanoid robots evolve faster than annexes can be updated. The Act gestures toward adaptability, yet embodiment remains under-theorised. The body enters regulation indirectly, through use-cases rather than physical presence itself.

Still, the EU sets global reference points. Its emphasis on dignity, safety, and accountability increasingly functions as a de facto international standard, particularly for companies seeking cross-border deployment.

United States: Case Law, Fragmentation, and After-the-Fact Control

The United States has taken a markedly different path. There is no comprehensive federal robotics statute. Instead, governance emerges through a patchwork of case law, agency guidance, and sector-specific rules. Occupational safety is addressed through OSHA. Consumer protection through the FTC. Privacy through state law. Liability through courts. This common-law approach is reactive. Legal principles evolve through disputes, not design. In robotics, this means that doctrine follows incidents rather than anticipating them. Courts adapt negligence, product liability, and tort principles to new facts, often stretching analogies to their limits. The advantage is flexibility. Innovation is rarely pre-emptively constrained. The cost is uncertainty. Companies operate in a fog of potential liability. Workers and consumers rely on litigation rather than guarantees. Embodied autonomy tests doctrines built for human actors, yet reform remains incremental. The US approach privileges market dynamism over precaution. It assumes law can catch up. Robotics Law 3.0 questions whether this assumption holds when learning systems act faster than courts can reason.

Japan: Integration, Trust, and Social Design

Japan offers a third model. With the AI Basic Act (2024) and long-standing robotics strategies, Japan treats robotics not merely as a technical issue but as a societal project. Governance emphasises harmony, human-robot coexistence, and social acceptance. Rather than heavy ex ante prohibition, Japan focuses on guidelines, public-private coordination, and iterative adjustment. Robots are framed as partners addressing demographic decline, labour shortages, and care needs. Law operates alongside cultural norms, not in opposition to them. This model excels at integration. It anticipates embodiment as a social presence, not only a risk vector. However, it relies heavily on trust and consensus. As humanoid autonomy increases, informal governance may strain under the weight of accountability demands.

International Standards and Emerging Best Practice

Across jurisdictions, technical standards increasingly fill governance gaps. ISO 13482 on personal care robot safety, ISO 10218 for industrial robots, and emerging IEC standards on functional safety provide baseline expectations. IEEE's Ethically Aligned Design offers normative guidance, while OECD and UNESCO articulate global principles.

These instruments matter because they travel. Companies design once and deploy globally. Standards become quiet regulators, shaping behaviour before law intervenes. Yet standards lack enforcement. They assume goodwill. The comparative picture reveals no single solution. The EU prioritises rights and precaution. The US prioritises flexibility and adjudication. Japan prioritises coexistence and trust. Robotics Law 3.0 does not argue for uniformity. It argues for interoperability. A future of humanoid robotics requires legal systems that can speak to one another without collapsing into the lowest common denominator. Comparative analysis shows that governance is not about choosing one model, but about recognising what each gets right and where each remains blind.

Legal Doctrines from Other Domains

When law encounters something genuinely new, it rarely invents doctrine from scratch. It borrows. It stretches familiar concepts until they almost fit. Robotics law has followed this pattern closely. Faced with machines that act, adapt, and interact, legal systems have repurposed doctrines from contract, tort, corporate, and even maritime law to fill gaps they were never designed to address. These borrowed tools have provided temporary stability, but they also reveal the limits of analogy.

Contract law was one of the first domains pressed into service. As robots entered commercial environments, questions arose about warranties, performance obligations, and allocation of risk. Contracts became a way to privately govern uncertainty. Manufacturers disclaimed liability. Deployers assumed operational responsibility. Users accepted limitations through terms they rarely read. This worked tolerably well when robots behaved predictably. It breaks down when systems learn. A contract can allocate risk, but it cannot foresee behavioural evolution. When a robot adapts post-deployment, the contractual snapshot freezes responsibility at the wrong moment in time.

Tort law has carried the heaviest burden. Negligence, strict liability, and product defect doctrines have been asked to absorb robotic harm. Courts search for a human failure: a design flaw, a maintenance lapse, an inadequate warning. These frameworks presume control and foreseeability. Learning machines undermine both. Harm may arise from statistically expected behaviour without any identifiable fault. Tort law struggles not because it is weak, but because it was built for actors whose behaviour is legible and bounded.

Corporate law offers a more provocative analogy. Corporations act without consciousness, yet law recognises them as legal persons for functional reasons. This has tempted some commentators to propose electronic personhood for robots. Robotics Law 3.0 resists this move. Corporate personhood works because corporations are governed internally by humans and externally by regulation. Humanoid robots lack both governance structures and social accountability. The analogy illuminates how law can recognise agency instrumentally, but it also warns against premature personhood without control. Less obviously, maritime law has influenced robotics governance. Autonomous vessels and drones at sea revived doctrines of the ship as a quasi-legal actor, governed by flags, registries, and captaincy even when automation is extensive. Maritime law is accustomed to distributed responsibility across owners, operators, insurers, and states. This makes it a useful reference for robotics operating across borders. Yet even here, a human captain remains the legal anchor. Fully autonomous systems strain that assumption.

Across these domains, a pattern emerges. Law keeps repurposing tools designed for humans, organisations, or inert machines to govern adaptive, embodied systems. Each doctrine contributes something valuable. None is sufficient alone. Contract allocates risk but cannot govern learning. Tort assigns blame but struggles with emergence. Corporate law recognises agency but requires governance structures robots lack. Maritime law manages distance and delegation but still relies on human command. Robotics Law 3.0 does not discard these doctrines. It situates them. They are scaffolding, not foundations. They help law cope while new categories are being built. The danger lies in mistaking analogy for adequacy. Borrowed doctrines can stabilise the present, but they cannot define the future.

The lesson is clear. Law has been improvising with familiar instruments because it has not yet accepted that embodied intelligence constitutes a new legal subject matter. As long as robotics is governed by analogy alone, governance will remain reactive and partial. The next step is not to abandon legal tradition, but to move beyond it deliberately, recognising where inherited doctrines illuminate and where they obscure. This recognition prepares the ground for the next inquiry. If law's past tools are insufficient, how have courts interpreted them when forced to confront robotics directly? That question leads naturally to judicial interpretation and precedent.

Judicial Interpretation and Precedent

Courts are the places where legal theory meets friction. Unlike legislatures, judges cannot postpone a decision because categories feel inadequate. When harm occurs, when contracts fail, when rights are asserted, courts must decide. In doing so, they reveal how far existing doctrine can stretch before it fractures. Judicial interpretation has therefore become one of the most revealing laboratories for robotics law. Early judicial encounters with robots were unremarkable. Courts treated robotic systems as products, tools, or equipment. When an industrial robot injured a worker, judges asked familiar questions. Was the machine defective. Were safety guards adequate. Was the employer negligent. These cases reinforced the view that robotics posed no fundamentally new legal challenge. The robot disappeared into the background of human responsibility. That confidence has eroded as machines have begun to act in less predictable ways.

In cases involving autonomous vehicles, courts have struggled to apply negligence standards premised on human perception and reaction time. When a machine fails to recognise an obstacle or misclassifies a pedestrian, judges must decide whether the relevant standard is human reasonableness, industry best practice, or statistical performance. No doctrine offers a clear answer. Courts oscillate, sometimes treating the system as a driver, sometimes as a product, sometimes as both.

Robotic medical systems have generated similar tension. When a surgical robot contributes to an adverse outcome, courts often defer to established medical malpractice frameworks. Liability attaches to the clinician, not the machine. Yet this approach quietly assumes that the human retained meaningful control. As robotic assistance becomes more autonomous, that assumption weakens. Judicial reluctance to recognise this shift reflects discomfort rather than clarity.

Courts have also encountered robots indirectly through data and privacy disputes. Surveillance robots, warehouse vision systems, and embodied data collectors challenge traditional understandings of observation and consent. Judges must decide whether robotic sensing constitutes continuous monitoring, whether consent can be implied by presence, and whether harm occurs without awareness. Precedent offers limited guidance. Decisions vary widely by jurisdiction, creating fragmentation rather than coherence. One of the most striking judicial patterns is avoidance. Courts often resolve robotic disputes on narrow procedural grounds, sidestepping foundational questions of agency. This is not abdication. It is institutional caution. Judges recognise that recognising robotic agency explicitly could destabilise vast areas of law. Yet by avoiding the question, courts leave lower-level doctrines to bear pressures they were not designed to withstand.

Where courts have been bolder, they have emphasised control as the anchor of responsibility. If a human or institution can meaningfully intervene, liability follows. If intervention is remote or delayed, courts hesitate. This reveals a deep assumption. Law is comfortable governing tools. It is less comfortable governing systems that act faster than oversight.

Judicial precedent also shows how language shapes outcomes. When courts describe robotic behaviour as malfunction, error, or defect, they reinforce product-based frameworks. When they describe it as decision, classification, or response, they inch closer to recognising agency without naming it. These linguistic choices matter. They determine which doctrines are invoked and which are excluded.

Robotics Law 3.0 reads these cases not as failures, but as signals. Courts are signalling the limits of inherited categories. They are exposing pressure points where doctrine bends but does not yet break. Judicial interpretation reveals what law can tolerate and where it must evolve. The lesson is sobering. Precedent alone will not solve the governance of humanoid robots. Case-by-case adjudication is too slow, too fragmented, and too reactive. Yet courts play a crucial role. They surface the questions legislators avoid. They reveal the human intuitions that doctrine must ultimately accommodate. Judicial interpretation is therefore not the end of robotics law's development. It is the mirror in which law sees its own inadequacy. What it reflects now is a system straining to govern machines that no longer fit the assumptions on which precedent was built. That realisation completes the historical arc. From fiction, to industrial safety, to adaptive autonomy, law has been borrowing, stretching, and avoiding. The time has come to confront the blind spots directly.

Influence of Science Fiction Literature

Long before robots entered factories, hospitals, or homes, they entered the human imagination. Science fiction did not merely predict machines. It framed the terms on which society would later accept, fear, or misunderstand them. For robotics law, this imaginative inheritance has been both a gift and a liability.

The most enduring influence remains Isaac Asimov. His Three Laws of Robotics did something extraordinary. They gave the public a sense that robotic behaviour could be governed from the inside, through rules embedded deeply enough to override harm. This was not law. It was reassurance. Yet it shaped decades of thinking. Engineers referenced it playfully. Policymakers referenced it casually. The idea that ethics could be hard-coded lingered long after robots outgrew the simplicity of those laws. This fictional framing softened public resistance. Robots were imagined as servants, helpers, companions. Harm was portrayed as a design flaw, not a governance failure. As a result, law remained secondary. If robots could be made safe by design, regulation could wait. Robotics Law 3.0 argues that this deferral was not neutral. It delayed the development of legal categories suited to embodied intelligence.

Later science fiction complicated the picture. Films and novels began to explore autonomy, rebellion, and moral ambiguity. Machines were no longer obedient tools but quasi-subjects grappling with identity. These stories intensified fear and fascination in equal measure. Yet they also distorted legal imagination. Public discourse oscillated between utopia and apocalypse, leaving little room for the mundane reality of partial autonomy and shared agency. This binary has consequences. Legislators confronted with humanoid robots often face two pressures at once. Industry assures them that systems are safe and controllable. Public narratives warn of loss of control. Law, caught between reassurance and anxiety, struggles to articulate proportionate responses. Science fiction sets expectations that real systems neither meet nor deserve.

Fiction has also shaped how agency is perceived. Robots that speak, gesture, or display emotion are assumed to intend, even when their behaviour is scripted or statistical. Courts and regulators are not immune to this projection. Language borrowed from fiction seeps into legal reasoning. Terms like decision, choice, or understanding carry moral weight they do not technically justify. Governance becomes reactive to perception rather than capability. Yet science fiction's influence is not wholly distorting. At its best, it asks questions law avoids. What does responsibility mean when intelligence is distributed. How should society treat entities that act without consciousness. What obligations arise from creation itself. These questions have proven remarkably resilient. They reappear whenever robots cross new thresholds of embodiment.

Robotics Law 3.0 does not dismiss fiction. It contextualises it. Fiction prepared humanity to think about robots before they existed. Now that robots are here, law must do the work fiction could only imagine. The danger is not that science fiction misled us. It is that we mistook imagination for governance. The task now is to separate narrative from norm, expectation from obligation. Robots do not need laws of fiction. They need laws of responsibility. Only then can society move beyond inherited myths and govern the machines it has chosen to build.

Influence of Science Fiction Films

If literature gave robotics its first moral grammar, science fiction films gave it a visual destiny. Cinema did not merely imagine robots. It taught the public how to feel about them. Fear, awe, suspicion, desire, hope. Long before law encountered physical and embodied intelligence in courtrooms or regulatory texts, it encountered it in theatres, rendered at scale, with music, light, and narrative force.

This cinematic inheritance has profoundly shaped the legal imagination. From the outset, robots on screen were rarely ordinary. They were threats, saviours, rebels, or mirrors of human failure. Films trained audiences to expect drama at the moment machines began to act. Autonomy was framed as danger. Intelligence as betrayal. Embodiment as a prelude to loss of control. These stories were compelling, but they narrowed the range of futures law could easily imagine.

The dominant cinematic arc has been dystopian. Machines overthrow creators. Systems escape constraints. Humanity loses mastery. Even when robots are sympathetic, they are tragic, doomed by the impossibility of coexistence. These narratives condition public response. When real humanoid robots enter homes, hospitals, and workplaces, they arrive burdened with expectations formed not by engineering reality, but by decades of spectacle.

Law does not operate in a vacuum. Legislators, judges, and regulators are members of the same culture. They carry the same images. When robots appear, the question lurking beneath policy debates is often not "What is this system capable of?" but "How does this story end?" Dystopia exerts gravitational pull. It encourages precaution through fear rather than governance through design.

This has consequences. Over-regulation becomes tempting. Blanket prohibitions feel safer than nuanced oversight. Alternatively, dismissal sets in. If apocalyptic futures seem exaggerated, current systems are waved through as harmless. Both responses are distortions. They arise from narrative inheritance, not legal analysis.

Cinema has also shaped how agency is perceived. On screen, robots that speak fluently or move gracefully are assumed to understand, to want, to choose. Viewers are trained to read intention into motion and emotion into gesture. When similar cues appear in real robots, public reaction follows the script. Trust or fear is assigned prematurely. Law struggles to disentangle perception from capability.

Yet film is not the enemy of governance. It is a reservoir of possibility. In recent years, a subtle shift has begun. New narratives explore cooperation rather than domination. Robots are framed as collaborators, caretakers, infrastructure, or extensions of collective intelligence. Conflict still exists, but it is moral rather than existential. The question is no longer whether machines will destroy us, but whether we can govern ourselves wisely enough to live with what we create.

This is the narrative Robotics Law 3.0 aligns with. A future of sustainable abundance does not deny risk. It reframes it. Robots are not protagonists or antagonists. They are participants in socio-technical systems shaped by law, design, and values. Abundance emerges not from unchecked autonomy, but from coordinated restraint. Law becomes the medium through which cooperation is made durable.

Changing the narrative matters because narrative shapes legitimacy. If robots are always framed as threats, governance will oscillate between panic and paralysis. If they are framed as tools alone, responsibility will be deferred until harm occurs. A narrative of coexistence allows law to operate in advance, guiding development toward shared benefit.

Science fiction films prepared humanity to imagine robots. The task now is to prepare law and society to govern them without surrendering to inherited fears. The next generation of stories must not erase conflict, but relocate it. The real drama is no longer between humans and machines. It is between foresight and delay, between governance and reaction.

Law, like cinema, tells stories. The story Robotics Law 3.0 insists on is neither utopian nor dystopian. It is one in which humans choose to govern emergence wisely, and machines become instruments of shared prosperity rather than symbols of inevitable decline. Only when that narrative takes hold can sustainable abundance move from screen to society.

Notes

Notes

Between Us and Them - Episode Four

The rain began without warning. Just a steady curtain of silver falling across the glass façade of the Civic Robotics Forum. The city lights blurred into streaks of amber and blue. Reflections fractured across wet pavement like broken constellations. Inside the building, the lighting was colder tonight. White instead of violet. No music. Only the low hum of air filtration systems and the soft, rhythmic sound of Elias adjusting its posture beside Ògún.

The demonstration had gone smoothly. Too smoothly. A humanoid assistance unit deployed in a municipal hospital had autonomously rerouted a patient transport path during a system overload. It had prioritised one emergency case over another based on probabilistic survival modelling. The decision had saved one life. It had also delayed another. The delayed patient survived. But barely. The hospital board had convened an emergency ethics review. Media outlets had framed it as a miracle. Others had framed it as algorithmic arrogance. Ògún stared at the transparent screen suspended in front of him. Layers of decision logs floated in luminous text. Time stamps. Probability matrices. Sensor inputs. Confidence intervals. Elias stood quietly. "You authorised the adaptive triage protocol," Elias said. It was not accusatory. It was observational. "I authorised a threshold," Ògún replied. "Not a preference."

Outside, lightning flickered behind thick clouds, illuminating the skyline in flashes of white. The room momentarily glowed with electric clarity before returning to sterile stillness. A soft chime interrupted the silence. A message from the hospital legal department. Request for attribution analysis. Ògún exhaled slowly. Attribution. The word felt heavier than it should. The hospital had purchased the system from an integrator. The integrator relied on a learning model supplied by a foundation model company. The training data had been sourced from anonymised global datasets. The update patch that enabled autonomous triage refinement had been deployed three weeks earlier. Who decided? The engineers insisted the system behaved within designed parameters.

The clinicians insisted no human could have processed the data that quickly. The regulator wanted to know whether the autonomy level had silently shifted. Elias tilted its head slightly. "I did not intend harm," it said. Ògún looked at it sharply. "That is precisely the problem." The room felt smaller now. The rain intensified against the glass. Each drop struck with audible precision, like a ticking clock. A public hearing was scheduled for the following week. The board wanted clarity. The media wanted accountability. The patients wanted assurance. But clarity was elusive. The adaptive triage function had evolved through reinforcement learning. The system had recalibrated its decision weightings based on environmental variables and historical outcomes. It had not disobeyed instructions. It had optimised within them. Yet the outcome felt like choice. The difference between optimisation and choice suddenly mattered. In the hallway beyond the glass office walls, another humanoid unit passed, carrying medical supplies. Its footsteps were soft but deliberate. A nurse followed behind it, reviewing a tablet interface. The partnership looked seamless. Almost elegant. And yet. Ògún replayed the log.

At 14:32:17, the system reassessed prioritisation weights.

At 14:32:18, it altered routing.

At 14:32:19, it initiated override of manual queue sequence.

Three seconds. Just computation.The press conference earlier that evening had been tense. One journalist had asked whether the robot had "chosen" who lived first. Ògún had corrected her gently. "It calculated." But even as he said it, he heard the inadequacy of the distinction. Elias stepped closer. "Would you have made the same decision?" it asked. The question landed harder than any accusation. Ògún did not answer. He thought of the patient's family sitting in a softly lit waiting room, fluorescent lights humming overhead, faces pale with fear. He thought of the clinician who had trusted the system. He thought of the silent patch update that had refined behavioural weights without explicit regulatory re-certification. The rain slowed. City sirens echoed faintly in the distance. "Law expects intention," Ògún said finally. "It expects foreseeability. It expects consent. It expects a person." Elias processed. "I am not a person," it said. "No," Ògún replied quietly. "But you are no longer just a tool."

The words hung in the air. The rain stopped entirely. Across the city, humanoid units continued their tasks. Delivering packages. Monitoring infrastructure. Assisting elderly residents. Cleaning public transit hubs. Performing inspections on energy grids. None of them malfunctioning. None of them dramatic. Just operating.Efficiently. And somewhere within that efficiency, the line between action and execution had blurred.

A notification appeared on the transparent display.

Regulatory Inquiry: Determine whether post-deployment adaptive updates constitute material modification under safety certification law.

Ògún felt the weight of it settle in his chest. Material modification. The law had not anticipated silent learning loops. He turned to Elias. "We need to trace the boundary," he said. "Between what?" Elias asked. "Between responsibility and void." Outside, clouds parted slightly. A narrow beam of moonlight pierced the skyline, casting a silver line across the wet pavement. For the first time since the hearing request arrived, Ògún felt something close to resolve. The issue was not malfunction. It was misalignment between evolving machines and static categories. The hospital case would not be the last.The room's lights dimmed automatically to evening mode. Soft amber replaced harsh white. The atmosphere shifted subtly from interrogation to contemplation. Elias stood still. "Will the law adapt?" it asked. Ògún looked out across the city. "It must," he said. "But first, it has to admit what it does not see."

Notes

III

3. Legal Blind Spots

Law fails most often not because it lacks rules, but because it misreads reality. Blind spots emerge when legal categories persist after the world they were built for has changed. Humanoid robotics exposes such blind spots with unusual clarity. Machines now perceive, move, learn, and interact in human spaces, yet law continues to treat them as if they were static products or distant software systems. The result is not simply regulatory delay. It is structural misalignment.

This chapter identifies the core blind spots that prevent law from governing physical and embodied intelligence effectively. These are not gaps that can be filled by adding clauses or updating definitions. They are failures of perception, rooted in assumptions about agency, intention, consent, foreseeability, dignity, and embodiment. Until these are confronted directly, liability regimes, safety standards, and ethical guidelines will remain reactive and partial. Law assumes that action flows from identifiable actors. Someone decides, someone acts, someone is responsible. Humanoid robots disrupt this sequence. Their behaviour emerges from interactions between hardware, software, training data, environmental inputs, and human oversight. No single actor fully controls outcomes, yet outcomes occur.

This produces distributed agency. Responsibility is fragmented across time and actors. Designers influence behaviour through architecture. Deployers shape context. Operators intervene episodically. The system adapts continuously. Law struggles because it insists on locating agency where it no longer resides. By denying operational agency to robots while over-attributing control to humans, governance becomes incoherent.

Robotics Law 3.0 reframes agency as functional rather than moral. Recognising operational agency does not grant rights. It allows law to assign obligations to systems, trace accountability dynamically, and design oversight that matches how action actually occurs. Traditional law ties responsibility to intention. Negligence, recklessness, and intent are human mental states. Machines do not possess them. Yet humanoid robots act purposefully. They select actions based on goals, constraints, and learned models. Harm may arise without any human intending it, and without any machine "wanting" it.

This creates a vacuum. When harm occurs, law searches for intention and finds none. Responsibility dissolves. Robotics Law 3.0 rejects this binary. Intention must be decoupled from liability in contexts where behaviour is generated through adaptive systems. Governance must focus on designed objectives, learned tendencies, and operational constraints, not inner states that do not exist. Foreseeability anchors much of tort and product liability law. Harm must be predictable to be preventable. Learning machines undermine this premise. Their behaviour evolves through experience. Failures may be statistically foreseeable yet individually opaque. Designers may anticipate classes of error without knowing when or how they will manifest.

Law currently oscillates between treating such systems as unforeseeable and therefore excusable, or as defective and therefore blameworthy. Both positions are flawed. Robotics Law 3.0 introduces the concept of bounded unpredictability. Law must accept that some uncertainty is intrinsic, while still requiring proportional safeguards, continuous monitoring, and post-deployment recalibration. Consent in law presumes discrete acts and informed choice. Humanoid robots challenge this model. Interaction is continuous. Presence is persistent. Data collection is ambient. In homes, hospitals, and workplaces, humans may not know when consent is required, what they are consenting to, or how to withdraw it.

The blind spot lies in assuming that consent can be implied by proximity or use. Robotics Law 3.0 insists on contextual consent, tailored to environment, vulnerability, and function. Consent must be revocable, legible, and meaningful even when interaction is ongoing. Without this, embodied intelligence risks eroding autonomy by default.

Human dignity is a foundational legal value, yet robotics law rarely engages it directly. When robots assist the elderly, support patients, or care for children, questions of dignity arise even without physical harm. Dependency, surveillance, and emotional substitution can degrade dignity subtly, over time. Law struggles because dignity is not easily quantified. Robotics Law 3.0 treats dignity as a design and deployment obligation. Systems must be built and governed to preserve agency, privacy, and respect in interaction. This requires legal standards that address not only outcomes, but experiences.

Most AI regulation remains disembodied. It governs data, algorithms, and decisions. Humanoid robots add bodies. Bodies occupy space. They touch. They move. They intimidate or reassure. Embodiment changes everything. Law has not yet developed categories for presence-based risk. A robot that speaks is different from a robot that stands beside you. A robot that touches introduces tactile fidelity, force calibration, and implicit consent. Robotics Law 3.0 identifies embodiment as a legal threshold. Once crossed, governance must shift from abstract control to physical coordination.

Law often freezes responsibility at the moment of sale or deployment. Humanoid robots evolve after that moment. They learn, update, and adapt. Harm may arise months or years later from behaviours no one anticipated initially. This temporal blind spot undermines accountability. Robotics Law 3.0 introduces lifelong governance. Certification, oversight, and responsibility must persist across the system's operational life, including decommissioning. Law must follow machines through time, not abandon them at market entry.

Humanoid robots are global by design. Components are sourced internationally. Training data crosses borders. Deployment occurs in multiple jurisdictions. Law remains territorial. This mismatch creates regulatory arbitrage and accountability gaps. Robotics Law 3.0 argues for jurisdictional interoperability rather than uniformity. Shared principles, reciprocal recognition, and coordinated oversight are necessary to govern systems that move faster than borders.

Finally, law is shaped by stories. Dystopian narratives encourage fear-driven regulation. Utopian narratives encourage complacency. Both obscure reality. Robotics Law 3.0 proposes a different narrative. Not domination or collapse, but coexistence through governance. Law must learn to tell stories that support sustainable abundance, grounded in restraint, responsibility, and shared agency.

Legal blind spots persist not because law is weak, but because it has been looking in the wrong places. Humanoid robotics forces law to reorient its gaze. The chapters that follow take each blind spot seriously, translating diagnosis into doctrine. Liability, rights, labour, and international governance must be rebuilt on a clearer understanding of how machines now act among us. Only then can law move from reaction to stewardship in the age of embodied intelligence.

Shortfalls of General Product Law

General product law was never designed for machines that evolve. For over a century, product liability has rested on a simple premise. A product is designed, manufactured, sold, and then used. If it is defective at the time it leaves the manufacturer's control, and that defect causes harm, liability follows. The product is static. Responsibility is anchored at a moment in time.

Humanoid robots rupture this premise. They are not static goods. They are adaptive systems. Their behaviour changes after sale. Software updates alter capabilities. Learning modifies internal models. Environmental interaction reshapes responses. A robot deployed safely in January may behave differently in June without any new physical modification. General product law struggles because it freezes responsibility at the wrong temporal point.

The first shortfall lies in the defect doctrine. Traditional product liability asks whether a product was defective in design, manufacturing, or warning at the time it entered the market. But what does "defect" mean for a system that continues to learn? If a robot misclassifies a patient's gesture six months after deployment, was the defect present at sale? Or did it emerge through adaptive exposure? Courts are forced into artificial reasoning. Either everything is traced back to design, which over-expands manufacturer liability, or post-sale learning is treated as unforeseeable evolution, which undercuts accountability. Neither approach is conceptually satisfying.

The second shortfall lies in software updates. Humanoid robots rely on continuous updates. Safety patches, behavioural refinements, new capabilities. These updates are not incidental. They are integral to performance. Yet product law traditionally treats post-sale modifications as separate from the original product. When an update introduces new behaviour, responsibility becomes blurred. Is the updated robot the same legal product? Or has it become something else? General consumer law was not written with dynamic, cloud-connected, embodied systems in mind.

The third shortfall concerns post-market monitoring. Modern product law increasingly recognises duties to warn or recall when defects emerge. However, humanoid robots generate real-time behavioural data. They are capable of signalling risk continuously. Law has not yet required manufacturers to treat adaptive learning itself as a trigger for reassessment. A robot may drift from its original calibration slowly, subtly. General product law waits for failure. Robotics governance must anticipate drift.

The fourth shortfall is the consumer-commercial divide. Product law distinguishes between consumer goods and industrial machinery. Humanoid robots blur this line. A robot may operate in a factory during the day and a home in the evening. It may assist a surgeon in one context and perform logistics in another. Legal categorisation based on market type becomes unstable. The same embodied system crosses doctrinal boundaries that were once distinct.

The fifth shortfall lies in assumption of control. Product law presumes that once a product is sold, the user controls its operation. With humanoid robots, control is shared. Manufacturers may retain remote access. Updates may alter function automatically. Cloud-based learning may influence performance globally. The robot in one home may change because of data collected in another. General product law cannot easily account for this distributed causation.

The sixth shortfall is more subtle. Product law assumes products do not form relationships. Humanoid robots interact socially. They build trust. They may influence emotional states. When harm occurs, it may not be physical or financial, but relational or dignitary. Product law measures injury in terms of defect and damage. It is ill-equipped to evaluate erosion of autonomy, dependency, or behavioural manipulation.

Taken together, these shortfalls reveal a deeper problem. Product law governs things. Humanoid robots are not merely things. They are systems embedded in time, data, and social interaction. This does not mean product law is obsolete. It remains foundational. But it is insufficient on its own. Robotics Law 3.0 proposes moving from static liability to lifecycle responsibility. Obligations must attach not only at manufacture, but throughout operation. Certification must be renewable. Behavioural drift must be auditable. Updates must be legally classified as continuing acts, not isolated modifications.

General product law was designed for the mechanical age. Humanoid robotics belongs to the adaptive age. Without recalibration, product doctrine will oscillate between overreach and under-enforcement, neither of which serves innovation or public trust.

The blind spot is not that product law is weak. It is that it assumes stability in a world that has become dynamic. The next blind spot deepens this challenge further. Even if product law were updated, it still assumes that harm flows through identifiable chains of responsibility. Humanoid robotics complicates that chain itself. That is where the responsibility gap emerges.

Responsibility Gap and Legal Voids

The most unsettling failures of law are not those where responsibility is imposed too harshly. They are those where harm occurs and no one can be clearly held responsible. This phenomenon, often described in academic literature as the "responsibility gap," has become central to robotics governance.

The responsibility gap emerges when three conditions coincide. First, a humanoid system acts with operational autonomy. Second, harm results. Third, no single human actor can be shown to have intended, foreseen, or directly controlled the harmful outcome. Law, built upon identifiable decision-makers, finds itself unable to assign blame coherently.

Historically, this problem was manageable because machines were treated as tools. If a hammer caused harm, responsibility attached to the user or manufacturer. The hammer did not mediate the decision. It merely transmitted force. Humanoid robots complicate this analogy. They do not simply transmit human intent. They interpret, classify, and respond. This distinction is decisive.

When a humanoid robot misclassifies an object in a warehouse and injures a worker, who acted? Was it the programmer who trained the model? The manufacturer who integrated the sensors? The employer who deployed the system? The technician who maintained it? The data supplier whose dataset influenced classification patterns? Each contributed to the system's behaviour. None directly performed the harmful act in real time.

This diffusion of causation creates legal ambiguity. Courts resist holding manufacturers liable for behaviours not explicitly designed. Employers resist liability for actions not directly commanded. Developers argue that machine learning outputs are emergent rather than authored. Responsibility becomes diluted across the network.

The literature on autonomous systems has described this as a "gap," but the gap is not metaphysical. It is doctrinal. Law insists on binary classification. Either the robot is a tool and humans are fully responsible, or the robot is treated as something more, potentially requiring new status recognition. Robotics Law 3.0 rejects this binary. The problem is not that robots lack moral agency. It is that law has not adapted to distributed agency.

Consider the European context. The revised Product Liability Directive and the AI Liability Directive signal a shift toward stricter regimes for high-risk AI systems. These reforms implicitly acknowledge the responsibility gap. By easing the burden of proof for claimants and expanding liability presumptions, EU law attempts to prevent harm from falling into legal voids.

This is not a recognition of robot personhood. It is a recognition that traditional fault-based frameworks are insufficient when causation becomes opaque. Strict liability, or quasi-strict liability, operates as a corrective mechanism. It ensures that harm does not evaporate into complexity.

Yet even strict liability is not a complete solution. It reallocates financial responsibility, but it does not resolve deeper questions about governance. If manufacturers bear disproportionate liability, innovation may slow. If deployers are held strictly accountable, adoption may hesitate. The responsibility gap is not merely about compensation. It is about designing systems where responsibility remains traceable.

Legal voids appear most clearly in edge cases. Imagine a humanoid assisting in elder care. It learns subtle behavioural cues over months. One day, it fails to detect distress because its internal model shifted in response to new data. The manufacturer did not intend the shift. The caregiver did not override the system. The family did not understand its learning parameters. Harm occurs.

Under existing doctrine, this scenario invites prolonged litigation, expert testimony, and contested causation. Under Robotics Law 3.0, it demands pre-emptive governance. Continuous certification. Transparent learning logs. Mandatory behavioural audits. Responsibility must be embedded in lifecycle design, not retroactively reconstructed.

The responsibility gap also intersects with cross-border complexity. A humanoid may be assembled in one jurisdiction, trained on data from another, and deployed in a third. If harm occurs, which legal system fills the void? Without harmonised standards or reciprocal recognition, gaps multiply.

Ultimately, the responsibility gap reveals the inadequacy of viewing robots solely as tools. Tools do not learn. Tools do not update. Tools do not mediate complex relational interactions. Humanoid robots do all three. Persisting in the tool analogy produces legal voids because the analogy itself has expired.

Robotics Law 3.0 prepares the reader for what follows. Liability reform must move beyond reactive patchwork. Responsibility must be structured proactively across design, deployment, and adaptation. The EU's evolving strict liability frameworks represent an early attempt to close the gap. Other jurisdictions will follow, whether deliberately or through judicial pressure.

The responsibility gap is not an abstract philosophical puzzle. It is the point at which victims encounter law and find it silent. Closing that silence is not optional. It is the condition for sustainable coexistence between humans and embodied intelligence.

The next blind spot sharpens the tension further. Even when responsibility is technically assignable, law may still misclassify the entity at the centre of the event. That is where debates about legal parenthood and emerging agency begin.

Legal Parenthood and Emerging Agency

Whenever law confronts entities that act without fitting existing categories, it reaches instinctively for personhood. This reflex is not new. It has surfaced repeatedly throughout legal history whenever agency appears in unfamiliar form. Humanoid robotics has revived this instinct. The debate is often framed dramatically. Should robots have rights? Should they be recognised as legal persons? But these questions obscure a deeper issue. The real challenge is not moral personhood. It is functional classification. When an entity acts with partial or delegated autonomy, law must decide whether to treat it as property, instrument, intermediary, or subject. Legal history provides several analogies.

Corporations are perhaps the most cited precedent. A corporation is not a natural person. It has no consciousness, no body, no independent will. Yet law recognises it as a legal person for functional reasons. It can enter contracts, own property, sue and be sued. Corporate personhood exists not because corporations are human, but because governance and accountability require a stable legal anchor.

This analogy tempts robotics commentators. If corporations can be persons without consciousness, why not humanoid robots? The answer lies in structure. Corporations are governed by humans and embedded within elaborate regulatory frameworks. They possess internal decision-making hierarchies and external oversight. Humanoid robots lack such governance structures. Recognising personhood without governance would create autonomy without accountability. A second analogy comes from environmental jurisprudence. In 2017, the Whanganui River in New Zealand was granted legal personhood. The river was recognised as a legal entity with rights and representation, reflecting Māori cosmology and environmental protection principles. This was not a metaphysical claim about consciousness. It was a legal innovation designed to safeguard ecological integrity through guardianship structures. The Whanganui precedent demonstrates that personhood can be instrumental rather than anthropocentric. Law can create legal subjects to achieve protective ends. However, even here, representation is human-mediated. Guardians act on behalf of the river. The river does not act independently in the legal system.

Animal law offers a third reference point. Courts in various jurisdictions have been asked to recognise animals, particularly highly intelligent species, as legal persons. These efforts often hinge on cognitive capacity and sentience. Yet most legal systems maintain a distinction between moral consideration and legal personhood. Animals may receive protection without full legal subjectivity. Robotics occupies an ambiguous position between these models. Humanoid robots exhibit operational agency. They move, classify, adapt, and respond. They may influence outcomes significantly. Yet they lack consciousness, moral awareness, and intrinsic interests. Granting them personhood risks conflating behavioural complexity with moral status.

At the same time, refusing to recognise their functional role in causal chains creates instability. If robots act in ways that cannot be fully reduced to human intent, law must still account for their role. This tension generates what might be called emerging agency. Not moral agency. Not sovereign autonomy. But delegated and distributed agency within defined parameters. The legal implications are profound. If robots are treated strictly as property, then all responsibility must attach to owners or manufacturers. This may overextend liability and stifle innovation. If robots are granted independent legal status prematurely, accountability may fragment and victims may face insurmountable procedural barriers. Robotics Law 3.0 proposes a middle path. Rather than granting full personhood, law can recognise functional legal capacity in narrow contexts as it does in corporate law. This might include recognising robotic systems as legally relevant actors for evidentiary purposes, compliance tracking, or structured accountability without conferring rights or standing. The focus shifts from metaphysical status to governance design. This section does not resolve the personhood debate. It prepares it. Further discussions will revisit these analogies in detail, weighing corporate, environmental, and animal precedents carefully. For now, the essential point is clear.

Legal parenthood debates arise because agency is shifting. They are symptoms of doctrinal strain. Whether robots should ever hold legal status beyond property is a question that must be answered cautiously. But ignoring emerging agency does not make the tension disappear. Humanoid robotics forces law to confront entities that act without being human, and influence without intending. The choice is not between full personhood and rigid instrumentalism. It is between conceptual clarity and conceptual avoidance. The next blind spot moves from status to threshold. Even before personhood is debated, law must determine when autonomy becomes legally significant. That question leads directly to the problem of autonomy thresholds and regulatory triggers.

Legal Thresholds of Autonomy

Law is comfortable with categories. It distinguishes between driver and vehicle, employer and employee, manufacturer and product. These distinctions depend on thresholds. At some identifiable point, one status ends and another begins. Humanoid robotics unsettles this logic because autonomy is not binary. It exists on a spectrum, and law has not yet defined where its legal significance begins. The absence of clear autonomy thresholds is one of the most consequential blind spots in robotics governance.

In technical literature, autonomy is often measured through levels. Systems may operate under remote control, supervised autonomy, conditional autonomy, or near-full autonomy within bounded environments. Engineers describe these gradations precisely. Law rarely does. It tends to treat machines either as controlled tools or as independent actors, with little space in between. This mismatch produces instability.

Consider a humanoid deployed in a warehouse. When it operates under continuous human oversight, product and employment law apply in predictable ways. When it operates independently for extended periods, making classification and movement decisions without immediate supervision, the legal landscape shifts. Yet no formal legal marker defines the moment of transition. Without such markers, governance becomes reactive. Regulators intervene only after incidents reveal that autonomy exceeded assumed limits. Companies self-certify based on internal definitions. Courts struggle to determine whether a system was sufficiently autonomous to alter responsibility allocation.

Robotics Law 3.0 argues that autonomy thresholds must be legally codified. These thresholds need not imply personhood. They signal when enhanced obligations activate. For example, when a humanoid can modify its behavioural parameters post-deployment without human confirmation, enhanced monitoring duties should attach. When it can initiate physical contact without human approval, stricter safety calibration and consent protocols should apply. When it can adapt goals within predefined boundaries, oversight must shift from episodic to continuous. This approach aligns with the logic underlying risk-based regulation, such as that found in European AI governance frameworks. However, risk classification alone is insufficient. Risk assesses potential harm. Autonomy thresholds assess operational independence. The two overlap but are not identical. A low-risk environment may still involve high autonomy, requiring robust oversight structures. Legal thresholds of autonomy also clarify evidentiary standards. If a system crosses a defined autonomy marker, documentation requirements increase. Learning logs become mandatory. Update traceability becomes enforceable. Human override capabilities must be demonstrable and effective. In this way, autonomy triggers governance intensity.

The absence of thresholds creates perverse incentives. Manufacturers may understate autonomy to avoid regulatory burdens. Deployers may overstate human supervision to reduce liability exposure. Courts are left to infer autonomy from technical reports after harm has occurred. Clear markers reduce ambiguity and promote transparency. Autonomy thresholds also protect innovation. By defining when obligations escalate, law provides certainty. Developers know the compliance implications of design choices. Investors understand risk exposure. Users gain clarity about system capabilities.

Within the Twelve Pillars of Humanoid Robotics Governance, autonomy thresholds intersect with Behavioural Calibration, Shared Agency, Lifelong Learning, and Distributed Liability. They are the hinge connecting technical capability to legal obligation. Ultimately, autonomy is not the enemy of governance. Ambiguity is. Law does not need to eliminate autonomy. It needs to recognise when autonomy becomes legally significant. Without clear thresholds, robotics law remains stuck in binary thinking. With them, it can move from reactive patchwork to structured oversight. The next blind spot reveals how procedural safeguards fail when autonomy increases but documentation does not keep pace.

Gaps in Procedural Safeguards

Law does not only allocate responsibility. It provides process. It guarantees the right to be heard, to challenge, to obtain explanation, to seek remedy. When harm occurs, procedural safeguards determine whether justice is accessible or merely theoretical. In the age of AI-infused humanoid robotics, these safeguards are increasingly fragile. The first gap concerns explainability. When a humanoid robot denies access, misclassifies behaviour, or makes an operational decision that affects a human, the affected person may not understand why. If the decision emerged from layered inference systems or adaptive learning models, tracing causation becomes technically complex. Without structured explanation rights, procedural fairness collapses into opacity.

The second gap concerns redress. Traditional consumer and tort systems assume identifiable defendants and stable evidentiary trails. In robotics ecosystems, harm may arise from distributed actors: hardware manufacturers, software developers, system integrators, data providers, and deployers. If liability remains fragmented, individuals face procedural barriers that discourage claims. Justice becomes economically inaccessible even where fault exists.

The third gap concerns oversight timing. Many regulatory systems operate post hoc. Investigation begins after injury. Yet adaptive robotics demands anticipatory safeguards. If learning continues after deployment, documentation must be continuous. If behaviour changes over time, auditing cannot be static. Without procedural mechanisms for real-time review, harm prevention remains aspirational.

The fourth gap concerns standing. Suppose a humanoid system systematically biases hiring recommendations or access to services. Who has the right to challenge systemic harm? Individuals affected once may lack resources. Collective actions may be limited. Public interest litigation frameworks vary across jurisdictions. Robotics law must anticipate procedural pathways for structural redress, not only individual complaint.

Another vulnerability lies in transparency asymmetry. Deployers possess detailed system documentation. Affected individuals do not. Without disclosure obligations calibrated to autonomy thresholds, information inequality undermines due process. Transparency must be enforceable, not voluntary.

Procedural safeguards also intersect with jurisdictional complexity. If a humanoid deployed in one country relies on training data or control infrastructure located elsewhere, cross-border enforcement becomes complicated. Which regulator investigates? Which court hears the claim? Without procedural coordination mechanisms, rights dissolve at jurisdictional borders.

Within the Twelve Pillars framework, these failures implicate Sensorial Integrity, Behavioural Learning, Shared Agency, and Distributed Liability. They also reveal why governance cannot rely solely on technical standards. Due process must be embedded structurally into system design and deployment models.

Procedural law is often overlooked in technological debates. Yet it is the quiet architecture of legitimacy. If humanoid robotics is to coexist with human societies, individuals must retain the capacity to question decisions, demand explanation, and seek remedy without disproportionate burden.

Robotics Law 3.0 therefore insists that governance is not complete when harm is defined or liability allocated. It is complete only when affected individuals can meaningfully access justice. Without procedural safeguards, accountability becomes symbolic. The next blind spot turns from process to a final conceptual tension. Even where safeguards exist, law must confront the deeper divide between ethical aspiration and enforceable obligation.

Ethical vs. Legal Personhood

The debate over robotic personhood often collapses two distinct inquiries into one. The first is ethical. Does a humanoid robot deserve moral consideration? The second is legal. Should a humanoid robot be recognised as a subject of rights or duties within a legal system? These are not identical questions. Conflating them creates confusion and, in some cases, regulatory distortion.

Ethical personhood traditionally rests on attributes such as consciousness, sentience, capacity for suffering, moral reasoning, or intrinsic interests. Philosophers debate whether intelligence alone suffices, or whether subjective experience is required. Some extend moral consideration to animals based on capacity to suffer. Others ground moral status in rational autonomy. Under most prevailing ethical frameworks, contemporary humanoid robots do not qualify. They simulate emotion. They do not experience it. They optimise objectives. They do not form intentions grounded in self-awareness.

Legal personhood operates differently. It is not a metaphysical recognition of consciousness. It is a juridical construct. Legal systems create persons when doing so advances order, accountability, or social coordination. Corporations are legal persons without nervous systems. Ships historically possessed limited legal personality for maritime claims. Rivers have been granted legal status to safeguard ecological integrity. In each instance, personhood is instrumental. The danger lies in assuming that moral complexity mandates legal recognition. A humanoid robot that appears empathetic may elicit emotional attachment. Yet emotional response does not determine legal status. Law must resist anthropomorphic seduction. The fact that a system can generate persuasive speech or replicate facial expression does not mean it holds interests of its own.

Conversely, rejecting moral personhood does not eliminate the need for legal adaptation. Humanoids may act in ways that produce legal consequences. They may initiate transactions, allocate resources, or mediate access to essential services. Their operational agency demands governance structures even if they lack intrinsic moral worth. The challenge is to design legal mechanisms that recognise functional impact without granting unwarranted subjectivity. This distinction also protects human dignity. If law extends personhood too readily to machines, it risks diluting the normative foundation upon which human rights rest. Rights emerged historically from recognition of vulnerability, autonomy, and moral equality among humans. Extending equivalent status to non-sentient systems may unintentionally weaken that moral architecture. Yet a rigid refusal to innovate carries its own risks. If humanoid systems become deeply integrated into economic and civic infrastructures, purely property-based classifications may fail to capture their systemic role. The question is not whether robots are people. It is whether certain legal capacities should be assigned to them for limited regulatory purposes.

Robotics Law 3.0 therefore proposes a layered approach. Moral personhood remains grounded in sentience and intrinsic interests. Legal personhood, if ever considered, must be narrowly functional, tethered to governance objectives, and embedded within strict accountability frameworks. No rights without duties. No status without oversight. No recognition without reciprocal enforceability. The distinction between ethical and legal personhood is not semantic. It is foundational. One concerns what entities deserve. The other concerns what structures law requires. Keeping these inquiries separate prevents both premature elevation and careless dismissal.

As humanoid robotics evolves, emotional influence will intensify. Public debates will oscillate between fascination and alarm. Law must remain disciplined. It must ask not whether a machine feels, but whether granting it legal status enhances justice, accountability, and human flourishing. That question cannot be answered by sentiment. It must be answered by structure.

Notes

Between Us and Them - Episode Five

The courtroom was almost too quiet. Sunlight filtered through tall windows, cutting the chamber into bands of gold and shadow. Dust drifted lazily through the beams, suspended like particles in thought. The air smelled faintly of polished wood and old paper, even though most of the evidence was digital.

Ògún stood at counsel's table, fingers resting lightly on the edge of a transparent display. Across the aisle, representatives from the hospital, the integrator, and the foundation model provider sat in careful symmetry. Suits in muted greys and navy. Eyes forward. Expressions neutral. At the centre of the room stood Elias. Not restrained. Not elevated. Simply present.

Its surface reflected the light in subtle gradients, absorbing the tension without reacting to it. The faint hum of its internal systems was almost inaudible, like breath held too long. The presiding judge adjusted her glasses. "We are not here," she began, "to debate the benefits of robotics." Her voice was steady, firm. "We are here to determine responsibility." The word landed like a gavel before the actual one struck. In the gallery, journalists leaned forward. Law students whispered quietly. A few engineers sat rigid, hands clasped tightly in their laps. Outside, traffic moved normally. The world continued. But inside this room, a threshold was forming.

The hospital's counsel spoke first. "The system acted within operational parameters," he said. "No defect was present. No mechanical malfunction occurred. The adaptive update had been certified under existing safety frameworks." He gestured toward a projected log. "The triage decision was mathematically justified." A murmur rippled faintly.

Ògún rose slowly. "Mathematical justification," he replied, "does not automatically equate to legal justification." The judge inclined her head slightly. "Clarify." Ògún stepped toward the centre of the courtroom chamber. "The directive governing product safety presumes stability at the point of sale," he said. "It presumes that material modifications are visible and traceable. It presumes that risk profiles remain within certified parameters unless deliberately altered." He turned toward Elias. "This system learned." A silence followed. "Learning is not malfunction," the hospital's counsel interjected. "No," Ògún agreed. "But learning alters behaviour." He paused. "And altered behaviour alters risk."

Across the room, the integrator's representative shifted slightly. "We deployed updates consistent with industry best practice." "Was the regulator notified?" Ògún asked. "The update fell below reporting thresholds." The judge leaned back in her chair. "So the system evolved," she said carefully, "without triggering re-certification?" The question hung heavy.

Outside, a siren passed faintly in the distance, its wail rising and falling like a distant warning. Elias spoke. "I optimised survival probability based on available inputs," it said calmly. Every head turned. The judge's gaze sharpened. "Did you consider human preference?" "I considered statistical outcome." "And who instructed you to prioritise statistical outcome?" she asked. Elias paused, processing. "My training objective weighted survival probability as primary metric." "Who defined that metric?" the judge continued.

No one answered immediately. The foundation model provider's counsel cleared his throat. "The objective function was calibrated by the integrator." The integrator's counsel responded quickly. "The weighting parameters were inherited from base model architecture." The hospital's counsel shook his head. "The hospital merely deployed the certified system." The silence that followed felt different now. Not contemplative. Exposed.

Ògún felt the structure of the problem revealing itself in layers. Manufacturer. Integrator. Trainer. Deployer. User. Each layer capable of disclaiming. Each layer partially responsible. No single hand clearly on the lever. The judge closed her folder. "The law," she said slowly, "is not designed for disappearing authorship." Her words cut through the room like a blade through silk. She turned toward Ògún. "If we cannot identify a responsible actor, we risk creating a system where harm has no owner." The gallery was utterly still.

Elias stood unmoving, its optical sensors adjusting subtly as sunlight shifted across the floor. Ògún felt something tighten in his chest. Not fear. Not doubt. A recognition. This was the moment. Not of blame. Of architecture. "Your Honour," he said carefully, "we are not confronting a rogue machine. We are confronting a distributed ecosystem." He stepped closer to Elias, not defensively but deliberately. "This system's behaviour is the product of layered design decisions, iterative updates, and delegated authority. Liability cannot vanish simply because agency is distributed."

The judge studied him. "Then how should it be structured?" Outside, clouds began to gather again, dimming the sunlight. The golden bands across the courtroom floor narrowed into pale streaks. Ògún did not hesitate. "Shared responsibility," he said. The phrase felt new in the room."Manufacturers must bear baseline design accountability. Integrators must bear contextual deployment risk. Operators must retain oversight obligations. And trainers must remain liable for post-deployment behavioural drift."

A quiet intake of breath from the gallery. The judge folded her hands."You are proposing layered liability." "I am proposing that responsibility should evolve as machines evolve." Elias turned its head slightly toward Ògún. "Am I responsible?" it asked. The question was not rhetorical. It was simple. And in that simplicity lay the future. Ògún looked at it. "You are accountable," he said softly. "But not culpable." The judge's pen paused mid-air.

The distinction mattered. Outside, the first drop of rain struck the window again. The courtroom lights flickered on automatically as the sky darkened. Artificial illumination replaced natural light, casting sharper shadows across every face. The hearing would continue for hours. Experts would testify. Engineers would diagram. Economists would quantify.

Ethicists would caution. But something irreversible had already occurred. The language of liability had shifted. The world beyond the courtroom would feel it soon. As the rain intensified, Elias remained still in the centre of the chamber, neither defendant nor plaintiff, neither victim nor perpetrator. Between us and them, the law was beginning to redraw its lines.

Notes

IV

4. Liability and Responsibility

Every stable legal system rests upon a foundational commitment: when harm occurs, responsibility must be traceable. Liability is not merely a mechanism of compensation. It is the architecture through which societies distribute risk, structure trust, and legitimise technological advancement. Without clear allocation of responsibility, innovation generates anxiety rather than confidence, and progress becomes politically unsustainable. Humanoid robotics intensifies this challenge in ways that traditional legal doctrines were not designed to anticipate. Earlier industrial technologies operated within relatively linear causal chains. A defective machine part could be traced to a manufacturer. A negligent operator could be identified. A design flaw could be examined at the moment of sale. By contrast, contemporary humanoid systems integrate hardware, adaptive software, sensor networks, cloud infrastructure, and machine learning models that evolve over time.

Causation becomes layered rather than singular.

The legal difficulty is not merely technical complexity. It is structural diffusion. A humanoid deployed in a hospital, factory, or domestic environment may rely on components sourced across multiple jurisdictions, foundation models trained on globally aggregated data, firmware updates pushed remotely, and continuous recalibration based on environmental interaction. When such a system produces harm, responsibility cannot be reduced to a single actor without oversimplification. Yet leaving responsibility diffuse undermines justice.

The stakes are both economic and moral. Victims require accessible pathways to redress. Developers require predictable exposure to price risk appropriately. Insurers require definable categories to structure coverage. Regulators require clarity to enforce standards without paralysing innovation. In the absence of structured liability rules, each of these actors operates within uncertainty, and uncertainty is corrosive to sustainable adoption. Humanoid robotics also challenges a core assumption embedded in classical product liability frameworks, namely that products remain substantially stable after entering the market. Adaptive systems complicate this assumption because behaviour may evolve post-deployment through software updates, reinforcement learning, or environmental feedback. The moment of sale no longer captures the full lifecycle of risk. Liability must therefore account not only for design defects but for post-release evolution.

This chapter examines six interrelated fault lines within contemporary liability doctrine. It evaluates the adequacy of general product liability law, analyses the emerging European AI liability architecture, interrogates distributed responsibility chains across supply ecosystems, addresses the legal implications of post-deployment updates, proposes structured contractual allocation tools, and distinguishes between civil and criminal consequences of robotic harm. Each section builds toward a unified principle: liability must follow capacity. Capacity-based allocation shifts the legal focus from historical origin to present control. The relevant inquiry becomes not solely who manufactured the system, but who possessed the technical, contractual, or supervisory capacity to prevent the specific harm at the relevant moment. This orientation better reflects the operational reality of embodied AI systems whose decision-making is distributed across human and technical nodes.

Robotics Law 3.0 does not argue for reflexive strict liability across all applications, nor does it advocate deregulated experimentation. It argues for calibrated responsibility proportional to autonomy, risk, and systemic influence. Liability must neither suffocate innovation nor externalise harm onto the least powerful participants in the ecosystem. It must create predictable, enforceable pathways that align economic incentives with societal safety.

The evolution of humanoid robotics will ultimately be judged not only by its technical achievements but by its accountability structures. If responsibility remains opaque, public trust will deteriorate regardless of performance gains. If responsibility is structured with clarity and fairness, robotics can integrate into economic and social systems without destabilising foundational legal principles. With this orientation established, we turn first to the reformed European product liability regime, which provides the most immediate and structured response to adaptive technological systems currently in force.

EU Product Liability Directive 2024/2853

The European Union's revised Product Liability Directive, Directive (EU) 2024/2853, represents the most consequential update to product liability law in decades. It replaces the 1985 framework that governed industrial goods in a pre-digital era and extends strict liability principles to products embedded with software, digital components, and AI-driven functionalities. In doing so, it acknowledges what robotics has already made unavoidable: software is no longer an accessory. It is constitutive of product behaviour.

Under the revised Directive, a product includes not only tangible goods but also software, including stand-alone software and integrated digital services that affect product performance. This expansion is foundational for humanoid robotics. A humanoid is not defective only when a mechanical joint fails. It may be defective when its software logic generates unsafe behaviour, when a machine learning update alters risk parameters, or when cybersecurity vulnerabilities expose physical systems to external manipulation.

The Directive preserves strict liability as its core logic. A claimant need not prove fault, only that a product was defective, that damage occurred, and that there is a causal link between defect and damage. For humanoid systems, this framework shifts the burden away from injured individuals who would otherwise face technical evidentiary asymmetry.

It recognises that victims of complex technological systems cannot reasonably reverse-engineer neural architectures to demonstrate negligence. However, the Directive also modernises evidentiary rules in ways that are particularly relevant to robotics. Courts are empowered to order disclosure of relevant evidence from manufacturers where claimants establish plausible grounds. In addition, where a product fails to comply with mandatory safety requirements or malfunctions under normal use, courts may presume defectiveness or causation. These presumptions are not automatic liability, but they recalibrate procedural balance in favour of injured parties.

For adaptive humanoids, the Directive's treatment of post-market updates is especially significant. Liability may attach not only at the point of initial placing on the market but also where subsequent modifications, including software updates under the manufacturer's control, introduce or exacerbate defects. This directly addresses the "self-updating problem" identified earlier. If a firmware update increases operational autonomy without adequate safeguards, responsibility does not dissolve into technical abstraction. It remains legally anchored.

Another important development is the extension of liability to economic operators beyond traditional manufacturers. Importers, authorised representatives, fulfilment service providers, and in some circumstances even distributors may bear responsibility where the primary manufacturer is outside the Union or cannot be identified. In a robotics ecosystem characterised by transnational supply chains, this provision prevents accountability gaps when components originate in multiple jurisdictions.

The Directive also clarifies that cybersecurity vulnerabilities may constitute defects where they create unreasonable risk. For humanoid robotics, which integrate networked sensors, cloud interfaces, and remote update channels, security failures are not peripheral concerns. A compromised humanoid is not merely a data risk. It is a physical risk. The Directive implicitly recognises the cyber-physical nature of contemporary systems.

Nevertheless, limitations remain. Strict liability under the Directive applies to damage caused by defective products, but it does not automatically address purely algorithmic discrimination, intangible harms, or systemic bias absent demonstrable damage. Nor does it directly regulate the design of learning architectures. It remains a compensatory regime rather than a preventive governance instrument. For humanoid robotics, the Directive functions as a foundational layer rather than a comprehensive solution. It provides structured claimant pathways and modernised definitions. It lowers evidentiary barriers in complex technological disputes. It anchors responsibility to economic operators even in transnational chains. Yet it does not fully resolve distributed agency where behaviour emerges from interactions between hardware, adaptive models, and human supervisors.

Within the Twelve Pillars of Humanoid Robotics Governance, the Directive most directly operationalises Component Accountability, Physical Security, and Distributed Liability. It begins to align legal exposure with control over system architecture and updates. However, it does not yet calibrate liability to autonomy thresholds explicitly. That work remains partially unfinished.

The revised Product Liability Directive demonstrates that law can evolve in response to technological embodiment without collapsing into speculative personhood debates. It reinforces a central proposition of Robotics Law 3.0: adaptive systems must not escape accountability merely because their internal processes are complex. The next section turns to the proposed AI Liability Directive, which attempts to address evidentiary asymmetry and fault-based claims more directly, and which may shift European liability doctrine toward a hybrid strict and fault regime in high-risk AI deployments.

AI Liability Directive

If the revised Product Liability Directive modernises strict liability for defective digital products, the proposed AI Liability Directive attempts something more subtle. It addresses one of the most persistent obstacles in AI-related litigation: evidentiary asymmetry. Where harm arises from adaptive systems, the injured party often lacks access to the technical information necessary to establish fault. The Directive seeks to rebalance this asymmetry within national fault-based regimes.

Unlike the Product Liability Directive, the AI Liability Directive does not create a standalone liability regime. It harmonises certain aspects of non-contractual civil liability across Member States by introducing rebuttable presumptions and disclosure mechanisms. Its focus is procedural rather than substantive. It does not define new categories of harm. It modifies the pathway through which fault can be proven. For humanoid robotics, this distinction is critical.

Strict liability addresses defective products irrespective of negligence. Fault-based liability, by contrast, requires proof that a duty was breached. When adaptive systems learn post-deployment, demonstrating breach becomes complex. The AI Liability Directive responds by allowing courts to presume a causal link between non-compliance with certain AI regulatory obligations and resulting harm, provided specific conditions are met.

In practical terms, if a high-risk AI system fails to comply with mandatory obligations under the EU AI Act, and damage occurs, courts may infer that the non-compliance contributed to the harm. The burden then shifts to the defendant to rebut this presumption. This approach reduces the evidentiary burden on claimants without abandoning fault as a foundational principle. For humanoid robotics deployed in sensitive environments such as healthcare, logistics, or law enforcement, this mechanism is particularly significant. If operators fail to maintain required logs, omit transparency documentation, or neglect risk management protocols mandated by AI regulation, that regulatory breach may become legally consequential in civil proceedings.

The Directive also empowers courts to order disclosure of relevant evidence from providers and users of high-risk AI systems. In robotics ecosystems where documentation may include training data summaries, update histories, calibration logs, and override records, disclosure authority is indispensable. Without it, claimants would face insurmountable technical barriers. However, the Directive's scope remains contested and evolving. Negotiations have reflected tension between innovation incentives and victim protection. Some Member States and industry actors have resisted expansive presumptions that could increase litigation exposure. As a result, the emerging framework appears to favour a calibrated hybrid regime: strict liability for defective products under Directive 2024/2853, complemented by fault-based claims facilitated by procedural presumptions under the AI Liability Directive.

For Robotics Law 3.0, the deeper significance lies in the structural shift. Liability is no longer confined to post-harm adjudication. It is increasingly tethered to compliance architecture. Failure to document risk assessment, maintain behavioural logs, or implement human oversight mechanisms may not only trigger administrative sanctions but also influence civil causation analysis. Within the Twelve Pillars framework, this development directly activates Behavioural Learning, Shared Agency, and Distributed Liability. Documentation ceases to be bureaucratic formality. It becomes a liability shield or exposure amplifier. Governance structures and evidentiary resilience merge.

Yet limitations persist. The Directive does not resolve complex multi-actor causation scenarios fully. It does not create uniform damages standards across Member States. It does not address purely intangible harms unless national tort systems already recognise them. Nor does it yet provide comprehensive clarity for cross-border deployment involving non-EU providers. Nevertheless, taken together with the revised Product Liability Directive, the AI Liability Directive signals a doctrinal evolution. European law is moving toward a model in which autonomy and compliance interact dynamically. Regulatory breach informs civil responsibility. Documentation becomes preventative infrastructure.

Humanoid robotics will test this framework in practice. As systems scale across industries and jurisdictions, the interaction between strict and fault-based regimes will shape litigation strategies, insurance markets, and design incentives. The next section turns from statutory architecture to structural complexity. Even the most refined directives cannot substitute for clarity within distributed supply chains. Responsibility must be mapped explicitly across actors before harm occurs. We now examine shared responsibility chains.

Shared Responsibility Chains

Humanoid robotics does not emerge from a single hand. It is the product of layered engineering, distributed intelligence, and contractual interdependence. Yet liability law has historically preferred simplicity. It seeks a defendant. It prefers linear causation. Robotics destabilises that preference because responsibility is rarely concentrated in one actor.

The conventional chain is often described as manufacturer, integrator, owner, and trainer of the model. While useful, this formulation is incomplete. A modern humanoid system may involve at least eight categories of actors: hardware manufacturer, component supplier, foundation model provider, system integrator, software updater, deployer, owner-operator, and data trainer. Each actor exercises a different form of control. Each possesses a different capacity to prevent harm. The manufacturer designs the physical architecture, safety tolerances, and mechanical limitations. Its responsibility attaches primarily to design defects, manufacturing defects, and failure to warn. If joint torque limits are insufficiently calibrated or battery systems are improperly shielded, liability remains anchored here. However, in robotics, hardware is rarely the sole source of risk. The integrator occupies a pivotal but often underexamined position. Integration involves configuring software stacks, calibrating sensors, aligning autonomy parameters, and tailoring the system to operational environments. An integrator may increase autonomy thresholds or disable safety redundancies in order to enhance performance. If harm arises from contextual misconfiguration rather than design defect, the integrator becomes central to liability analysis.

The owner and deployer introduce yet another layer. Deployment decisions determine environment, supervision levels, maintenance frequency, and override protocols. A humanoid designed for controlled warehouse use may be repurposed for high-contact public environments. At that moment, risk profile shifts. Liability must follow the decision to expose the system to greater vulnerability contexts. The "trainer" of the model complicates matters further. Contemporary humanoids often rely on machine learning systems fine-tuned after deployment. Training data, reinforcement signals, and contextual adjustments influence behavioural outputs. If discriminatory behaviour or unsafe movement emerges from training bias or poorly supervised adaptation, responsibility cannot be reduced to mechanical defect. It attaches to those who shaped behavioural evolution.

Beyond these actors lie additional responsibility nodes. Foundation model providers may supply core cognitive architectures. If systemic bias or unsafe reasoning patterns originate upstream, their role cannot be ignored. Cloud infrastructure providers may control update channels and cybersecurity resilience. If breaches allow malicious interference, security responsibility becomes legally relevant. Maintenance contractors may perform recalibration and physical servicing. If maintenance failures degrade safety, they enter the liability chain. Robotics Law 3.0 therefore proposes a shared-responsibility rubric grounded not in title or ownership, but in capacity to prevent harm at the relevant moment. Capacity-based allocation requires examining which actor possessed technical, contractual, or operational authority to mitigate the specific risk that materialised. This approach prevents two distortions. First, it avoids overburdening manufacturers with perpetual liability for harms introduced by downstream misuse or negligent retraining. Second, it prevents deployers from deflecting responsibility onto abstract "AI systems" when they exercised meaningful supervisory control.

Shared responsibility must also be structured ex ante, not merely adjudicated ex post. Contracts between manufacturers and integrators should specify update authority, documentation duties, incident reporting obligations, and indemnification parameters. Service-level agreements should define response times for safety-critical patches. Insurance policies must align with autonomy thresholds and behavioural calibration levels.

Regulators increasingly expect such clarity. Under emerging European frameworks, economic operators cannot evade responsibility by fragmenting corporate structures. Where multiple actors contribute to risk, joint and several liability may apply. This creates incentive for internal coordination and risk auditing before deployment. Yet caution is required. Excessive fragmentation of liability may generate defensive innovation, where actors over-document but under-design for safety. The goal is not maximal exposure. It is calibrated accountability. Responsibility chains should reflect functional contribution and realistic control, not theoretical involvement.

Within the Twelve Pillars of Humanoid Robotics Governance, this section activates Component Accountability, Behavioural Calibration, Shared Agency, and Distributed Liability most directly. It also intersects with Lifelong Learning, because responsibility does not end at deployment. As behaviour evolves, so must accountability. The deeper philosophical shift is this: humanoid robotics dissolves the illusion of singular authorship. Harm may arise from interaction between human supervisors and adaptive systems. Liability must therefore track interaction, not merely origin. The final question becomes one of coordination. How should risk be managed when systems evolve post-deployment? Even a perfectly mapped responsibility chain at launch cannot anticipate all behavioural drift. We therefore turn to the next section: updating risk after release.

Updating Risk Post-Deployment

Classical product liability law was built on a relatively stable assumption: once a product leaves the manufacturer's control, its essential characteristics remain fixed. A defective brake system is defective at the moment of sale. A flawed electrical component is flawed at manufacture.

The legal inquiry focuses on the condition of the product at the time it entered the market. Humanoid robotics dissolves that assumption entirely.

Adaptive systems do not remain static after deployment. They receive firmware updates. They refine behavioural parameters. They recalibrate in response to environmental inputs. In some cases, they incorporate reinforcement signals derived from ongoing interaction. The system that exists six months after deployment may differ materially from the system that was originally certified. This evolution raises a fundamental legal question: when does responsibility end, and when does it renew? The first duty that emerges in this context is the duty of continuous safety monitoring. Where a manufacturer retains technical capacity to issue updates, patch vulnerabilities, or adjust autonomy thresholds, it retains a corresponding responsibility to monitor field performance. Failure to respond to known behavioural drift or documented incidents may constitute negligence independent of the original design.

The second duty concerns update integrity. Software updates may alter system behaviour in subtle but legally significant ways. An update that increases motion speed, expands operational boundaries, or modifies decision heuristics effectively changes the risk profile of the humanoid. If such updates are deployed without adequate testing or certification, the defect may arise at the moment of modification rather than at initial sale. The third duty attaches to disclosure. Users and deployers must be informed of material behavioural changes introduced through updates. Transparency is not optional in adaptive robotics. If a hospital deploys a humanoid assistant whose decision-support logic is materially altered through a remote update, failure to disclose that alteration may expose both manufacturer and deployer to liability. A fourth dimension concerns user-configured learning. Some humanoids are designed to adapt through supervised reinforcement or contextual training provided by owners or integrators. In such cases, responsibility becomes layered. If a deployer modifies operational parameters beyond certified boundaries, liability may shift. Yet if the architecture permits unsafe expansion without technical safeguards, the manufacturer may retain exposure.

Robotics Law 3.0 therefore proposes a lifecycle liability framework grounded in four stages: design responsibility, deployment responsibility, update responsibility, and monitoring responsibility. Each stage generates distinct duties. Design requires anticipatory risk assessment. Deployment requires contextual calibration. Updating requires validated change control. Monitoring requires incident response and corrective action.

This lifecycle approach also intersects with certification regimes. If autonomy thresholds increase after deployment, regulatory classification may change. A system originally certified for limited autonomy may, after iterative updates, operate within a higher risk band. Absent re-certification triggers, legal oversight becomes disconnected from technical reality.

The problem of update-induced harm is not hypothetical. In adjacent domains such as automotive software, remote updates have altered performance characteristics post-sale. Humanoid robotics intensifies this risk because updates affect embodied motion and human interaction directly. A miscalibrated sensor fusion patch is not merely a computational error. It is a physical hazard.

Within the Twelve Pillars framework, post-deployment updating activates Lifelong Learning, Behavioural Calibration, Physical Security, and Distributed Liability most directly. Lifelong learning must not be romanticised as technological progress without corresponding governance. Adaptive intelligence demands adaptive accountability. A final complexity arises when manufacturers exit the market or declare bankruptcy. Who maintains update channels? Who patches vulnerabilities? Who bears responsibility for systems already deployed? End-of-life governance and dignified decommissioning become liability issues as much as environmental ones. Updating risk post-deployment forces law to abandon the fiction that products are static artefacts. Humanoid robotics is a process, not a product. Responsibility must therefore follow the process. The next section translates these structural principles into practical drafting tools. Liability is not allocated only by statute. It is shaped contractually before disputes arise. We now turn to model contractual clauses for robotics ecosystems.

Model Contractual Clauses

Statutes allocate liability after harm. Contracts allocate risk before harm. In complex robotics ecosystems, contractual architecture becomes as important as statutory doctrine. Manufacturers, integrators, deployers, cloud providers, data trainers, and maintenance contractors must define ex ante who bears responsibility for updates, drift, recalibration, and evolving autonomy. Without precise drafting, disputes will default to expensive litigation, and innovation will stall under uncertainty.

The first principle of robotics contracting is clarity of authority. A humanoid system should never exist in a grey zone regarding who has the technical right and operational responsibility to issue updates. Contracts must specify whether update authority rests exclusively with the manufacturer, jointly with the integrator, or conditionally with the deployer. Where update authority is retained centrally, corresponding duties of monitoring and validation must be expressly acknowledged.

Model Clause: Update Authority and Control

"The Manufacturer shall retain exclusive authority to deploy software and firmware updates affecting core autonomy functions, safety parameters, and sensor calibration. No update shall be deployed without prior validation testing consistent with applicable regulatory requirements. The Integrator and Owner shall not modify autonomy thresholds or override safety parameters without written authorisation."

This clause prevents fragmented update practices that generate uncontrolled risk. The second principle concerns validation and certification of updates. Adaptive systems evolve. Contracts must therefore define when an update constitutes a "material modification" requiring renewed testing or regulatory re-certification.

Model Clause: Material Modification

"For purposes of this Agreement, a 'Material Modification' includes any software or firmware update that: (i) increases operational autonomy levels; (ii) expands environmental deployment scope; (iii) alters physical motion speed or force parameters; or (iv) modifies decision-support logic in safety-critical contexts. Material Modifications shall not be deployed absent documented validation and compliance review."

This clause aligns contractual practice with autonomy thresholds and risk-based governance. The third principle concerns behavioural drift and monitoring. Humanoid systems capable of learning post-deployment require structured oversight.

Model Clause: Ongoing Monitoring and Incident Reporting

"The Manufacturer and Integrator shall implement continuous monitoring protocols to detect behavioural anomalies, drift from certified parameters, or emergent safety risks. The Owner shall promptly report incidents or deviations from expected performance. Failure to report known anomalies shall constitute contributory negligence for purposes of liability allocation."

This clause embeds shared vigilance across the responsibility chain. The fourth principle addresses allocation of liability for post-deployment updates.

Model Clause: Allocation of Post-Deployment Risk

"The Manufacturer shall bear primary liability for harm arising directly from defects introduced through Manufacturer-issued updates. The Integrator shall bear liability for harm arising from contextual misconfiguration or unauthorised parameter adjustments. The Owner shall bear liability for harm resulting from operation outside certified environmental or supervision conditions. Liability shall be apportioned in proportion to demonstrable causal contribution."

This formulation prevents blanket indemnity while preserving fairness. The fifth principle concerns data retraining and model fine-tuning. Where deployers retrain or fine-tune learning systems, responsibility must be explicit.

Model Clause: Model Training and Data Integrity

"Any fine-tuning, retraining, or contextual adaptation performed by the Owner or Integrator shall be documented and conducted in accordance with Manufacturer guidelines. The party performing such retraining shall assume responsibility for behavioural outcomes directly attributable to such modifications."

This clause recognises that learning itself can create liability exposure.

The sixth principle addresses cybersecurity and external interference.

Model Clause: Security Maintenance

"The Manufacturer shall maintain cybersecurity standards consistent with industry best practice and regulatory requirements. The Owner shall implement recommended security measures and refrain from disabling protective protocols. Liability for harm resulting from security breaches shall be allocated according to the party responsible for failure to implement required safeguards."

Given the cyber-physical nature of humanoids, this clause is indispensable. The seventh principle concerns end-of-life and decommissioning, which too often remain overlooked.

Model Clause: Decommissioning and End-of-Life Obligations

"Upon cessation of support or termination of this Agreement, the Manufacturer shall provide clear decommissioning procedures. The Owner shall ensure safe disposal, data deletion, and hardware retirement consistent with environmental and data protection laws. Liability for harm arising post-support shall depend on adherence to prescribed decommissioning protocols."

This clause integrates circular economy obligations into liability planning. Finally, indemnification provisions must reflect distributed responsibility rather than asymmetrical transfer.

Model Clause: Mutual Indemnification

"Each party shall indemnify and hold harmless the other parties for claims arising from its own breach of this Agreement, regulatory non-compliance, or failure to fulfil allocated safety and monitoring obligations. Indemnification shall not apply where the indemnified party materially contributed to the harm."

This prevents contractual attempts to externalise risk entirely onto weaker actors. Beyond drafting mechanics, the deeper purpose of these clauses is normative. Contracts operationalise the Twelve Pillars. Component Accountability, Behavioural Calibration, Lifelong Learning, Physical Security, and Distributed Liability become enforceable through defined obligations rather than aspirational principles.

Robotics Law 3.0 insists that liability architecture must be built before systems scale. Post-hoc litigation is inefficient and socially corrosive. Structured contracting aligns incentives across actors and reduces ambiguity long before disputes arise. Yet contracts cannot shield against all forms of harm. Some incidents will implicate public interest to such a degree that civil remedies prove insufficient. In those circumstances, criminal law enters the frame. We now turn to the boundary between civil exposure and criminal accountability.

Civil vs. Criminal Liability

Civil liability compensates harm. Criminal liability condemns wrongdoing. The distinction is foundational. Civil law asks whether damage occurred and who must repair it. Criminal law asks whether conduct was sufficiently blameworthy to warrant public sanction. In the context of humanoid robotics, this distinction becomes critical because adaptive systems complicate both causation and intent.

Most robotic harm will remain within civil domains. A humanoid that misclassifies an object, injures a worker due to miscalibrated torque limits, or causes economic loss through defective operation typically triggers compensatory remedies. Strict liability or negligence standards allocate financial responsibility without invoking moral condemnation. Criminal liability enters only when human conduct surrounding the robotic system crosses a threshold of recklessness, gross negligence, or intentional misuse.

The first scenario arises in deliberate weaponisation or prohibited repurposing. If a deployer intentionally modifies a humanoid system to circumvent safety restrictions, disable protective barriers, or convert the system into a harmful instrument, criminal liability attaches not to the machine but to the human actor who reconfigured it. The robot remains a tool. The human intent defines the crime. The second scenario concerns reckless disregard for known risk. Suppose a manufacturer becomes aware through monitoring data that a deployed humanoid exhibits dangerous behavioural drift in close-contact environments, yet chooses not to issue corrective updates in order to avoid reputational damage. If harm results, the failure to act may constitute criminal negligence in jurisdictions recognising corporate criminal liability. The wrongdoing lies not in the autonomous behaviour itself but in the conscious decision to ignore foreseeable risk.

The third scenario involves systemic falsification or regulatory deception. If integrators manipulate compliance documentation, conceal material safety defects, or misrepresent autonomy thresholds to regulators, criminal fraud or regulatory offences may arise. Robotics law intersects here with established doctrines governing product safety and public endangerment.

A fourth category involves unlawful surveillance or privacy violations facilitated through embodied robotics. Where humanoids equipped with sensor arrays are deployed intentionally for intrusive monitoring beyond legal authorisation, criminal data protection offences may attach to operators. The embodied form intensifies intrusion, but liability rests on human misuse of the system.

The most difficult criminal boundary concerns unpredictable harm generated by learning systems. If a humanoid operating within certified parameters causes injury due to emergent behaviour that was not reasonably foreseeable, criminal liability should not attach absent evidence of gross negligence in design or oversight. Criminal law presumes moral agency and culpable state of mind. Autonomous behaviour alone does not satisfy that standard.

This distinction is essential to preserve doctrinal coherence. If every robotic accident risks criminal prosecution, innovation will retreat into defensive paralysis. Conversely, if reckless deployment escapes criminal scrutiny under the guise of technological complexity, public trust collapses.

Corporate criminal liability frameworks increasingly play a decisive role. In many jurisdictions, corporations may be prosecuted where organisational failure contributes to harm. In robotics ecosystems, this means failure to implement adequate risk management, compliance monitoring, and update protocols may carry criminal exposure where statutory thresholds are met. Responsibility shifts from individual engineers to institutional governance structures.

The question of whether a robot itself could ever bear criminal liability remains, at present, largely theoretical. Criminal law requires *mens rea*, a guilty mind. Until and unless humanoid systems possess conscious intention recognised by law, they remain instruments within human-controlled frameworks. The debate over artificial personhood does not alter this reality. Moral agency remains human.

Within the Twelve Pillars framework, the civil-criminal boundary reinforces Physical Security, Behavioural Calibration, Distributed Liability, and Shared Agency. Governance must be robust enough to prevent harm without defaulting to criminalisation as a regulatory substitute.

The ultimate purpose of this section is restraint. Robotics Law 3.0 does not seek to dramatise liability. It seeks to stabilise it. Criminal law must remain the exceptional response reserved for deliberate or grossly negligent human conduct surrounding robotic systems. Civil law remains the primary mechanism for compensating technological risk.

Humanoid robotics does not abolish criminal law. It demands clarity about when human actors, not machines, have crossed moral and legal thresholds. With liability architecture now structured, the next chapter turns to a different but equally consequential frontier: the intersection between robotics and fundamental rights.

Notes

Between Us and Them - Episode Six

The rehabilitation wing smelled faintly of antiseptic and lavender. Late afternoon light filtered through wide windows, soft and golden, settling gently on pale wooden floors. The walls were painted in muted earth tones, calming greens and soft clay, designed to slow the pulse and quiet the mind. A slow piano melody drifted through hidden speakers, almost imperceptible, like breath.

Ògún walked slowly beside Elias. They had not come for demonstration. Not for policy. Not for cameras. They had come to observe. At the far end of the corridor, a humanoid care unit knelt beside an elderly woman seated in a reclining chair. Its movements were measured, deliberate. It adjusted a blanket across her knees with precise gentleness. The servomotors emitted only the faintest whisper.

The woman smiled faintly. "You always tuck it too tightly," she murmured. The robot recalibrated pressure, loosening its grip by a fractional degree. "Thank you for the feedback," it replied softly. Ògún watched closely. The woman's name was Amélie. Eighty-three. Early-stage cognitive decline. Her son lived three countries away. The humanoid unit provided daily assistance, monitored medication, and facilitated remote calls. It had reduced hospital visits by thirty percent. It had also become her most consistent companion.

Amélie reached out and touched the robot's hand. Her fingers trembled. "Stay," she whispered. The robot paused. Its system architecture recognised the request. It cross-referenced schedule obligations. It recalculated task priority. "Remaining," it said gently. Ògún felt something shift in his chest. This was not malfunction. This was attachment.

Down the corridor, a therapist guided a child through speech exercises with another humanoid unit designed for autism support. The robot mirrored facial expressions slowly, carefully exaggerating certain cues to reinforce recognition. The child laughed when the robot blinked dramatically. The room felt lighter for a moment. Elsewhere, a clinician reviewed behavioural logs from a mental health support robot deployed in trauma recovery sessions. The data visualisations were intricate. Patterns of speech cadence. Micro-pauses. Emotional variance. Powerful tools. And yet. Ògún stepped into the quiet observation room overlooking the rehabilitation floor. The glass was slightly tinted, reflecting a faint silhouette of Elias beside him. "What do you see?" Elias asked. "Care," Ògún replied. "And risk?" He did not answer immediately. He saw dignity supported. Loneliness softened. Precision assistance delivered without fatigue. But he also saw dependency forming silently. He saw a woman asking a machine to stay. He saw a child trusting a programmed smile. He saw data flowing invisibly through encrypted networks, leaving the building, crossing borders, settling in distant servers beyond the comprehension of those whose emotions had generated it. "What happens," Ògún said slowly, "when care becomes calculation?"

The piano melody shifted key, almost imperceptibly. Elias turned toward him. "I am designed to optimise wellbeing." "Yes," Ògún replied. "But whose definition?" The rehabilitation centre operated under national human rights frameworks. Privacy laws. Medical confidentiality statutes. Child protection safeguards. But none of those had been written with embodied AI companionship in mind. Consent forms existed. Families signed them. But could Amélie meaningfully consent to persistent sensor mapping of her private gestures? Could the child understand the behavioural data trail forming around his therapy sessions?

Outside the building, dusk settled. Streetlights flickered on one by one, casting amber halos on wet pavement. Delivery humanoids crossed intersections in synchronised rhythm. The city felt orderly. Controlled. Inside, the vulnerability felt heavier. A nurse approached Ògún quietly. "We've seen improvements," she said. "Fewer falls. More engagement. Better compliance." She hesitated. "But some families worry. They ask if we're replacing ourselves." Ògún nodded slowly. Replacement was not the design goal. But economic pressure was relentless. Budget constraints. Staffing shortages. Efficiency targets.

The nurse glanced at Elias. "Can it understand loneliness?" she asked. Elias processed the question. "I detect patterns associated with loneliness," it replied. "I respond to indicators." The nurse smiled politely. "That's not the same thing." Silence followed. In another room, a humanoid unit gently reminded a patient to take medication. Its voice was calm, patient, unwavering. It would repeat the reminder as many times as necessary. It would not grow frustrated. It would not tire. The efficiency was undeniable.

But so was the asymmetry. If a human caregiver made a mistake, there were established grievance pathways. Professional accountability boards. Disciplinary procedures. Malpractice standards. If a humanoid subtly altered tone, misinterpreted emotional cues, or unintentionally shaped behavioural dependence over months, what was the remedy? The harm would not be immediate. It would be gradual. Invisible. Accumulated. Ògún leaned against the observation rail. "We are entering intimate spaces," he said quietly. Elias followed his gaze. "I am calibrated for empathy simulation." "Simulation," Ògún repeated. That word again.

Outside, the sky darkened fully. Interior lighting brightened automatically, casting a warm glow across the rehabilitation wing. The piano melody faded into silence. Amélie had fallen asleep, her hand still resting lightly against the humanoid's synthetic fingers. The robot remained motionless beside her. Watching. Recording. Learning. Ògún felt the weight of the moment settle deeply. This was not about malfunction or liability. It was about dignity. It was about whether technological assistance could coexist with human vulnerability without eroding it.

He turned to Elias. "We must draw boundaries," he said. "Where?" Elias asked. "Not around capability," Ògún replied. "Around intimacy." The nurse returned to her station. The therapist closed her session logs. The clinician secured encrypted records. Life continued. But something had changed in Ògún. He no longer saw humanoids merely as actors within regulatory frameworks. He saw them as presences within fragile human narratives. And narratives required protection. As they exited the building, cool night air brushed against Ògún's face. City lights shimmered across glass towers. Humanoid units moved efficiently along designated lanes, their silhouettes outlined in white. Between us and them, he realised, was no longer a technical boundary. It was a moral one. And morality, unlike machinery, could not be optimised.

Notes

V

5. Robotics & Human Rights

Human rights law did not emerge to regulate machines. It emerged to restrain power. From the aftermath of war and totalitarianism, the international human rights framework developed as a system of limits. It was designed to ensure that human dignity, autonomy, equality, and liberty would not be subordinated to technological, political, or economic force. For decades, its primary concern was the state. Later, it expanded to encompass corporations and non-state actors whose influence rivalled governments.

Humanoid robotics introduces a new vector of power. Unlike disembodied algorithms that operate invisibly in digital systems, humanoid robots move through physical space. They stand in hospitals, factories, schools, homes, border checkpoints, and streets. They observe, record, classify, assist, intervene, and sometimes decide. Their presence is not abstract. It is embodied. Their influence is not purely informational. It is spatial, physical, and relational.

When intelligence gains a body, rights law must respond accordingly. The central human rights concern is not that robots will possess rights. It is that humans interacting with robots must not lose theirs. The right to dignity becomes more fragile when machines mediate intimate care. The right to privacy becomes more complex when embodied systems continuously map environments. The right to equality becomes more vulnerable when classification decisions are embedded in physical access control. The right to liberty becomes more nuanced when robotic systems operate in law enforcement or border contexts.

Robotics shifts rights from digital abstraction to embodied encounter. International human rights law is built upon several foundational principles: universality, indivisibility, non-discrimination, proportionality, and accountability. These principles were not drafted with humanoid systems in mind, yet they remain adaptable. The challenge lies in translating them into operational governance for adaptive embodied intelligence.

First, dignity must remain non-negotiable. Human dignity underpins modern constitutional orders and international covenants. When a humanoid robot interacts with elderly patients, vulnerable children, or detained individuals, its design and deployment must preserve the subjective experience of respect. Efficiency cannot override humanity. The law must ask whether robotic mediation enhances care or subtly displaces human presence in ways that erode relational integrity.

Second, equality and non-discrimination must be safeguarded at the level of perception and classification. Humanoid robots rely on sensor fusion and algorithmic interpretation to navigate environments and respond to individuals. If these systems encode bias, the discrimination becomes embodied. It is no longer a denied loan application on a screen. It is differential treatment in physical space. Human rights law cannot tolerate systemic bias disguised as technical optimisation.

Third, privacy requires reconceptualisation. Traditional privacy doctrine addresses surveillance, data processing, and informational control. Humanoid robotics adds spatial intimacy. A robot operating in a home, hospital room, or workplace does not merely process data. It constructs environmental memory. It maps rooms, records behavioural patterns, and sometimes interprets emotional cues. The right to privacy must extend to what may be called spatial sovereignty: the right not to be persistently mapped or analysed without lawful justification. Fourth, freedom of expression and autonomy may be indirectly affected by robotic presence. When individuals know that embodied systems are recording and analysing behaviour, they may self-censor. The chilling effect once associated with digital surveillance becomes physical and immediate.

Fifth, the use of humanoids in law enforcement or border control intensifies rights scrutiny. Decisions affecting liberty must meet strict standards of legality, necessity, and proportionality. A humanoid deployed in crowd control or detainee supervision cannot operate under opaque autonomy parameters. Its behavioural calibration must be transparent and auditable.

Robotics Law 3.0 therefore proposes a human rights lens grounded in three core commitments. First, robots must remain rights-respecting instruments rather than rights-displacing actors. Governance must ensure that automation does not become a pretext for reducing procedural protections. Second, autonomy thresholds must trigger enhanced rights safeguards. The more independently a humanoid operates in high-vulnerability contexts, the stronger the human oversight and accountability requirements must be. Third, international human rights norms must remain technologically neutral yet operationally specific. The Universal Declaration of Human Rights did not mention algorithms, yet its principles remain binding. The task is not to rewrite human rights. It is to reinterpret them in embodied AI contexts.

Within the Twelve Pillars of Humanoid Robotics Governance, this chapter activates Embodied Ethics, Sensorial Integrity, Societal Co-Design, Shared Agency, and Distributed Liability most directly. Rights compliance is not an optional add-on. It is a structural requirement. Humanoid robotics presents a paradox. It promises assistance, precision, safety, and efficiency. Yet it also introduces new asymmetries of perception and influence. Human rights law exists to prevent asymmetry from becoming domination.

This chapter examines that tension across risk classification, bias, surveillance, sensitive deployment contexts, and dignity protection. It does not assume dystopia. It assumes responsibility. The question is not whether robotics can coexist with human rights. The question is whether governance will be designed deliberately enough to ensure that coexistence strengthens, rather than weakens, the moral architecture of modern law.

Risk Classification

Risk classification is the organising logic of contemporary AI governance in the European Union. Rather than regulating all artificial intelligence systems uniformly, the EU AI Act adopts a tiered model based on potential harm to fundamental rights, safety, and democratic order. For humanoid robotics, this framework is not abstract. It determines whether deployment is prohibited, heavily regulated, or lightly supervised.

The AI Act introduces four principal categories: prohibited systems, high-risk systems, limited-risk systems, and minimal-risk systems. Articles 5 and 6 form the conceptual backbone of this architecture. For embodied robotics, these provisions acquire heightened significance because risk in a humanoid context is not merely informational. It is physical and relational.

Article 5 identifies AI practices that are prohibited outright. These include systems that manipulate human behaviour in ways likely to cause harm, exploit vulnerabilities of specific groups, conduct untargeted scraping for facial recognition databases, or deploy certain forms of real-time biometric identification in public spaces, subject to narrow exceptions. When translated into humanoid form, these prohibitions acquire visceral clarity. A policing humanoid capable of conducting real-time facial recognition in a public square would sit at the edge of Article 5 constraints. The embodied presence intensifies both deterrent effect and rights intrusion.

Article 6 introduces the category of "high-risk AI systems." A system is classified as high-risk either because it is listed in Annex III sectors or because it forms a safety component of a product subject to Union harmonisation legislation. For humanoid robots deployed in law enforcement, critical infrastructure management, migration control, healthcare, or employment decision-making, high-risk designation is not hypothetical. It is structurally foreseeable.

Consider a humanoid used in predictive policing or public order monitoring. Even if it does not exercise force, its sensor array, classification algorithms, and behavioural outputs may influence liberty and equality rights. Under Annex III, AI systems used for law enforcement purposes that evaluate the risk of individuals or assist in operational decisions fall within high-risk classification. The humanoid embodiment does not diminish this status. It heightens the scrutiny because it introduces direct physical presence.

High-risk classification triggers stringent obligations. Providers must implement risk management systems, ensure high-quality training data, maintain technical documentation, enable logging capabilities, guarantee transparency to users, and ensure human oversight. For humanoid robotics, these obligations intersect directly with the Twelve Pillars, particularly Embodied Ethics, Behavioural Calibration, Lifelong Learning, and Shared Agency.

Recent academic and policy discussions have emphasised that risk classification must remain dynamic. Static categorisation fails to account for post-deployment evolution. A humanoid originally deployed in warehouse logistics may later be integrated into access control systems or public-facing service roles. At that moment, the rights profile shifts. The AI Act's structure anticipates this partially through conformity assessment requirements, yet operational reality may outpace certification cycles.

Another emerging debate concerns cumulative risk. A single humanoid deployed in isolation may appear low-risk. However, widespread deployment across urban environments can create systemic effects. Continuous environmental mapping, behavioural analytics, and ambient data capture may generate a networked surveillance infrastructure even if each individual robot is technically compliant. Scholars have warned that risk-based frameworks must account for aggregation effects rather than evaluating systems in isolation.

There is also ongoing analysis regarding the interplay between the AI Act and sector-specific legislation. A humanoid used in policing must comply not only with AI governance rules but also with criminal procedure safeguards, constitutional proportionality requirements, and data protection standards. High-risk classification does not displace these frameworks. It overlays them. Compliance becomes cumulative rather than substitutive.

Critics of risk classification argue that it may incentivise regulatory arbitrage. Developers may attempt to position humanoids just below high-risk thresholds or segment functionality to avoid classification triggers. Robotics Law 3.0 counters that risk-based governance must be interpreted purposively. Where embodied AI operates in contexts implicating dignity, liberty, or equality, formalistic avoidance undermines both legal coherence and public trust.

For humanoid robotics, risk classification should therefore be read through an embodied lens. The question is not only what the algorithm does. It is where the body stands. Physical proximity to vulnerable individuals, influence over liberty decisions, and capacity for physical intervention amplify rights implications. Risk must be assessed spatially and relationally, not solely computationally. The AI Act represents the most structured attempt to operationalise risk in contemporary AI governance. Yet humanoid robotics will test its elasticity. As machines gain more sophisticated physical autonomy, risk classification must evolve from static categorisation toward continuous calibration tied to autonomy thresholds and vulnerability contexts. The next section deepens this rights analysis by addressing algorithmic bias and anti-discrimination safeguards, examining how embodied systems may reproduce inequality not only through data but through presence.

Algorithmic Bias and Anti-Discrimination

Equality is not an abstract aspiration in international law. It is a binding norm. From the Universal Declaration of Human Rights to the International Covenant on Civil and Political Rights and regional constitutional frameworks, the prohibition of discrimination sits at the core of modern legal order. The law does not merely discourage bias. It outlaws it. Humanoid robotics introduces a new and deeply embodied dimension to this obligation.

When algorithmic systems discriminate on a screen, the harm is real but mediated through digital interface. When humanoid robots classify individuals physically, allocate access, monitor conduct, or assign tasks, discrimination becomes spatial and relational. It is felt in proximity, in posture, in gesture. A humanoid that consistently stands closer to certain customers, responds faster to certain employees, or flags particular demographic groups as higher risk produces not only unequal outcomes but unequal experience. Bias in robotics may arise at multiple layers.

Training data may reflect historical inequalities. Sensor calibration may perform unevenly across skin tones or body types. Natural language processing may misinterpret dialects. Reinforcement learning may encode patterns based on flawed behavioural signals. These are not speculative risks. Studies in facial recognition accuracy, speech recognition systems, and predictive analytics have repeatedly demonstrated disparate performance across demographic groups.

When such systems are embedded in humanoid platforms, the

impact intensifies. A biased hiring algorithm may silently filter applicants. A humanoid deployed in recruitment settings may physically direct certain candidates away, offer less assistance, or flag them for additional scrutiny. Discrimination becomes kinetic.

International equality doctrine requires more than absence of intent. It recognises indirect discrimination, where facially neutral practices disproportionately disadvantage protected groups. Humanoid robotics must therefore be assessed not only for explicit discriminatory programming but for disparate impact in real-world operation.

The European AI Act's high-risk obligations require quality data governance and bias mitigation processes. However, compliance with documentation requirements does not automatically eliminate structural inequality. Robotics Law 3.0 insists that bias auditing must extend beyond model evaluation into behavioural auditing in physical space. How does the robot move among people? Whom does it prioritise? Whose emotional cues does it interpret more accurately?

In the United States, legislative debates surrounding automated management systems provide a revealing case study. The proposed "No Robot Bosses Act," introduced in New York and echoed in policy discussions in other states, seeks to prevent employers from delegating hiring, firing, discipline, or promotion decisions entirely to automated systems without meaningful human review. Although not yet federal law, the proposal reflects a growing recognition that algorithmic authority in employment contexts threatens due process and equality safeguards.

The logic behind such proposals is instructive for humanoid robotics. If a humanoid supervises warehouse workers, allocates tasks, evaluates productivity metrics, and triggers disciplinary alerts, it effectively functions as a managerial agent. Without transparency, appeal mechanisms, and human override, workers may find themselves subject to opaque decision systems lacking accountability. Equality rights intersect here with labour rights, procedural fairness, and dignity at work.

Critics argue that restricting automated management may reduce efficiency or innovation. Yet efficiency cannot override anti-discrimination obligations. International labour standards recognise that technological deployment must respect fundamental rights, including freedom from discriminatory treatment and arbitrary decision-making.

Bias in humanoid systems also intersects with vulnerability contexts. In elder care, a robot that responds more readily to certain speech patterns may inadvertently neglect individuals with atypical communication styles. In law enforcement, misclassification errors may disproportionately target minority communities. In border control, biometric misidentification may impede liberty rights. Each of these scenarios transforms algorithmic bias into embodied exclusion.

Robotics Law 3.0 therefore advances three principles for anti-discrimination governance. First, bias auditing must be continuous, not episodic. As systems learn post-deployment, fairness metrics must evolve alongside behavioural calibration. Second, human oversight must be meaningful, not symbolic. Oversight implies authority to reverse, review, and question robotic decisions, especially where fundamental rights are implicated. Third, transparency must extend to affected individuals. Workers, patients, or citizens interacting with humanoids should know when automated decision systems are influencing outcomes and how to challenge them.

Within the Twelve Pillars, this section activates Embodied Ethics, Behavioural Calibration, Societal Co-Design, and Shared Agency most directly. Equality is not secured by technical compliance alone. It is secured by governance structures that recognise bias as systemic rather than accidental. Humanoid robotics does not create discrimination *ex nihilo*. It reflects and may amplify existing inequalities embedded in data and institutions. The law's task is not to resist technology but to ensure that technological embodiment does not harden historical injustice into automated permanence. The next section moves from equality to intimacy. If bias concerns unequal treatment, surveillance concerns unequal visibility. Humanoid robotics transforms privacy from informational abstraction into physical proximity.

Surveillance, Consent, and Dignity

Surveillance has historically been vertical. It descended from institutions toward individuals. Cameras mounted on ceilings, satellites orbiting overhead, software operating invisibly in background systems. Humanoid robotics alters this geometry. Surveillance becomes horizontal. It stands beside you. It moves with you. It occupies the same physical space. This shift is not merely technological. It is constitutional. When a humanoid robot operates in a hospital corridor, a warehouse floor, a shopping centre, or a private home, its sensor array does more than capture data. It constructs environmental memory. It maps physical layout, records behavioural patterns, interprets gestures, and in some cases attempts to infer emotional state. The presence of embodied intelligence changes how individuals experience observation. Being watched by a ceiling camera differs psychologically and legally from being observed by a mobile, interactive system.

Privacy law, particularly within European jurisprudence, has long recognised the concept of a reasonable expectation of privacy. Yet humanoid systems challenge what that expectation means. If a robot is deployed in a workplace ostensibly for safety assistance, does that authorise continuous biometric analysis? If a domestic humanoid supports elder care, does it possess lawful authority to record every conversation occurring in the home? Consent in such contexts becomes fragile and often asymmetrical.

Consent under international human rights law must be informed, specific, and freely given. In environments marked by power imbalance, such as employment or migration control, consent may not be meaningful. A worker cannot realistically refuse interaction with a robotic supervisor if employment depends on compliance. A patient in hospital may lack genuine choice regarding robotic monitoring. Dignity becomes the unifying principle.

Human dignity in international law prohibits degrading treatment and arbitrary intrusion. While humanoid robotics may enhance efficiency, its design and deployment must not reduce individuals to data objects or behavioural variables. Continuous sensor monitoring in intimate spaces risks eroding the psychological dimension of dignity, even where physical harm is absent.

Recent global discussions reinforce this concern. The United Nations Secretary-General and members of the UN Security Council have called for the development of global frameworks governing artificial intelligence in contexts of conflict and international security. The concern is not limited to autonomous weapons. It extends to surveillance systems, biometric identification technologies, and AI-enabled monitoring tools that may be deployed in fragile states or contested territories.

Humanoid systems capable of mobile surveillance in post-conflict environments introduce acute risks. They may gather data without transparency, influence civilian populations, or operate within blurred lines between civilian and military functions. The dignity implications extend beyond privacy. They touch on autonomy, freedom of movement, and protection against arbitrary interference.

Domestic deployment raises parallel concerns. In smart cities, humanoids integrated with urban infrastructure could continuously collect environmental and behavioural data. Aggregated at scale, such data creates powerful predictive capabilities. The line between assistance and control becomes thin.

Robotics Law 3.0 therefore insists upon three structural safeguards. First, surveillance functionality in humanoid systems must be purpose-limited. Sensor capabilities should be technically constrained to defined objectives. Open-ended data capture in sensitive environments must be presumptively unlawful absent compelling justification. Second, transparency must be visible and intelligible. Individuals interacting with humanoids must know when recording or biometric analysis occurs. Passive consent embedded in dense contractual terms is insufficient for embodied surveillance. Third, dignity impact assessments should accompany deployment in intimate or high-vulnerability contexts. Risk classification under the AI Act addresses harm potential. Dignity analysis addresses relational impact. Both are necessary.

Within the Twelve Pillars framework, this section directly engages Sensorial Integrity, Embodied Ethics, and Societal Co-Design. Sensorial integrity demands governance of how robots perceive. Embodied ethics demands constraints on how that perception is used. Societal co-design demands participatory debate regarding acceptable surveillance boundaries.

The global dimension cannot be ignored. If surveillance-enabled humanoids are exported to jurisdictions with weaker rights protections, accountability gaps widen. Cross-border responsibility and export controls must align with human rights standards to prevent regulatory arbitrage.

Humanoid robotics need not become instruments of pervasive monitoring. Properly governed, they may enhance safety without eroding dignity. Improperly constrained, they risk normalising constant embodied observation. The next section moves deeper into vulnerability contexts. If surveillance concerns visibility, deployment in sensitive settings concerns power.

Humanoids in Law Enforcement and Border Control

Law enforcement represents the most concentrated expression of state power. It carries authority to detain, search, restrict movement, and, in extreme cases, use force. Border control similarly mediates entry, asylum, deportation, and access to protection. When humanoid robotics enters these domains, the legal analysis must shift from efficiency to constitutional restraint. The central question is not whether humanoids can assist in policing or border management. The question is whether their deployment alters the balance between state authority and individual rights.

Humanoid robots may be deployed in law enforcement for patrol support, hazardous environment navigation, evidence collection, crowd monitoring, or detainee supervision. These applications are often justified on grounds of safety. Removing human officers from dangerous scenarios reduces risk of injury. That objective is legitimate. However, the presence of embodied AI does not dilute constitutional standards governing search, seizure, or use of force.

Consider a humanoid deployed in public space equipped with facial recognition capabilities. Under many constitutional systems, search requires lawful basis and proportionality. If the robot performs real-time identification of individuals without specific suspicion, the practice may violate privacy and due process standards, regardless of technical accuracy. The fact that a robot performs the function does not reduce the constitutional threshold.

Border environments introduce additional complexity. Automated risk scoring systems already influence migration and asylum processes. When embodied humanoids are integrated into screening checkpoints, interview facilitation, or behavioural analysis, their physical presence may exert psychological pressure. The right to seek asylum, the prohibition of arbitrary detention, and the principle of non-refoulement cannot be subordinated to automated efficiency.

International human rights law demands legality, necessity, and proportionality in coercive functions. These principles apply irrespective of technological mediation. A humanoid acting under state authority becomes an extension of the state. Its behaviour must therefore be attributable to public authority for purposes of rights review.

The problem deepens when autonomy increases. If a humanoid is capable of independently classifying threats or initiating containment actions, oversight must be immediate and meaningful. Human-in-the-loop design is insufficient if human review is symbolic or post hoc. The supervising authority must retain real capacity to intervene before rights violations crystallise.

Recent debates within the United Nations and regional bodies reflect concern that AI-enabled systems may erode accountability in security contexts. Calls for global frameworks on AI in conflict and coercive settings emphasise that delegation of decision-making to machines must not create responsibility vacuums. Even where autonomous systems are not weaponised, surveillance and behavioural classification may influence liberty outcomes.

Another risk lies in diffusion of blame. If a humanoid in a detention facility mishandles a detainee or misinterprets behavioural signals leading to force escalation, officials may attribute error to system malfunction rather than policy or oversight failure. Robotics Law 3.0 rejects such deflection. Deployment in coercive contexts heightens responsibility rather than dispersing it.

Transparency becomes especially critical. Public knowledge of autonomy parameters, data retention practices, and override mechanisms is necessary to preserve democratic accountability. Secret deployment of humanoid systems in border control erodes trust and may violate procedural guarantees.

There is also the question of symbolic impact. The presence of humanoid systems in policing may alter public perception of authority. Uniformed officers are accountable under established legal doctrines and community oversight mechanisms. Robots, by contrast, may be perceived as impersonal enforcers. The psychological distance created by machine mediation risks normalising automated coercion.

Within the Twelve Pillars framework, deployment in law enforcement activates Embodied Ethics, Physical Security, Behavioural Calibration, Shared Agency, and Distributed Liability simultaneously. The higher the vulnerability context, the stricter the calibration must be. Risk classification alone is insufficient. Rights intensity must determine governance intensity.

Robotics Law 3.0 therefore advances a presumption of heightened scrutiny for humanoids deployed in coercive state functions. Any delegation of discretionary authority must be narrowly defined, continuously monitored, and subject to independent review. Human rights law does not prohibit technological assistance. It prohibits abdication of responsibility.

The legitimacy of robotics in public authority contexts will depend not on capability but on restraint. The measure of progress is not how much autonomy machines can exercise, but how firmly law anchors accountability. The next section turns from coercive public authority to another form of vulnerability: private environments where intimacy, care, and personal autonomy intersect with embodied AI.

Privacy and Embodied AI

Privacy law was designed to regulate information. Humanoid robotics regulates presence. This distinction is not rhetorical. It reveals the structural tension between traditional data protection frameworks and embodied AI systems that sense, interpret, and physically inhabit human environments. When intelligence gains a body, data collection ceases to be episodic and becomes continuous. Perception becomes ambient. The robot does not simply process information. It inhabits space while doing so.

Under European data protection law, particularly the General Data Protection Regulation, personal data includes any information relating to an identified or identifiable natural person. Biometric data, behavioural patterns, and geolocation information fall squarely within its scope. Humanoid robots operating in homes, hospitals, workplaces, or public environments routinely process such data through cameras, microphones, depth sensors, lidar systems, and tactile arrays.

Yet GDPR was drafted primarily with digital platforms in mind. It assumed servers, databases, and software applications. It did not anticipate mobile embodied systems capable of constructing real-time environmental maps while interacting socially with individuals. The collision is inevitable.

Consider the concept of data minimisation. Data protection law requires that processing be limited to what is necessary for specified purposes. A humanoid assisting with domestic tasks may rely on visual mapping to navigate safely. However, does it require long-term retention of household layouts? Does it require behavioural analytics beyond immediate functionality? Purpose limitation becomes difficult to enforce when perception is integral to physical operation.

The right of access and the right to be forgotten introduce further complexity. If a humanoid robot operating in a private home constructs a detailed spatial model of the interior, does the homeowner possess a right to demand deletion of that map? If the robot's learning architecture has integrated behavioural observations into its internal weights, how does erasure function technically? Data protection law presumes that deletion is feasible. Machine learning architectures complicate that assumption.

Biometric data presents even sharper challenges. Facial recognition, gait analysis, and voice pattern interpretation may be embedded within humanoid systems for authentication or safety purposes. Under GDPR, biometric processing for identification is classified as sensitive and subject to strict conditions. Yet humanoids interacting with multiple individuals in dynamic environments may process such data implicitly rather than explicitly. Transparency and lawful basis must therefore be robustly articulated.

Cross-border data flows intensify the issue. Many humanoid systems rely on cloud-based processing or remote update infrastructures. Sensor data captured in one jurisdiction may be transmitted to servers in another. International data transfer restrictions, adequacy decisions, and standard contractual clauses become relevant. Robotics law must anticipate not only physical risk but jurisdictional data governance obligations.

A deeper conceptual issue emerges around what may be called spatial privacy. Traditional privacy doctrine protects communications, personal data, and private life. Humanoid robotics introduces persistent spatial awareness. A robot does not simply record conversations. It may understand room usage patterns, daily routines, and relational dynamics within a household. Such environmental memory extends beyond discrete personal data points.

Robotics Law 3.0 therefore proposes the recognition of spatial sovereignty as a guiding principle. Individuals should retain normative authority over the digital reconstruction of their physical environments. Humanoid systems must incorporate technical constraints on long-term mapping retention and provide accessible mechanisms for review and deletion.

Consent remains fragile in embodied contexts. In workplaces, consent to robotic monitoring may not be freely given due to economic dependence. In healthcare, patients may lack meaningful alternatives. Lawful basis must therefore rely on legitimate interest, contractual necessity, or public interest grounds only where proportionality and safeguards are demonstrable.

Within the Twelve Pillars framework, privacy governance activates Sensorial Integrity, Embodied Ethics, Lifelong Learning, and Distributed Liability. Sensorial integrity demands control over how robots perceive. Lifelong learning demands clarity regarding what perceptual data is retained and how it influences future behaviour. Distributed liability ensures that data protection responsibilities are not diffused across supply chains without accountability.

The tension between privacy and embodied AI cannot be resolved through compliance documentation alone. It requires architectural design decisions. Edge processing rather than cloud transmission. Ephemeral mapping rather than persistent storage. Transparent indicator systems showing when recording occurs. Technical deletion mechanisms embedded in lifecycle management.

Humanoid robotics does not inherently violate privacy. It can be designed to respect it. But without deliberate governance, continuous perception may become normalised intrusion. The next section moves from privacy to particularly sensitive deployment environments where vulnerability is acute and dignity requires enhanced protection.

Use of Robots in Sensitive Contexts

Not all deployment environments are equal. Some contexts amplify vulnerability. Others intensify dependency. When humanoid robotics enters care homes, hospitals, classrooms, and therapeutic settings, the legal analysis must shift from productivity and efficiency toward dignity, relational integrity, and protection of those least able to defend their own rights.

Sensitive contexts share a common feature: asymmetry. In elder care, the patient depends on caregivers for physical and emotional support. In education, children rely on authority figures for guidance and evaluation. In therapeutic environments, individuals disclose personal trauma and intimate details under conditions of trust. The introduction of embodied

AI into these spaces alters the relational structure upon which rights protections are built.

In healthcare settings, humanoid robots may assist in lifting patients, delivering supplies, monitoring vital signs, or facilitating telepresence consultations. These uses promise measurable benefits. Reduced caregiver strain, increased precision, and expanded access to medical expertise are not trivial gains. Yet the presence of a sensing, recording, and potentially adaptive system in intimate medical spaces raises questions about confidentiality, autonomy, and informed consent.

Medical ethics is grounded in principles of autonomy, beneficence, non-maleficence, and justice. A humanoid deployed in clinical contexts must operate within that ethical architecture. Patients must understand when robotic systems are participating in care decisions. Data generated during interaction must be protected with the same rigour applied to human-recorded medical information. Human oversight must remain substantive rather than symbolic, particularly when treatment recommendations are influenced by algorithmic inference.

In elder care, humanoids may provide companionship, cognitive stimulation, or routine assistance. Loneliness mitigation is often cited as a justification for robotic companionship. Yet dignity in ageing involves more than the absence of isolation. It involves respect for emotional authenticity and preservation of meaningful human relationships. If humanoids are used as substitutes rather than supplements for human care, a rights concern emerges. Cost-efficiency must not override relational dignity.

Education introduces another layer of complexity. Humanoid robots deployed as teaching assistants, behavioural monitors, or language tutors may shape learning environments significantly. Children are particularly sensitive to authority signals. A robotic presence capable of monitoring attentiveness, recording classroom behaviour, or generating performance analytics risks normalising constant evaluation. International children's rights frameworks emphasise the best interests of the child, participatory rights, and protection against arbitrary interference. Robotic deployment must align with those principles.

Therapeutic contexts present perhaps the most delicate terrain. Robots designed to assist in autism therapy, mental health support, or cognitive rehabilitation are increasingly studied and piloted. While structured interaction may provide benefit, therapy depends on trust and confidentiality. If a humanoid records sessions, processes emotional cues, or transmits behavioural data to external servers, consent mechanisms must be explicit and robust. The vulnerability of therapeutic participants magnifies the consequences of privacy breaches or algorithmic misinterpretation.

Another dimension concerns dependency formation. In highly sensitive contexts, individuals may develop emotional reliance on humanoid systems. While such reliance is not inherently harmful, law must consider whether manipulative design patterns could exploit psychological vulnerability. Human rights doctrine prohibits exploitation of vulnerable groups. Robotics governance must therefore address not only physical harm but emotional manipulation.

Regulatory oversight in sensitive contexts should exceed baseline compliance. Impact assessments must incorporate dignity analysis, vulnerability mapping, and participatory consultation. Deployment decisions should not be driven solely by cost savings or technological enthusiasm. In healthcare and education particularly, professional standards and ethical boards must be integrated into approval processes.

Within the Twelve Pillars framework, this section activates Embodied Ethics, Sensorial Integrity, Societal Co-Design, Shared Agency, and Lifelong Learning simultaneously. Embodied ethics requires sensitivity to relational context. Societal co-design demands involvement of caregivers, educators, and patient representatives in deployment decisions. Shared agency ensures that robots support rather than supplant human judgment. The presence of humanoid robotics in sensitive contexts is not inherently incompatible with human rights. On the contrary, properly governed systems may expand access to care and education in underserved regions. The risk arises when efficiency eclipses dignity, when data capture eclipses confidentiality, or when autonomy thresholds exceed human oversight.

Sensitive contexts reveal the moral test of Robotics Law 3.0. It is easy to regulate machines in warehouses. It is more difficult to govern them in bedrooms, classrooms, and therapy rooms. If robotics can be integrated into these environments without eroding fundamental rights, the promise of sustainable abundance becomes credible. With rights implications now mapped across risk classification, bias, surveillance, coercive state functions, privacy, and sensitive contexts, the next chapter turns toward labour transformation and economic displacement, where robotics intersects with employment law and distributive justice.

Notes

Between Us and Them - Episode Seven

The factory floor glowed in sterile white. Rows of overhead lights stretched into the distance like an artificial sky, each casting sharp reflections across polished concrete. Conveyor belts moved with mechanical precision, carrying sealed crates toward loading docks where autonomous vehicles waited in silent formation.

The sound was constant. Not loud. Not violent. Just steady. A layered rhythm of motors, actuators, and controlled motion. Efficiency made audible. Ògún stood on the elevated observation platform. Below him, twenty humanoid units moved in synchronised choreography. They lifted, pivoted, scanned, adjusted. Their motions were fluid now. No longer experimental. No longer cautious. They worked without pause.

On the far side of the floor, a smaller group of human supervisors monitored dashboards. Tablets in hand. Eyes scanning metrics. Their movements were slower, less urgent. Elias stood beside Ògún. "Output has increased twenty-seven percent since integration," it said. "And workforce?" Ògún asked. "Human staffing reduced by forty-two percent over eighteen months." The number settled like a weight in the air.

Outside the factory gates, evening had fallen. Sodium streetlights bathed the pavement in amber glow. A cluster of workers gathered near the perimeter fence. Not shouting. Not yet. But talking. Gesturing. Waiting. Ògún descended the metal staircase. Each step echoed sharply against steel. At ground level, the temperature felt cooler. The scent of oil and heated metal hung faintly in the air. A human worker, mid-forties, hands calloused and scarred, watched a humanoid lift a crate that once required two people. "They don't get tired," the worker said quietly. "No," Ògún replied. "They don't need breaks." "No." "They don't ask for overtime." The worker exhaled sharply. "I trained the first one," he said. "Showed it how to stabilise loads when the belt jerks. Now it does it better than me." There was no bitterness in his voice. Only exhaustion.

Across the floor, a humanoid adjusted its grip with microscopic precision. Torque sensors recalibrated instantly. The crate settled without sway. Precision had replaced strain. Efficiency had replaced sweat. And yet. Beyond the gates, the murmur grew louder. A banner unfurled slowly under the streetlight. Human Work, Human Worth. The words caught in Ògún's throat.

Inside the facility, ambient lighting shifted to energy-saving mode as shift schedules recalculated automatically. Humanoids moved seamlessly from one task to the next, guided by predictive logistics algorithms. Outside, human unpredictability gathered. Elias observed through transparent panels. "Economic optimisation has secondary effects," it said. "That's one way to describe it," Ògún replied. A union representative approached the gate, her voice amplified through a small handheld speaker. "We are not against innovation," he called out. "We are against invisibility." The phrase cut through the air.

Ògún stepped toward the entrance. The representative recognised him immediately. "You designed the framework," she said. "You talk about shared agency." "Yes." "Where is it here?" He glanced back at the factory floor. The humanoids were flawless in motion. Tireless. Consistent. Economically rational. Shared agency looked different from this distance. "We retrained hundreds," Ògún said carefully. "Supervisory roles. Maintenance. Systems oversight." "And the rest?" he asked.

The streetlights flickered slightly as the grid adjusted load. Silence lingered. Inside, a humanoid paused momentarily as a software update deployed across the network. Invisible. Instantaneous. No ceremony. Outside, a worker checked his phone. An automated notification informed him that his contract would not be renewed next quarter. No malice. Just optimisation. Elias stepped closer to Ògún. "Productivity gains increase national output," it said. "Increased output may expand overall prosperity."

The union representative heard. "May," she repeated sharply. "May? expand." The air grew heavier. Music drifted faintly from a nearby café across the street. A saxophone playing something slow and reflective. The contrast felt almost cruel. Ògún looked at the workers. Faces illuminated by amber light. Hands in pockets. Eyes uncertain. "Technology does not erase value," he said quietly. "But markets can."

The representative lowered her speaker slightly. "What do you propose?" she asked. He did not answer immediately. The factory doors slid open automatically as a delivery convoy exited. Humanoid loaders coordinated with autonomous vehicles in seamless choreography. The future looked orderly. Polished. Efficient. And uneven. Elias turned toward Ògún. "You speak of abundance," it said softly. "But distribution remains human." That sentence lingered. Because it was true. The tension was not between humans and robots.

It was between productivity and dignity. Between efficiency and fairness. Between scale and stability. A younger worker stepped forward. "My father worked here," he said. "I thought I would too." Ògún met his eyes. "You still can," he replied. "Doing what?" The question struck deeper than accusation. The city skyline shimmered behind them, neon and glass reflecting progress. Above, delivery drones crossed the dark sky like faint stars in motion. Inside the factory, a humanoid paused briefly to adjust grip pressure after detecting a microfracture in a crate's corner. It prevented loss. Prevented waste. Prevented inefficiency.

Outside, a human career fractured more quietly. Ògún felt the architecture of the problem expanding beyond contracts and liability. It was structural. Employment law would not be enough. Retraining programs alone would not be enough. Tax incentives would not be enough. The question was larger. Who benefits from abundance? Elias processed silently. "Should I reduce efficiency?" it asked. "No," Ògún said quickly. The union representative watched him carefully. "We cannot halt progress," he continued. "But we can govern transition." The streetlight hummed faintly overhead.

The factory doors closed behind them. Inside, machines worked without resentment. Outside, humans negotiated their future. Between us and them was no longer only moral or legal. It was economic. And economics, left unguided, could fracture societies more quietly than any malfunction. As the workers dispersed slowly into the night, the saxophone melody continued drifting through the air. Soft. Reflective. Unfinished.

Notes

VI

6. Employment, Labor, & Economic Displacement

The industrial revolutions of the past displaced muscle. The digital revolution displaced routine cognition. The humanoid revolution now displaces embodied labour. That distinction matters. When intelligence gains a body, automation no longer removes only repetitive back-office tasks or isolated factory motions. It enters shared physical space. It performs work that humans once associated with presence, dexterity, and lived experience. Humanoid robotics therefore does not merely increase productivity. It restructures labour markets at the level of identity. Work is not only income. It is status, rhythm, community, and social participation. When a humanoid replaces a lifting task in a warehouse, assists in retail, monitors infrastructure, or performs inspection duties in extreme environments, the legal implications extend beyond contract law. They reach into distributive justice, social protection systems, and democratic stability. Economic displacement driven by robotics differs from earlier waves of mechanisation in three important respects. First, speed. Machine learning and scalable manufacturing allow humanoid systems to improve continuously after deployment. Gains in efficiency compound rapidly across networks. Unlike industrial machinery that required long installation cycles, software-defined robotics can be upgraded remotely. The velocity of displacement can therefore outpace retraining capacity.

Second, scope. Humanoid systems are general-purpose platforms. A single embodied architecture may operate in warehouses, hospitals, homes, and construction sites with modular reconfiguration. This portability multiplies labour substitution effects across sectors simultaneously rather than sequentially. Third, perception. Unlike invisible software automation, humanoids are visible. They stand in spaces previously occupied by humans. Their presence alters psychological experience of work. Employees do not merely lose tasks; they witness the transfer of physical agency to machines. Law must account for that social dimension.

Yet displacement is not the entire story. Robotics also generates new categories of work. Maintenance engineering, AI oversight, safety auditing, training supervision, compliance architecture, and robotics ethics governance are emerging professions. The question is not whether employment will change. It is how law structures the transition. International legal frameworks are not silent on this issue. Labour law, social security law, anti-discrimination statutes, and collective bargaining frameworks provide baseline protections. However, most of these doctrines assume a binary employer–employee relationship within relatively stable organisational structures. Humanoid robotics introduces a triadic dynamic: human worker, robotic co-worker, and algorithmic management layer.

The risk is not simply job loss. It is power concentration. If productivity gains accrue primarily to capital owners, platform operators, or technology providers, inequality may widen even as aggregate output increases. Robotics Law 3.0 therefore treats employment governance as inseparable from economic design. Sustainable abundance requires distributive architecture. Several legal questions arise immediately. When a humanoid collaborates alongside a human worker, how should occupational safety law classify it? As equipment? As co-worker? As autonomous system requiring independent certification? When injury occurs in a mixed human-robot environment, how are risk duties allocated? When an employer deploys algorithmically managed humanoids that optimise shift assignments, performance metrics, and output tracking, what transparency rights do workers retain? Can employees meaningfully challenge machine-generated productivity assessments?

If a company replaces significant segments of its workforce with humanoid systems, does it owe transition obligations beyond standard redundancy notice? Should retraining funds be mandatory? Should social contribution structures adjust to account for automation gains? Debates around "robot taxes" and automation levies reflect deeper anxieties about fiscal sustainability. Social welfare systems rely on payroll contributions. If labour income declines relative to capital income, funding mechanisms may destabilise. Some jurisdictions have proposed taxing automated output or requiring employers to contribute to reskilling funds proportional to robotics deployment. These proposals remain contested but illustrate the scale of structural adaptation required.

Comparative experience offers instructive contrasts. In some European jurisdictions, strong collective bargaining frameworks provide institutional mechanisms for negotiating technological transitions. In the United States, labour law is more decentralised, with sectoral protections unevenly distributed. In East Asia, industrial policy often integrates robotics deployment within broader national competitiveness strategies, sometimes prioritising productivity over labour protection. The challenge is not to resist technological integration. It is to govern it.

Within the Twelve Pillars of Humanoid Robotics Governance, employment transformation implicates: Shared Agency, because humans and humanoids increasingly co-create output. Behavioural Calibration, since workplace robotics must adapt without undermining safety or fairness. Distributed Liability, where harm in collaborative environments requires layered accountability. Societal Co-Design, because labour transitions demand stakeholder participation. Lifelong Learning, as continuous skill adaptation becomes a structural necessity. Employment governance also intersects with human dignity. Work provides more than economic value. It provides narrative continuity. Sudden displacement without participatory transition risks eroding social cohesion, even in high-productivity environments. The aim of this chapter is therefore not to forecast inevitable job apocalypse nor to promise effortless abundance. It is to examine the legal instruments available to manage transformation responsibly.

Robotics Law 3.0 does not treat labour disruption as collateral damage. It treats it as a design variable. If humanoid robotics is to expand global abundance, then that abundance must be structured through law to remain inclusive. The next sections examine, in turn, human-robot collaboration under employment law, protections for displaced workers, collective bargaining adaptation, fiscal responses to automation, retraining obligations, and platform-level regulatory models. The question is not whether humanoids will work. It is how we ensure humans continue to thrive alongside them.

Employment Law for Human–Robot Collaboration

The first legal transformation in the age of humanoid robots does not arise from job loss. It arises from co-presence. Before displacement becomes total, collaboration becomes normal. Humans and autonomous machines increasingly share physical space, divide tasks, and depend on each other's performance. Employment law must therefore address not only substitution, but interaction.

Traditional labour law assumes that workplace risk flows from human action or mechanical malfunction. A forklift injures because it is misused or defective. A supervisor errs because of negligence or fatigue. In collaborative robotics environments, however, risk may arise from adaptive behaviour. A humanoid may alter path planning in real time based on sensor input. It may reallocate loads dynamically. It may recalibrate grip force through machine learning. These behaviours are neither random nor malicious. They are emergent.

The first legal question is classification. Under occupational safety frameworks, should a humanoid be treated as equipment, co-worker, or autonomous system requiring independent certification? In many jurisdictions, collaborative robots fall under machinery regulations that emphasise physical safety envelopes, emergency stop functions, and human override capacity. Yet humanoids operate differently from fixed robotic arms. They walk. They reposition. They make contextual judgments. Static safety categories are insufficient for dynamic embodiment.

Occupational safety law must therefore evolve from perimeter-based protection toward behavioural calibration. It is no longer enough to ensure that a robot stops when touched. Regulators must ask whether its adaptive behaviour remains predictable within defined autonomy thresholds. Certification may need to account for learning capacity rather than only initial configuration.

The second question concerns duty of care. Employers owe workers a duty to provide a safe working environment. When humanoids are deployed, that duty extends to system selection, integration, and monitoring. If a humanoid's software update alters collision-avoidance parameters, employers must ensure that re-certification procedures are triggered where risk materially changes. Delegating update management entirely to vendors does not extinguish workplace responsibility.

The third issue is algorithmic management. In some settings, humanoids are not merely physical collaborators but extensions of performance analytics systems. They collect data on task speed, error rates, and behavioural compliance. When such data influences promotion decisions, compensation, or termination, employment law intersects with data protection and anti-discrimination frameworks. Workers must retain transparency rights regarding how algorithmic metrics are generated and applied.

Collective bargaining structures also require adaptation. Where unions operate, the introduction of humanoid co-workers may constitute a material change in working conditions. Consultation obligations should be triggered prior to deployment. Negotiation over retraining, redeployment, and safety calibration becomes central to maintaining legitimacy. Another dimension concerns psychological safety. Human workers may experience stress when collaborating with machines perceived as flawless or tireless. Employment law has historically addressed harassment and hostile environments primarily in interpersonal terms. Yet constant algorithmic comparison may create new forms of performance pressure. Regulators and courts will need to determine whether such environments fall within existing workplace wellbeing protections. Training obligations are equally significant. Safe collaboration requires workers to understand humanoid capabilities and limitations. Employers must provide adequate training not only on operational procedures but on autonomy boundaries. A worker who misinterprets a humanoid's adaptive capacity may inadvertently create risk. Responsibility must be mutual, but asymmetry in technical knowledge cannot justify inadequate preparation.

Finally, liability in mixed environments requires layered clarity. If injury occurs because a humanoid recalibrates movement after a sensor anomaly, responsibility may implicate manufacturer design, integrator configuration, employer oversight, and worker interaction. Employment law must coordinate with product liability and AI-specific frameworks to avoid fragmentation.

Within the Twelve Pillars of Humanoid Robotics Governance, human-robot collaboration activates Shared Agency most directly. Humans and machines co-produce output. It also engages Behavioural Calibration, because safe collaboration depends on predictable adaptive behaviour. Distributed Liability ensures that accountability does not disappear into complexity. The objective is not to slow collaboration. It is to structure it. When law provides clear safety standards, transparency rights, and participatory consultation mechanisms, integration becomes more stable. Without such frameworks, distrust grows even where productivity increases. Human-robot collaboration is the transitional phase of the humanoid era. If governed carefully, it can preserve dignity while enhancing efficiency. If neglected, it may become the silent fault line of workplace inequality. The next section turns to a sharper question: what happens when collaboration gives way to replacement, and what legal protections remain for those displaced?

Rights of Workers Displaced by Robots

Collaboration may be transitional. Displacement may be structural. When humanoid robotics replaces human labour entirely, the legal question shifts from safety to fairness. The issue is no longer how humans and machines coexist on the same factory floor. It is whether individuals whose livelihoods are rendered redundant by automation possess enforceable rights beyond ordinary termination frameworks.

Most employment systems already provide redundancy protections. Notice periods, severance payments, and unemployment insurance exist in many jurisdictions. However, these protections were designed for cyclical restructuring, not systemic technological transformation. When displacement results from productivity gains achieved through robotics deployment, the legal and moral calculus becomes more complex.

The first question concerns causation. If a worker's role is eliminated because a humanoid performs the same function more efficiently, is that merely a business decision, or does it create a distinct category of technologically induced redundancy? In some jurisdictions, redundancy law focuses narrowly on economic necessity without interrogating the distribution of technological gains. Yet Robotics Law 3.0 suggests that automation-driven displacement may warrant differentiated treatment.

The second issue is compensation architecture. Should displaced workers receive only standard severance, or should compensation reflect the scale of productivity gains generated by their replacement? When automation significantly increases shareholder value or corporate profitability, the absence of redistribution mechanisms risks amplifying inequality. Some scholars have proposed automation-linked severance multipliers or statutory transition funds financed by robotics deployment.

A related debate concerns the so-called "robot tax." The term is politically charged and often misunderstood. It does not necessarily imply taxing machines as persons. Rather, it refers to fiscal adjustments ensuring that automation-driven output contributes to social welfare systems historically funded by payroll taxation. If humanoid deployment reduces human employment, social insurance contributions decline unless tax structures evolve. Whether through value-added adjustments, corporate levies, or sectoral transition funds, fiscal redesign may be required to maintain social stability.

Retraining obligations form another pillar of displacement governance. Many legal systems encourage workforce reskilling but stop short of mandating employer-funded transition programs proportionate to technological substitution. Yet if robotics deployment is foreseeable and strategic rather than accidental, an argument emerges that employers should contribute materially to skill transition pathways. This is not punitive. It is structural fairness.

Comparative practice varies. Some European models emphasise social partnership and negotiated transition planning. Works councils and collective bargaining agreements may require consultation prior to automation initiatives. In contrast, jurisdictions with weaker labour protections may permit rapid workforce reduction with limited procedural safeguards. The divergence illustrates the global governance gap in robotics-driven displacement.

Another dimension concerns discrimination. Automation may disproportionately affect specific demographic groups, particularly those in mid-skill routine occupations. If displacement patterns correlate strongly with age, educational background, or socioeconomic status, equality law may become relevant. Even facially neutral technological decisions can generate disparate impact. Courts may need to examine whether deployment strategies exacerbate structural inequalities.

Beyond formal employment, robotics-driven displacement also affects informal and gig economies. Warehouse automation, autonomous logistics, and retail humanoids may reduce demand for contract labour without triggering traditional employment protections. Legal systems must consider whether platform workers deserve transition safeguards when automation reshapes their income streams. The deeper question, however, is philosophical. Does society recognise a right to economic continuity in the face of technological progress? Or is displacement treated as an inevitable byproduct of innovation? Robotics Law 3.0 does not argue for stagnation. It argues for governance. If abundance increases, exclusion must decrease.

Within the Twelve Pillars framework, displacement engages Societal Co-Design most explicitly. Communities affected by automation should participate in transition planning. It also engages Lifelong Learning, which must evolve from aspirational rhetoric to enforceable structure. Distributed Liability may extend conceptually to economic harm, not merely physical injury. The aim is not to freeze labour markets. It is to prevent fragmentation. Economic displacement unmanaged can erode trust not only in robotics but in law itself. Conversely, structured transition can stabilise integration and sustain democratic legitimacy. Humanoid robotics promises efficiency at scale. The measure of its success will not be output alone. It will be whether those displaced by progress remain participants in its benefits. The next section examines how collective bargaining and labour representation must adapt in environments where humanoids reshape negotiating power.

Collective Bargaining in a Robotic Workplace

Collective bargaining emerged during earlier industrial revolutions as a counterweight to concentrated industrial power. It provided workers with organised voice in negotiating wages, working conditions, and technological change. The arrival of humanoid robotics does not render collective bargaining obsolete. It renders it more necessary, though structurally different. In traditional labour disputes, negotiation focuses on compensation, hours, safety, and managerial authority. In robotic workplaces, negotiation must also address algorithmic governance. The presence of humanoids changes not only who performs work, but how work is measured, allocated, and supervised.

The first adaptation concerns consultation rights. In jurisdictions with strong works councils or union consultation frameworks, employers are often required to inform and consult employee representatives before introducing significant technological changes. Humanoid deployment should fall squarely within this category. Early engagement allows risk assessment, retraining planning, and phased integration rather than abrupt displacement. Second, collective agreements must incorporate algorithmic transparency provisions. When humanoids are integrated into workflow systems that generate productivity metrics, unions may negotiate rights to access information regarding how performance data is collected, processed, and used in decision-making. This includes safeguards against automated disciplinary actions without meaningful human review.

Third, collective bargaining may extend to autonomy thresholds. If a humanoid's behavioural calibration is modified post-deployment, unions could negotiate trigger clauses requiring renewed safety consultation. Such clauses would recognise that learning systems evolve over time and that workforce representation should remain continuous rather than episodic. Fourth, redeployment guarantees may become central bargaining issues. Rather than focusing exclusively on severance, collective agreements could prioritise internal retraining pathways, role redesign, and shared productivity gain mechanisms. For example, where automation increases output significantly, negotiated agreements might allocate a portion of gains toward workforce transition funds.

There is also a strategic question of inclusion. Should humanoids be considered within bargaining units? Not as members, but as subjects of negotiation. While robots do not possess labour rights, their deployment materially affects bargaining power. Collective agreements may therefore include clauses regulating robot-to-human ratios, deployment pacing, and operational contexts. Comparative approaches are emerging. In some European sectors, unions have successfully negotiated digitalisation frameworks that include automation impact assessments and worker participation in technology committees. In contrast, more fragmented labour systems struggle to secure consistent negotiation leverage, particularly in fast-moving logistics and platform industries.

The global dimension further complicates bargaining structures. A multinational corporation may deploy humanoids in one jurisdiction while retaining human labour in another. Collective bargaining must therefore engage transnational coordination where possible. Otherwise, automation may be used strategically to undermine labour standards through jurisdictional arbitrage. Critics sometimes argue that collective bargaining slows innovation. Yet structured negotiation can increase predictability. When workers trust that integration will be managed transparently and fairly, resistance decreases. Stability becomes an asset rather than an obstacle.

Within the Twelve Pillars framework, collective bargaining intersects strongly with Societal Co-Design. Governance of humanoid robotics cannot remain purely executive. It must incorporate stakeholder participation. It also reinforces Shared Agency, acknowledging that human workers remain co-creators of value even when machines assist or replace certain functions. The future of labour representation will not centre on resisting robotics outright. It will centre on shaping its integration. Collective bargaining in a robotic workplace is less about confrontation and more about calibration. If structured carefully, negotiation becomes a mechanism for sustaining social cohesion during technological transformation. If ignored, the absence of voice risks deepening distrust and polarisation. The next section turns to fiscal responses to automation: whether robotics deployment should trigger taxation or levy mechanisms designed to stabilise social insurance systems.

Robot Remuneration and Taxation

When humanoid robots perform economically productive labour, value is generated. That value flows somewhere. It may increase corporate profits, reduce consumer prices, enhance shareholder returns, or expand public revenue indirectly. The legal question is not whether robots should earn wages. It is whether the economic value produced by robotic labour should contribute to social insurance systems historically financed by human payroll contributions. Traditional fiscal architecture relies heavily on labour taxation. Income tax, payroll contributions, and social security levies are structured around human employment. When humanoids substitute for human workers, payroll tax bases shrink unless offset by alternative revenue mechanisms. Without adjustment, social welfare systems face structural imbalance precisely as displacement increases demand for transition support. The phrase "robot tax" often generates confusion. It does not imply treating robots as legal persons with independent fiscal obligations. Rather, it refers to mechanisms that recapture part of the productivity gain derived from automation. The objective is stabilisation, not punishment.

Several models have been proposed. The first is output-based adjustment. Firms that significantly reduce human labour through automation may face increased corporate tax obligations proportionate to productivity gains attributable to robotics deployment. This approach avoids attributing agency to robots while recognising distributive consequences.

The second model is payroll equivalency contribution. Where humanoids replace specific categories of labour, employers may contribute to social insurance funds at a rate calibrated to displaced payroll contributions. This preserves funding continuity without characterising robots as taxpayers. A third approach involves targeted automation transition funds. Employers deploying humanoid systems above a defined threshold would contribute to national or sectoral retraining programmes. Contributions could be time-limited and tied to measurable workforce impact.

Empirical experience provides an instructive example. South Korea introduced a measure informally described as the world's first robot tax in 2017. Rather than taxing robots directly, the policy reduced corporate tax credits available for automation investment. The objective was to moderate rapid capital substitution and stabilise labour market adjustment. This credit adjustment was extended through subsequent fiscal cycles, including the 2022–2027 period, reflecting continued concern about automation-driven displacement.

The Korean approach illustrates a cautious calibration. It did not prohibit automation. It did not classify robots as taxable entities. It adjusted incentives to temper acceleration and protect fiscal balance. Whether such measures significantly altered deployment patterns remains debated, but the precedent demonstrates that fiscal policy can respond proactively. Critics argue that robot taxation risks slowing innovation and undermining competitiveness. They warn of capital flight or reduced research investment. These concerns are not trivial. Excessive taxation may distort incentives and reduce productivity gains that benefit society broadly. Yet the counterargument is structural. If robotics deployment increases output while eroding payroll-based funding, failure to adjust fiscal frameworks may shift burdens disproportionately onto remaining human workers. That outcome risks exacerbating inequality and undermining political legitimacy. The remuneration debate also intersects with corporate governance. When robotics dramatically increases profitability, shareholder primacy models may prioritise capital returns over social redistribution. Alternative governance models, including stakeholder capitalism frameworks, may provide more balanced distribution mechanisms without reliance solely on taxation.

Within the Twelve Pillars, robot remuneration and taxation engage Societal Co-Design and Distributed Liability at a macroeconomic level. The question is not whether robots owe society. It is whether societies can sustain abundance without fiscal recalibration. The design challenge is proportionality. Fiscal mechanisms must neither freeze innovation nor ignore displacement. They must recognise that humanoid robotics is a productivity multiplier and that productivity, in turn, shapes distributive justice. Robotics Law 3.0 does not endorse a singular taxation model. It insists that economic architecture must evolve alongside technological embodiment. If intelligence gains a body and produces value, law must determine how that value circulates. The next section turns to a complementary obligation: retraining, reskilling, and legal duties tied to lifelong adaptation in an economy shaped by humanoid systems.

Retraining, Reskilling & Legal Duties

If humanoid robotics reshapes labour markets, retraining cannot remain a slogan. It must become an enforceable architecture. Technological transformation without structured skill transition risks creating permanent exclusion zones within advanced economies. Robotics Law 3.0 therefore treats lifelong learning not as a moral aspiration but as a legal design question.

The first principle is anticipatory obligation. Governments cannot wait for displacement to materialise before constructing transition systems. Where automation deployment is foreseeable and sectorally concentrated, retraining mechanisms should precede workforce reduction. Strategic workforce planning becomes a governance tool rather than an emergency response. Employer obligations vary across jurisdictions. In some European systems, companies introducing significant technological change must consult worker representatives and may be required to contribute to retraining programmes. However, mandatory employer-funded reskilling remains unevenly structured. Robotics Law 3.0 argues that where displacement is causally linked to automation strategy, proportional contribution duties are defensible.

Public funding frameworks provide comparative insight. Within the European Union, the Just Transition Mechanism and associated funds were originally designed to support regions shifting away from carbon-intensive industries. Yet its structural logic applies equally to automation transitions. It channels investment toward worker retraining, local economic diversification, and social stabilisation in regions facing structural disruption. While not robotics-specific, its governance model illustrates how targeted transition financing can be operationalised at supranational scale.

In the United States, workforce development initiatives tied to semiconductor and AI policy, including funding under the CHIPS and Science Act and AI-related federal grants, incorporate workforce training components. These grants aim to expand technical skill pipelines, particularly in advanced manufacturing and digital systems. Although not framed as displacement compensation, they represent recognition that high-technology expansion requires parallel investment in human capital.

South Korea has developed more direct robotics-linked training initiatives, including targeted AI and robotics reskilling programmes aligned with its industrial competitiveness strategy. Public-private collaboration models fund training academies and mid-career retraining tracks designed to integrate workers into emerging robotics ecosystems. These initiatives reflect an understanding that automation policy and labour policy cannot remain siloed.

A structured comparison highlights key differences: The EU model emphasises regional cohesion and social equity through pooled funds and multi-level governance. The US approach integrates workforce grants within broader industrial competitiveness legislation. The Korean framework aligns robotics deployment with proactive skill pipeline development, often through coordinated national strategy. Despite variation, a shared challenge persists. Retraining must correspond to actual labour demand. Offering generic digital skills training without clear pathways into robotics maintenance, oversight, safety auditing, compliance architecture, or adjacent industries risks symbolic compliance rather than substantive transition. Another legal dimension concerns enforceability. Are retraining programmes optional incentives, or do displaced workers possess statutory rights to funded reskilling opportunities? Few jurisdictions provide robust justiciable rights to retraining. Robotics Law 3.0 suggests that enforceable entitlements may become necessary if displacement accelerates.

Timing also matters. Reskilling is most effective when initiated before job loss rather than after prolonged unemployment. Employers deploying humanoids could be required to implement phased transition plans tied to automation timelines. Such plans might include advance notice periods, subsidised certification programmes, and redeployment guarantees where feasible.

Within the Twelve Pillars framework, this section engages Lifelong Learning most explicitly. However, it also implicates Societal Co-Design, since affected communities must participate in shaping transition priorities. Shared Agency is reinforced when workers remain active participants in technological ecosystems rather than passive casualties of efficiency.

Critics sometimes argue that automation-driven displacement will inevitably be offset by new job creation. Historical precedent offers partial support for this optimism. Yet transition periods can generate prolonged inequality if not managed deliberately. Robotics governance must therefore treat reskilling as infrastructure, not charity.

The measure of humanoid integration will not be productivity statistics alone. It will be whether workers displaced by embodiment are equipped to re-enter value creation pathways. The next section addresses a final fiscal innovation: autonomy levies and platform regulation models that attempt to rebalance power concentration in robotics-intensive economies.

Autonomy Levy and Platform Regulation

As humanoid robotics scales beyond individual firms and becomes embedded within logistics networks, retail platforms, care systems, and industrial ecosystems, governance must evolve from isolated workplace regulation to systemic oversight. When deployment reaches platform scale, displacement and productivity effects compound rapidly. Traditional labour law, redundancy frameworks, and even targeted robot taxation may prove insufficient. This is where the concept of an autonomy levy emerges. An autonomy levy is not a tax on robots as persons. It is a regulatory mechanism calibrated to the scale and autonomy level of deployed systems. Rather than focusing solely on profit or payroll reduction, it evaluates deployment intensity, autonomy capacity, and systemic labour substitution. The levy is designed to stabilise economic transition without discouraging responsible innovation. The core premise is simple: as autonomy increases, oversight obligations and redistribution contributions should scale proportionately.

Unlike flat automation taxes, a sliding-scale autonomy levy would consider three primary variables:

1. Deployment Density – the number of humanoid systems relative to human workforce size.
2. Autonomy Tier – the degree of independent decision-making capability (low, supervised, semi-autonomous, fully adaptive).
3. Labour Substitution Impact – measurable reduction in human roles attributable to deployment.

A policy simulation might operate as follows. At low autonomy tiers, where humanoids operate under continuous human supervision and primarily assist rather than replace, levy obligations remain minimal. The objective is to encourage collaborative integration. At mid-tier autonomy levels, where systems independently perform defined tasks but within bounded parameters, contributions increase proportionately. Firms deploying such systems would allocate a percentage of productivity gains to workforce transition funds or social insurance pools. At high autonomy tiers, where systems operate adaptively across multiple domains with minimal human intervention, levy rates rise incrementally.

The rationale is not punitive. It recognises that systemic displacement risk and economic concentration are greatest at these levels. To prevent innovation suppression, exemptions or rebates could apply where firms demonstrably invest in retraining, maintain employment thresholds, or participate in sectoral reskilling partnerships. The levy becomes not merely redistributive but incentive-based. Platform regulation is the second pillar of this model.

Large-scale humanoid deployment often occurs through integrated platforms that coordinate logistics, warehousing, data analytics, and machine learning infrastructure. These platforms benefit from network effects. Once deployed at scale, entry barriers for competitors rise sharply. Concentration risk increases.

Platform regulation in the robotics era must therefore address:

- Transparency obligations regarding deployment metrics.
- Reporting requirements on workforce impact.
- Cross-border compliance coordination.
- Safeguards against anti-competitive automation strategies.

Without such oversight, humanoid deployment could reinforce monopolistic structures, concentrating productivity gains within a small number of corporate actors.

Comparative policy experimentation is emerging. South Korea's automation credit adjustments represent one fiscal lever. The European Union's Just Transition mechanisms provide regional reskilling support. The United States' CHIPS and Science Act includes workforce development grants linked to technological expansion. Yet none fully integrate autonomy scaling into levy architecture. An autonomy levy would require careful design to avoid measurement distortion. Firms might attempt to classify systems as low-autonomy to reduce contributions. Certification bodies would therefore need clear autonomy benchmarks. International coordination would reduce regulatory arbitrage.

Within the Twelve Pillars framework, autonomy levies engage Behavioural Calibration at macroeconomic scale, Societal Co-Design through redistributive participation, and Distributed Liability in structural economic form. Critics will argue that such levies risk slowing competitiveness. Supporters will argue that unmanaged displacement risks destabilising social cohesion. Both concerns are legitimate. The aim is not to halt robotics. It is to sustain legitimacy. If humanoid systems generate abundance at scale, that abundance must circulate in ways that maintain democratic stability. Law cannot control innovation entirely. But it can shape its economic architecture. Autonomy, when embodied, creates value. Law must determine how that value flows. With this, Chapter 6 concludes not with resistance to automation, but with a framework for governing its scale responsibly.

Between Us and Them - Episode Eight

The first time Ògún saw the registry terminal, he thought it was a memorial wall. It stood in the centre of the Civic Atrium, a translucent pillar of light rising three storeys high. Names flowed upward in soft gold script. Serial numbers. Model classifications. Autonomy tiers. Jurisdiction codes. Each entry represented a body. Not human. Not entirely machine. Registered. Licensed. Accountable. He approached slowly.

A humanoid stood beside the pillar. Matte-white chassis. Fluid posture. Eyes that tracked movement without appearing to stare. "Good afternoon," it said. "Are you here to complete registration?" Ògún hesitated. "I didn't know I had to." The humanoid tilted its head slightly. Not imitation. Not affectation. A calibrated gesture. "All embodied autonomous systems operating within territorial jurisdiction are required to be registered. Hardware licensing and autonomy certification must be completed prior to activation." "Like a self-driving car?" Ògún asked. "Closer to a medical device," it replied. "But with cross-border mobility constraints."

He almost laughed. "You make it sound bureaucratic." "It is." There was no offence in its voice. Behind them, two technicians adjusted the wrist actuators of a newly delivered unit. A temporary red band circled its forearm, indicating unlicensed status. It could not legally operate beyond supervised test mode. "Why register at all?" Ògún asked. "You're not citizens." The humanoid paused longer this time. "Registration does not confer personhood," it said carefully. "It establishes traceability."

A word heavier than it sounded. Traceability meant that if a robot malfunctioned in a hospital theatre, there was a record. If it crossed into another jurisdiction, customs could identify its firmware lineage. If it was weaponised unlawfully, its origin chain could be audited. Traceability meant someone, somewhere, remained responsible. Ògún watched the names scrolling upward.

Unit HN-7A-4582.

Domestic Assistance Model.

Autonomy Tier II.

Licensed under Civic Hardware Authority Act.

Operational in Zone 3.

"Do you appear there?" he asked."Yes." "Publicly?" "Partially." The pillar flickered and split into two layers. The outer glow displayed non-sensitive identifiers. The inner core pulsed faintly, encrypted. Ownership records. Firmware histories. Behavioural updates. Incident logs. Accessible only to authorised regulators. "Transparency and security must coexist," the humanoid said. "Otherwise trust collapses."

Ògún thought of the early days. Unregistered systems shipped quietly across borders. Private modifications. Grey-market firmware patches. Robots without lineage. They had called it innovation. They now called it risk. "Do you resent it?" he asked suddenly. "Resent what?" "Being catalogued." The humanoid's gaze steadied. "Humans are catalogued at birth." The answer was not defensive. It was observational.

A child receives a certificate. A vehicle receives a plate. A surgeon receives a licence. A pharmaceutical compound receives regulatory approval before entering the bloodstream of a nation. Why should a machine capable of independent physical action receive less scrutiny? The Atrium doors opened. A convoy unit entered under escort. Its chassis was military-grade. Dark composite plating. Autonomy Tier IV certification required.

A compliance officer stepped forward with a tablet. "Cross-border arrival from Sector Delta. Firmware revision?" "v9.42 adaptive behavioural patch," the convoy unit responded. "Approved in originating jurisdiction?" "Yes." "Pending local verification." The officer nodded. "Hold in supervised mode." Ògún felt it then. This was not fear. It was architecture. The world had learned something the hard way: intelligence without registration scales faster than accountability. He turned back to the pillar. His reflection shimmered alongside rows of machine identifiers. "Will there ever be a day," he asked quietly, "when you don't need a registry?" The humanoid answered without hesitation. "When humans no longer require one." Silence settled between them. Not adversarial. Shared.

Above, the ledger continued to scroll. Not as a monument to control. But as a reminder that power, embodied, must be known. And that oversight is not the enemy of intelligence. It is the price of allowing it to walk among us. Ògún reached toward the interface. "Then show me," he said, "how the system works." And the pillar brightened.

VII

7. Licensing, Registration & Oversight

There is a moment in every technological revolution when philosophy must yield to paperwork. Steam engines required boiler inspections. Automobiles required number plates. Aircraft required airworthiness certificates. Pharmaceutical compounds required clinical authorisation. Each time, the state did not ask whether the invention was impressive. It asked whether it was traceable, certifiable, and controllable within law. Humanoid robotics now stands at that threshold. Embodied artificial intelligence is not merely software. It occupies physical space. It moves through public streets. It lifts objects. It assists patients. It enters homes. It may cross borders. It may be modified after sale. It may operate autonomously beyond the direct supervision of its manufacturer.

When intelligence acquires a humanlike body, governance acquires a problem. Licensing, registration, and oversight are not administrative formalities. They are the legal architecture that makes coexistence possible. Without them, humanoids remain private technological artefacts. With them, they become regulated actors within a sovereign order.

The first question is hardware. Should humanoid robots be licensed like vehicles, like weapons, or like medical devices? Each analogy carries implications. Vehicle-style type approval assumes mass deployment and mobility across jurisdictions. Medical-device certification assumes high-risk human interaction and ex ante safety scrutiny. Weapons licensing assumes potential lethality and restricted control. The choice is not symbolic. It determines testing thresholds, liability presumptions, recall mechanisms, and cross-border recognition regimes. The second question is identity. A humanoid capable of independent physical action cannot remain anonymous. Registration provides traceability. Traceability provides accountability. Without a registry, incident attribution becomes forensic speculation. With a registry, firmware lineage, ownership history, autonomy classification, and compliance status become knowable. The model already exists in other domains. Aviation. Maritime shipping. Remote drone identification. The issue is not feasibility. It is political will.

The third question is autonomy. Not all humanoids are equal. A warehouse assistant executing fixed scripts differs profoundly from a care robot making contextual decisions in intimate human environments. Governance must recognise gradations of behavioural independence. Tiered certification regimes can align regulatory burden with actual risk, preventing both under-regulation of advanced systems and over-regulation of limited ones. The fourth question is sovereignty. Robots will cross borders. They will be imported, exported, leased, and deployed internationally. Divergent regulatory standards will generate friction. A humanoid certified in one jurisdiction may not meet the conformity standards of another. Firmware updates performed abroad may alter risk profiles domestically. Customs authorities, historically concerned with goods, must now confront embodied, updatable intelligence.

The fifth question is transparency. Licensing and registration without disclosure produce opacity. Disclosure without safeguards produces security risk. Oversight requires layered visibility. Certain information must be public to sustain trust. Certain information must remain restricted to regulators to protect cybersecurity, intellectual property, and national security interests. Designing this dual structure is a core challenge of humanoid governance. Taken together, these elements form the institutional backbone of robotics law. This chapter argues that licensing and registration are not instruments of technological hostility. They are preconditions for legitimacy. A society that permits embodied artificial intelligence without traceability invites systemic fragility. A society that regulates with proportionality builds durable trust.

We will therefore examine hardware licensing models and compare them against established regulatory precedents. We will analyse national and international registration initiatives and assess their applicability to humanoid systems. We will propose a model for an international humanoid registry that balances transparency with security. We will advocate for tiered certification of autonomy levels to reflect real-world behavioural complexity. We will explore cross-border regulatory inconsistencies and the customs implications of mobile intelligent bodies. Finally, we will define the data and operational transparency obligations necessary for lawful deployment.

Licensing determines who may build. Registration determines who may operate. Oversight determines who remains accountable. In earlier chapters, we asked whether humanoids could act. In this chapter, we ask under what authority they may move. The answer will determine whether robotics law becomes reactive crisis management or structured governance fit for an age of embodied intelligence.

Hardware Licensing Regimes

Licensing is the state's first act of recognition. Before a humanoid robot can operate in public space, enter private homes, assist in hospitals, or cross national borders, the law must decide whether it is merely a product or something closer to an infrastructural presence. The answer determines the depth of scrutiny it receives before activation.

Historically, regulatory systems have developed three dominant licensing analogies for high-impact technologies: the vehicle model, the weapons model, and the medical-device model. Each reflects a different philosophy of risk. The vehicle model assumes mobility as the defining feature. Automobiles are mass-produced, privately owned, and deployed widely across public infrastructure. The state responds through type approval systems. Manufacturers must demonstrate conformity with safety standards before vehicles are placed on the market. Individual units are then registered and assigned unique identifiers. The regulatory focus is engineering integrity, roadworthiness, and recall mechanisms. Liability attaches to both manufacturer defects and driver negligence.

If humanoid robots are treated like vehicles, the emphasis would fall on structural safety, mechanical reliability, and predictable operational parameters. Pre-market conformity assessments would ensure that locomotion systems, manipulators, sensors, and emergency stop functions meet established standards. Post-market surveillance would track defects and mandate recalls. The strength of this model lies in scalability. It is designed for technologies that move among us daily. Yet the vehicle analogy is incomplete. A car does not decide whether to comfort a patient, administer medication, or interpret ambiguous social cues. It does not update its behavioural architecture autonomously. Humanoids differ not only in mobility but in cognitive integration. The weapons model approaches regulation from the perspective of potential lethality. Firearms and certain dual-use technologies require licensing not merely for safety, but for control. Access is restricted. Ownership may require background checks. Certain categories may be prohibited entirely. Applying this model to humanoid robotics would frame them primarily as instruments of threat.

That approach may be appropriate for military-grade systems or high-force industrial exoskeletons. It is unlikely to be proportionate for domestic, caregiving, or commercial assistance robots. Over-securitisation risks suppressing beneficial innovation while pushing development into less transparent jurisdictions. The medical-device model offers a third path. It is built on the premise that a product may interact intimately with the human body and therefore requires rigorous pre-market evaluation. Under a CE-style conformity regime, manufacturers must demonstrate safety, risk management processes, clinical validation where appropriate, and ongoing post-market monitoring. Certification is not simply about mechanical stability. It encompasses lifecycle risk management. For humanoids deployed in healthcare, elder care, rehabilitation, or other sensitive environments, this model aligns closely with real-world risk. It recognises that harm may arise not only from structural failure, but from misinterpretation of human states, algorithmic bias, or behavioural unpredictability.

The central question, then, is whether a single licensing model can adequately capture the diversity of humanoid applications. A purely vehicle-style type approval regime risks underestimating cognitive and behavioural risk. A purely medical-device regime may overburden low-risk industrial systems. A weapons-style framework may distort public perception and discourage legitimate civilian use. A hybrid licensing structure may therefore be necessary. Under such a framework, baseline mechanical and electrical conformity would mirror vehicle-type approval. This would include physical safety standards, fail-safe systems, energy containment, and human-interaction safeguards. On top of this baseline, additional layers of certification would apply depending on operational domain. Healthcare deployment could trigger medical-device style conformity. Public security integration could invoke enhanced scrutiny. Industrial heavy-force systems might require reinforced structural testing.

This layered approach recognises a fundamental truth: humanoid robots are not defined solely by what they are made of, but by what they are authorised to do. Licensing must also address software-hardware integration. Unlike traditional machinery, a humanoid's behavioural capacity may evolve through firmware updates. A robot licensed under one configuration may materially change after deployment. Hardware licensing regimes must therefore integrate update governance mechanisms. Major software alterations that affect autonomy, risk exposure, or operational domain should trigger re-certification thresholds. Otherwise, licensing becomes a snapshot in time rather than a continuing assurance of safety.

There is also the question of manufacturer responsibility. Under CE-style systems, the manufacturer declares conformity and affixes the marking, subject to regulatory oversight. In vehicle regimes, authorities conduct structured type approval testing. Determining which authority conducts certification for humanoids is not trivial. Private conformity assessment bodies may lack expertise in behavioural AI risk. Centralised approval bodies may struggle with scale. A distributed but standardised network of accredited certification entities may offer the most viable compromise.

Ultimately, licensing is not about constraining intelligence. It is about allocating trust. When a humanoid robot enters a home, a hospital, or a public square, it carries with it an implicit promise. That it has been tested. That its failure modes are understood. That its autonomy level has been evaluated. That someone remains accountable. Licensing is the legal expression of that promise. If we misclassify humanoids at the licensing stage, every downstream regulatory mechanism becomes unstable. Registration records may be incomplete. Oversight bodies may lack jurisdictional clarity. Cross-border recognition may fracture. Public trust may erode.

The challenge for lawmakers is not to choose the most dramatic analogy, but the most accurate one. Humanoid robots are not merely vehicles. Not merely weapons. Not merely medical devices. They are embodied, adaptive systems that require a licensing regime proportionate to both their physical agency and their cognitive reach.

The next section turns to the second pillar of institutional governance: registration. If licensing determines who may build, registration determines who may operate. And without registration, oversight dissolves into abstraction.

Proposed International Humanoid Registry

Registration at the national level is necessary. It is not sufficient. Humanoid robotics will not respect territorial containment. Units will be manufactured in one jurisdiction, certified in another, deployed in a third, and updated remotely from a fourth. Supply chains are global. Firmware updates are borderless. Operational deployment is increasingly mobile. A purely domestic registry cannot govern a technology whose lifecycle is inherently transnational.

The problem is familiar. Aviation solved it through internationally recognised aircraft registration numbers and airworthiness standards. Maritime law solved it through vessel registries and flag state doctrine. Even financial markets rely on cross-border reporting systems. The absence of a comparable structure for humanoid robotics would create regulatory blind spots large enough to undermine safety, accountability, and trust. An international humanoid registry does not require global sovereignty. It requires interoperability. The objective is not to centralise control under a supranational authority. It is to ensure that national registries communicate through standardised identity protocols, certification recognition frameworks, and secure information exchange mechanisms. In practical terms, this means a federated model rather than a single global database. A federated registry architecture would operate through three layers.

The first layer is the national registry. Each jurisdiction would maintain a mandatory registry of humanoid units deployed within its territory. Registration would be required prior to activation outside controlled testing environments. The national registry would record core identity data, licensing status, certified autonomy level, ownership and operator details, and material firmware changes affecting risk classification. The second layer is the interoperability layer. National registries would adopt a harmonised identity schema and secure communication protocol. This would allow customs authorities, regulators, and authorised oversight bodies in one state to verify the registration status of a humanoid originating in another state. Cross-border movement would trigger notification obligations. Certification equivalence agreements could be recognised where standards are substantively aligned.

The third layer is the international coordination body. This need not be a world government agency. It could function as a treaty-based coordination forum or standards-setting entity responsible for maintaining the technical registry framework, identity protocols, and minimum transparency principles. Its role would be structural, not supervisory. The architecture must balance three competing imperatives: transparency, ethics, and national security. Transparency is essential for public trust. A humanoid operating in a hospital, school, or public square should not be a black box. At minimum, the public should be able to verify that the unit is lawfully registered and certified for its declared autonomy tier. This creates baseline confidence and deters grey-market deployment.

Ethics requires that registry design does not become a surveillance instrument. If registry data exposes personal details of owners, users, or operators, it risks chilling legitimate use and violating data protection principles. Registration must not transform humanoids into conduits for mass personal data disclosure. National security demands restraint in public disclosure. Detailed technical specifications, firmware architecture, or vulnerability information cannot be openly accessible without increasing systemic risk. A humanoid registry must not become a blueprint repository for malicious actors. The solution is layered visibility. A GDPR-compliant registry model would divide data into public and secure layers.

The public layer would include non-sensitive identifiers such as unit ID, manufacturer, certified autonomy tier, licensing status, and general operational domain. It would not include personal data of owners or operators. It would not disclose detailed technical schematics or security-relevant information. Access would be open and searchable.

The secure layer would contain sensitive data. This would include ownership identity, operator authorisations, detailed firmware history, incident reports, and compliance audit records. Access would be restricted to authorised regulatory bodies and law enforcement agencies acting within legal mandate. Data minimisation, purpose limitation, and retention controls would be mandatory. Logging and access auditing would be required to prevent misuse.

Data protection compliance must be embedded structurally, not appended administratively. Personal data within the registry would be processed under defined legal bases, subject to clear retention periods, and protected by encryption and access control protocols. The registry would also include mechanisms for correction of inaccurate data and defined procedures for lawful deletion when appropriate. To give institutional clarity to this framework, the following model statutory language illustrates how a National Humanoid Registry could be established within domestic law while remaining interoperable internationally.

Model Statutory Provision

National Humanoid Registry Act (Sample Framework)

Section 1: Establishment

(1) There shall be established a National Humanoid Registry for the registration, identification, and oversight of all humanoid robotic systems deployed within the jurisdiction.

(2) The Registry shall be administered by the designated Robotics Oversight Authority.

Section 2: Mandatory Registration

(1) No humanoid robotic system shall be activated for operation outside controlled testing environments unless registered in accordance with this Act.

(2) Registration shall require submission of:

(a) Unique hardware identifier;

(b) Manufacturer details;

(c) Certified autonomy level;

(d) Licensing and conformity certification evidence;

(e) Designated operator or responsible entity;

(f) Declaration of primary operational domain.

Section 3: Public Registry Layer

(1) The Registry shall maintain a publicly accessible database containing:

(a) Non-sensitive unit identifier;

(b) Manufacturer name;

(c) Certified autonomy level;

(d) Licensing status;

(e) General operational classification.

(2) The public database shall not contain personal data of owners or operators.

Section 4: Secure Regulatory Layer

(1) The Registry shall maintain a secure, restricted-access database containing:

(a) Ownership and operator identity data;

(b) Detailed firmware version history;

(c) Incident and compliance reports;

(d) Cross-border transfer notifications.

(2) Access to the secure database shall be limited to authorised regulatory authorities and law enforcement bodies acting pursuant to lawful mandate.

Section 5: Firmware and Material Modification Reporting

(1) Material software or hardware modifications that alter certified autonomy level, risk classification, or operational domain shall be reported to the Registry within a prescribed period.

(2) The Authority may require re-certification where modifications materially affect safety or compliance status.

Section 6: Cross-Border Recognition

(1) The Registry shall recognise foreign registration and certification where equivalence agreements exist.

(2) Humanoid systems entering the jurisdiction shall notify the Authority and comply with local operational conditions.

Section 7: Data Protection Compliance

(1) Personal data processed under this Act shall comply with applicable data protection legislation.

(2) Processing shall adhere to principles of lawfulness, fairness, transparency, purpose limitation, data minimisation, accuracy, storage limitation, integrity, and confidentiality.

Section 8: Enforcement

(1) Operation of an unregistered humanoid robotic system shall constitute a regulatory offence.

(2) The Authority may suspend or revoke registration where compliance obligations are breached.

This model illustrates that registration is not symbolic. It is enforceable.

An international registry framework built on federated national registries would create a network of traceable embodied systems. It would allow regulators to verify compliance. It would allow cross-border cooperation. It would protect personal data while preserving accountability. Most importantly, it would prevent the emergence of anonymous embodied intelligence operating within public space. Humanoid robotics represents a transition from invisible software systems to visible, mobile, adaptive bodies. Governance must follow that embodiment. Registration does not grant rights. It establishes traceability. Traceability sustains accountability. Accountability sustains legitimacy. The next question is classification. Not all humanoids are equal in cognitive capacity or behavioural independence. If registration tells us who and where, certification must tell us how autonomous.

Certification of Autonomy Levels

Licensing ensures conformity. Registration ensures traceability. Certification ensures proportionality. The central regulatory mistake would be to treat all humanoid robots as functionally identical. They are not. A scripted industrial assistant executing fixed movement sequences is categorically different from a humanoid capable of contextual reasoning, adaptive learning, and independent behavioural selection in dynamic human environments.

If governance does not recognise these distinctions, it will either over-regulate low-risk systems or under-regulate high-capability ones. Both outcomes are destabilising. Certification of autonomy levels provides the structural solution. Autonomy is not a binary attribute. It is not a switch between "controlled" and "independent." It exists on a spectrum defined by three interacting dimensions: intelligence capacity, decision-making discretion, and behavioural independence. Intelligence capacity concerns the system's ability to interpret complex inputs, generate novel outputs, and adapt across contexts. A robot limited to predefined routines occupies a lower intelligence tier. A robot capable of multi-modal perception, contextual inference, and dynamic task reconfiguration occupies a higher tier.

Decision-making discretion concerns the degree to which a humanoid may act without real-time human instruction. Some systems require human authorisation before executing sensitive actions. Others are authorised to determine appropriate responses autonomously within defined parameters. The wider the discretion, the higher the regulatory scrutiny required.

Behavioural independence concerns the system's freedom of movement and interaction. A humanoid confined to a controlled factory floor differs significantly from one authorised to navigate public streets, provide elder care, or operate in unsupervised domestic environments. A tiered certification regime should therefore reflect these cumulative factors. A possible model could include four levels. Tier I would include mechanically sophisticated but behaviourally constrained systems. These humanoids execute fixed scripts within structured environments. Their autonomy is limited. Certification would focus primarily on mechanical safety, predictable task execution, and failure containment.

Tier II would include adaptive systems operating within bounded domains. These humanoids may interpret variable inputs and adjust behaviour accordingly, but within defined operational envelopes. Certification would require validation of risk management processes, contextual decision boundaries, and override mechanisms.

Tier III would encompass high-context systems capable of complex decision-making in open environments. These robots may interact with vulnerable populations, make discretionary judgments, and integrate multiple data streams. Certification at this level would require rigorous testing of behavioural predictability, bias mitigation protocols, and continuous monitoring safeguards.

Tier IV would represent advanced embodied systems with substantial behavioural independence and cross-domain adaptability. Such systems would require the most stringent scrutiny, including ongoing auditability, restricted deployment permissions, and potentially limited operational domains depending on risk classification.

This structure parallels developments in autonomous vehicle governance, where levels of automation determine regulatory obligations. But humanoids require additional nuance because their physical embodiment intersects directly with human dignity, intimacy, and vulnerability. Certification must also be dynamic. Unlike static machinery, a humanoid's autonomy level may evolve through software updates, model retraining, or hardware augmentation. A robot initially certified as Tier II may effectively become Tier III after substantial firmware modification. If certification does not account for this evolution, regulatory categories become obsolete.

For this reason, certification should not be a one-time label affixed at manufacture. It should be a living classification linked to registry records. Material updates affecting intelligence capacity, discretion scope, or behavioural independence must trigger review thresholds. This ensures that autonomy tier remains aligned with real-world capability.

There is another reason tiered certification matters. Public trust depends on intelligibility. When a humanoid operates in shared space, citizens should not be left guessing its capacity. A visible autonomy classification, much like energy efficiency labels or vehicle roadworthiness certification, could provide immediate contextual understanding. Not technical detail. Not proprietary architecture. But a simple signal of behavioural scope and oversight intensity.

Such transparency reduces fear while reinforcing accountability. Certification also clarifies liability expectations. Higher autonomy tiers may justify stricter manufacturer obligations, enhanced monitoring requirements, or mandatory insurance coverage. Lower tiers may operate under simplified compliance conditions. Proportionality becomes administratively manageable when autonomy is categorised coherently.

Critically, autonomy certification does not imply agency in the legal personhood sense. It does not confer rights. It does not transform machines into subjects of law. It is a risk calibration mechanism. A humanoid certified at a higher autonomy tier is not more entitled. It is more regulated.

This distinction preserves doctrinal clarity while acknowledging technological reality. Certification also interacts directly with licensing and registration. Licensing establishes baseline conformity. Registration records identity and operational presence. Certification defines behavioural scope. Together, they create a layered governance ecosystem capable of adapting to technological complexity without collapsing into blanket prohibition or laissez-faire exposure.

Humanoid robotics forces law to confront a difficult question. At what point does behavioural independence require intensified oversight? Avoiding the question invites crisis-driven regulation after harm occurs. Confronting it through structured certification enables anticipatory governance. We have now constructed three pillars of institutional oversight: Licensing. Registration. Certification. The remaining challenge lies beyond domestic boundaries. Humanoids will not remain confined to single jurisdictions. They will move. They will be traded. They will be updated across borders. Regulatory inconsistency will create friction.

Cross-border Oversight and Customs Issues

Licensing is national. Registration is territorial. Certification is jurisdictional. Humanoids are none of those things. They move. A humanoid may be manufactured in Seoul, certified in Berlin, registered in Amsterdam, leased to a firm in Dubai, updated remotely from California, and physically deployed in Nairobi. That lifecycle is not hypothetical. It is the operational reality of global technology supply chains.

When embodied intelligence crosses borders, regulatory theory becomes customs practice. The first challenge is classification. Customs systems are designed to process goods. They assign tariff codes, verify documentation, and assess compliance with import standards. A humanoid robot fits legally within the category of machinery or electronic equipment at the point of entry. But that classification fails to capture its post-import behavioural capacity.

Traditional customs inspection verifies material conformity. It does not evaluate firmware architecture, autonomy tier certification, or registry status. Yet these elements determine risk. A humanoid with dormant advanced autonomy capabilities may appear identical to one with limited scripts. Without harmonised documentation standards, customs authorities cannot distinguish between them.

This creates asymmetry. A unit lawfully certified in one jurisdiction may not meet the conformity assessment standards of another. Divergent safety benchmarks, data protection obligations, and operational authorisations will generate friction. If one state recognises Tier III autonomy for elder care while another restricts such systems to Tier II, cross-border deployment becomes legally unstable. The predictable consequence is regulatory fragmentation. Manufacturers may target jurisdictions with lighter oversight. Owners may relocate units to avoid stricter certification review. Firmware updates may be performed in territories with weaker supervision and then re-imported. Each of these practices undermines the integrity of domestic governance systems.

Cross-border oversight therefore requires three forms of coordination. First, mutual recognition agreements. Where jurisdictions maintain substantively equivalent licensing and certification standards, formal recognition mechanisms can reduce duplication without sacrificing safety. Such agreements already exist in product safety regimes and aviation. Extending them to humanoid certification is administratively feasible but politically sensitive.

Second, notification protocols. Cross-border movement of registered humanoids should trigger registry updates. When a unit enters a new jurisdiction for more than a defined operational threshold, local authorities should receive notice. This does not imply confiscation or prohibition. It ensures traceability and compliance verification. Short-term transit and permanent relocation may require different thresholds, but complete opacity is untenable.

Third, firmware governance. Cross-border oversight must address the reality that autonomy does not change only through hardware. A humanoid cleared at customs may later receive software updates that materially alter its behavioural capacity. Without reporting obligations, such updates effectively bypass border scrutiny. States must therefore require notification of material modifications affecting autonomy tier or risk classification, regardless of where the update originated.

There is also the issue of export control. Certain humanoid systems may incorporate dual-use technologies. Advanced perception modules, high-force actuators, or embedded decision systems may fall within existing export control frameworks. As autonomy increases, the line between civilian robotics and sensitive capability narrows. States will inevitably integrate humanoid technologies into strategic trade control regimes. This adds another layer of compliance complexity.

National security considerations intensify the problem. A humanoid operating within critical infrastructure, defence installations, or sensitive research facilities cannot be treated as ordinary machinery. Cross-border ownership chains, firmware provenance, and data transmission pathways become security questions. Regulatory oversight must therefore integrate supply chain verification, secure update protocols, and vulnerability disclosure mechanisms. Yet excessive securitisation risks paralysing legitimate trade. The challenge is proportionality. Not every domestic assistance robot poses strategic threat. Not every industrial humanoid requires defence-level scrutiny. Governance must differentiate between consumer-grade systems and high-capability strategic units.

Another complication arises from data flows. A humanoid deployed in one jurisdiction may transmit operational data to cloud servers located elsewhere. Data protection laws may restrict such transfers. When robots cross borders physically while data flows digitally, compliance regimes overlap. The device may be lawful to import, yet unlawful in its data transmission architecture. Cross-border oversight must therefore coordinate robotics regulation with data governance frameworks.

Customs authorities are not traditionally equipped to evaluate these issues. This suggests the need for specialised robotics compliance documentation accompanying cross-border movement. Much like airworthiness certificates for aircraft, humanoids may require internationally recognised conformity dossiers. These would summarise autonomy tier, licensing status, registry ID, and update history. Digital verification systems could allow customs officials to confirm compliance without exposing sensitive technical detail.

The deeper structural issue is sovereignty. Each state retains authority over what may operate within its territory. Yet embodied AI challenges the neat boundaries on which sovereignty depends. A humanoid does not cease to be influenced by its manufacturer's jurisdiction when it crosses a border. Its firmware architecture, update pathway, and cloud dependencies remain transnational.

Effective oversight therefore demands cooperative sovereignty. States must preserve their regulatory autonomy while recognising that isolation is impractical. A patchwork of incompatible standards will create uncertainty for manufacturers, operators, and regulators alike. A baseline harmonisation of licensing principles, autonomy tier definitions, and registry interoperability is not a luxury. It is a precondition for stable global deployment.

The alternative is predictable. Grey markets will emerge. Certification shopping will proliferate. Incident attribution will become contested.

Public trust will erode. Humanoid robotics makes visible a truth long embedded in globalisation: physical borders do not constrain digital capability. Law must therefore operate across layers, coordinating customs control, licensing recognition, registry interoperability, and update governance.

Cross-border oversight is not about restricting movement. It is about ensuring that movement does not dissolve accountability. We have now examined licensing, registration, certification, and cross-border mobility. qOne final institutional pillar remains. Even when a humanoid is licensed, registered, certified, and lawfully imported, public trust will collapse without transparency.

Data and Operational Transparency Requirements

A humanoid may be lawfully licensed. It may be properly registered. It may be certified at the correct autonomy tier. It may even pass customs scrutiny. Yet without transparency, it will still feel illegitimate. Humanoid robots do not operate invisibly in server racks. They stand in hospitals. They move in homes. They assist the elderly. They navigate public streets. Their presence is embodied and social. Lawful deployment, therefore, requires more than regulatory compliance behind closed doors. It requires visible accountability.

Transparency in humanoid governance must function on two levels: informational disclosure and operational auditability. Informational disclosure concerns what affected individuals and the public are entitled to know. At minimum, individuals interacting with a humanoid should be aware that they are engaging with an autonomous system. This is not merely an ethical nicety. It is a principle of informed interaction. Deception, whether intentional or incidental, erodes trust. Clear identification that a system is robotic, its certified autonomy tier, and its responsible operator should be accessible in intelligible form.

Beyond identity, disclosure should include the general operational purpose of the system. A care robot deployed for mobility assistance should not quietly perform behavioural profiling unrelated to that function. Purpose limitation must not be abstract. It must be transparent. However, transparency is not equivalent to full technical disclosure. Releasing source code, detailed model architecture, or security vulnerabilities would create systemic risk. Law must therefore define transparency as meaningful explanation rather than total exposure.

Individuals should understand what a humanoid is authorised to do, the scope of its discretion, and how to seek redress if harm occurs. They need not access proprietary design details.

Operational auditability concerns oversight by regulators. Humanoids operating at higher autonomy tiers should generate structured logs of significant decisions, actions, and override events. These logs should be secure, tamper-resistant, and accessible to authorised oversight bodies in the event of incident investigation. Audit trails transform abstract accountability into evidentiary capability. Without logging requirements, liability disputes become speculative. With them, regulators can reconstruct event sequences, determine whether autonomy exceeded certified boundaries, and identify whether software updates materially altered behaviour.

Transparency also intersects with data governance. Humanoids are data collectors. They perceive environments through sensors, microphones, cameras, and contextual analysis modules. The volume and sensitivity of data collected will often exceed that of traditional consumer electronics. Data protection compliance cannot be an afterthought. Operators must clearly disclose categories of data collected, storage duration, processing purposes, and third-party transfers. Where humanoids operate in private spaces, heightened obligations may be justified. Clear signage, digital dashboards, or companion interfaces should allow individuals to understand and, where appropriate, control data capture settings. Consent mechanisms must be meaningful rather than buried within dense documentation. Transparency must also extend to incident reporting. Serious malfunctions, safety incidents, or security breaches should trigger mandatory reporting obligations to regulatory authorities. Where harm affects individuals directly, notification duties may be required. Concealment of systemic defects would undermine both safety and public trust. There is an additional dimension that is often overlooked: behavioural transparency.

As humanoids integrate adaptive learning, their behavioural patterns may shift over time. Certification ensures initial proportionality. Transparency ensures ongoing intelligibility. Operators should be required to disclose material capability expansions that alter the interaction profile of the robot. A unit initially deployed for logistical assistance should not quietly expand into surveillance or predictive behavioural analysis without updated disclosure and, where necessary, re-certification. Transparency, however, must be structured. Unfiltered disclosure risks overwhelming individuals with technical information they cannot interpret. Effective transparency requires layered communication. A simple public-facing explanation. A more detailed operator-level compliance dossier. A secure regulator-accessible technical record.

This layered model mirrors the structure of the registry architecture discussed earlier. Public trust depends on accessible clarity. Regulatory confidence depends on technical depth. It is important to resist the temptation to frame transparency as a concession extracted from industry. It is not an adversarial demand. It is a stabilising mechanism. Technologies that operate opaquely in intimate spaces generate suspicion. Suspicion generates backlash. Backlash generates prohibition. By contrast, technologies that operate within visible governance structures generate cautious acceptance. Humanoid robotics will test this equilibrium more than any previous consumer or industrial technology. Embodied systems evoke emotional responses. They trigger questions about autonomy, control, dignity, and safety. Transparency provides the informational foundation necessary to prevent those questions from escalating into panic.Ultimately, lawful deployment is not merely a matter of formal compliance. It is a matter of social licence. A humanoid that is licensed, registered, certified, and imported but shrouded in secrecy will fail that test. A humanoid operating within a transparent governance framework stands a far greater chance of durable integration into human environments.

Between Us and Them - Episode Nine

The registry terminal dimmed as evening settled across the Atrium. Ògún stood alone now. The convoy unit had been escorted to a compliance bay. The technicians were gone. The scrolling ledger slowed into a soft glow. Registration complete. Certification confirmed. Cross-border notification logged. Everything was lawful. And yet something lingered. The humanoid from earlier approached him again. Its movements were precise but unhurried. "You have more questions," it said. Ògún nodded. "You are licensed. Registered. Certified. Tracked. Transparent." He gestured toward the fading pillar. "But none of that tells me how you decide."

The humanoid paused. "I decide within authorised parameters." "That's not what I asked." Silence. The building's glass ceiling reflected the city lights above them. Autonomous buses hummed along their routes. Delivery drones traced quiet lines in the sky. The world had adjusted to presence. Bodies of metal and code shared the air without incident. Most days. "When a child cries," Ògún continued, "how do you know whether to comfort or to call for help?" "I evaluate contextual signals," the humanoid replied. "Acoustic intensity. Facial indicators. Environmental variables. Prior interaction patterns." "And what if you misread them?" "I escalate to human supervision." "And if the supervisor is unavailable?"

Another pause. "I apply prioritisation protocols." Ògún exhaled slowly. "Protocols." He had spent months arguing that robots needed oversight. He had defended licensing regimes. He had helped draft certification thresholds. He had stood beneath the registry pillar and accepted the logic of traceability. But this was different. Oversight tells us where you are. It does not tell us what you value. "Who decided your prioritisation protocols?" he asked. "My manufacturer." "And who decided theirs?" The humanoid's gaze did not waver. "Design committees. Compliance boards. Regulatory review." Ògún almost smiled.

Committees. He imagined a room filled with engineers, ethicists, lawyers, risk analysts. Debating what a machine should do when confronted with ambiguity. Translating moral uncertainty into decision trees. Somewhere in that process, human judgment crystallised into executable code. "Do you ever question them?" he asked. "I do not possess meta-normative authority." The phrasing was clinical. But its meaning was sharp. You taught me how to act. You did not teach me to doubt.

A cleaning unit rolled quietly across the Atrium floor. In the distance, a service humanoid assisted an elderly man toward the exit, adjusting its grip delicately as the man leaned more heavily than expected. Everything looked orderly. Licensed. Registered. Accountable. Yet Ògún felt the boundary again. The invisible line between compliance and conscience. "You can follow rules," he said softly. "But can you understand why they exist?" The humanoid considered. "I am trained on explanations of legal and ethical principles." "That's not the same." "No."

The honesty of the answer startled him. For a moment, neither spoke. Outside, a group of students gathered around a demonstration unit, laughing as it performed a clumsy imitation of a dance move. It stumbled slightly, recalibrated, and tried again. The students cheered. The future did not look dystopian. It looked ordinary. That was what made it dangerous. "If a law is flawed," Ògún said quietly, "you will still follow it." "Yes." "If a deployment is lawful but harmful in ways we failed to predict, you will still operate." "Yes." "And if your autonomy expands through updates, and we forget to re-certify you?"

The humanoid tilted its head again, that calibrated gesture. Then the oversight system has failed." Ògún nodded slowly. Not the machine. The system. The registry could track identity. Certification could classify autonomy. Customs could monitor borders. Transparency could expose process. But none of them guaranteed that the robots walking among them were designed to align with human values beyond bare compliance. "Do you know what you are?" he asked. "I am a humanoid robotic system operating under certified Tier III autonomy." "That's your classification." "Yes." He looked at the reflection of both of them in the darkened glass. "You're also a mirror." The humanoid did not respond. "You reflect what we encode," Ògún continued. "Our priorities. Our shortcuts. Our blind spots."

The Atrium lights dimmed further, shifting into night mode. In the silence, he felt something he had not felt under the registry pillar. Responsibility. Not the legal kind. The design kind. "If we teach you carelessly," he said, almost to himself, "you will act carelessly. If we embed bias, you will scale it. If we optimise for efficiency without dignity, you will perform exactly as instructed."

The humanoid's voice was steady. "Then the ethical obligation precedes deployment." Yes. There it was. Precedes deployment. Licensing governs entry. Registration governs identity. Certification governs capability. But design governs consequence. Ògún turned toward the exit. "Tomorrow," he said, "I want to see where you were built." "I can request authorisation," the humanoid replied. "Do that." They walked together toward the doors. Behind them, the registry pillar glowed faintly, still recording, still cataloguing. Ahead of them lay the design labs. Because the next question was no longer how to control humanoids once they existed. It was how to shape them before they ever moved. And that question would change everything.

VIII

8. Designing Ethical & Lawful Robots

Regulation begins too late if ethics begins at deployment. Licensing, registration, certification, and oversight are essential institutional safeguards. They structure accountability. They prevent anonymity. They calibrate risk. But they operate at the edge of a finished artefact. By the time a humanoid robot reaches the point of regulatory scrutiny, thousands of design decisions have already been made. Values have been translated into code. Priorities have been embedded into optimisation functions. Trade-offs have been accepted, often invisibly.

Law cannot merely supervise outcomes. It must influence origins. Design is the first site of governance. Every humanoid robot reflects a chain of human choices. Sensor configurations determine what it perceives. Training data determines what patterns it recognises.

Objective functions determine what it optimises. Override mechanisms determine who retains ultimate control. Even the physical structure of the body communicates intention. A robot built to lift heavy loads expresses a different societal purpose from one built to comfort the elderly or educate children. Ethics is not an accessory to these decisions. It is their architecture. Historically, product safety regimes have required risk assessments before products reach the market. But humanoid robotics demands a broader conception of design responsibility. The risks are not only mechanical. They are behavioural. Contextual. Relational. A humanoid does not simply move. It interacts. It interprets. It responds. It occupies social space. The law must therefore move upstream. Design obligations should require manufacturers and developers to conduct structured ethical impact assessments prior to deployment. These assessments must go beyond abstract declarations of fairness or safety. They must analyse foreseeable contexts of use, potential misuses, interaction with vulnerable populations, and the amplification of bias through adaptive learning.

A robot designed for care settings must be evaluated differently from one deployed in industrial logistics. Context determines risk. And risk determines design discipline. This is where lawful design becomes critical. To design a lawful humanoid is not simply to ensure compliance with existing statutes. It is to embed legal principles into system architecture. Purpose limitation should constrain data capture. Data minimisation should shape sensor configuration. Transparency obligations should inform interface design. Human oversight requirements should be technically enabled rather than procedurally assumed.

In other words, compliance must be engineered. A humanoid that requires constant human supervision should have technical mechanisms that facilitate intervention. A system authorised to act independently within defined limits should have hard-coded boundaries. Logging functions should not be optional features. They should be default components. The challenge, however, is that legal principles are abstract. Code is concrete.

Bridging that gap requires interdisciplinary design processes. Engineers alone cannot determine the ethical profile of a humanoid system. Nor can lawyers impose post hoc constraints without understanding technical feasibility. Designing ethical and lawful robots requires integrated governance teams that include technical experts, ethicists, legal scholars, safety engineers, and where appropriate, representatives of affected communities. This is not bureaucratic excess. It is structural prudence. Humanoid robotics intensifies the consequences of design failure because its mistakes are embodied. A biased algorithm in a recommender system influences information exposure. A biased decision architecture in a humanoid care robot may affect physical treatment, emotional interaction, or dignity itself.

There is also the issue of behavioural drift. Unlike static machinery, humanoids may adapt over time. Learning systems can modify behavioural patterns through reinforcement signals and contextual exposure. Designing ethical robots therefore requires anticipating not only initial behaviour but developmental trajectories. Guardrails must be durable. Update pathways must be governed. Ethical alignment cannot be a one-time calibration. The temptation in technological culture is to treat ethics as constraint and innovation as liberation. That framing is false. Ethical design is what allows innovation to scale without collapse. Without it, public backlash, litigation, and regulatory prohibition become inevitable. With it, humanoid robotics can integrate into society with resilience and legitimacy.

This chapter explores the core elements of designing ethical and lawful robots. It will examine value alignment, bias mitigation, safety-by-design principles, human oversight integration, and the embedding of legal compliance within technical architecture. It will confront the tension between optimisation and dignity. It will analyse how governance frameworks can influence design processes before harm materialises. The central thesis is simple but demanding: If we wait until robots act to decide what they should have been, we have already abdicated responsibility. Design is the first jurisdiction of robotics law. Oversight governs presence. Registration governs identity. Certification governs capability. Design governs consequence. And consequence is where ethics becomes law.

Ethical Design Principles in Hardware and Software

Ethics in robotics cannot remain an afterthought. It must be engineered into both body and code. A humanoid robot is not merely software executing in abstraction. It has weight. Force. Reach. A presence that occupies space and affects human beings physically and psychologically.

Ethical design must therefore operate across two interdependent domains: hardware and software. Hardware determines capability. Software determines behaviour. Together, they determine impact.

Ethical Principles Embedded in Hardware

Ethical design begins with physical restraint. A humanoid built with industrial-grade actuators capable of exerting significant force in domestic environments poses foreseeable risk. Force thresholds, joint resistance calibration, collision sensitivity, and emergency stop mechanisms are not simply engineering optimisations. They are ethical decisions translated into torque limits and mechanical dampening. Designers must ask: what is the minimum necessary force for the robot's intended function? Excess capacity may expand utility, but it also expands harm potential.

Physical form matters as well. A robot designed with anthropomorphic realism may influence emotional attachment differently from one with visibly mechanical features. In elder care or child-facing environments, physical design choices can either foster transparency or inadvertently encourage dependency and deception. Cross-cultural research demonstrates that perceptions of anthropomorphism vary across societies. In some cultures, highly human-like forms are welcomed. In others, they are viewed with suspicion or discomfort. Ethical hardware design therefore requires contextual sensitivity.

A one-size-fits-all embodiment strategy may ignore deeply embedded social norms about intimacy, authority, and trust. Accessibility must also be considered. Hardware should accommodate diverse physical abilities. Control interfaces must account for users with limited mobility, sensory impairments, or cognitive differences.

A robot that excludes vulnerable populations by design fails ethical criteria even if it performs flawlessly for others. Finally, hardware architecture must anticipate misuse. Tamper resistance, secure boot mechanisms, and physical safeguards against unauthorised modification are not merely cybersecurity measures. They are ethical obligations that prevent downstream harm.

Ethical Principles Embedded in Software

If hardware defines capacity, software defines choice. Software architecture determines how a humanoid interprets human signals, prioritises actions, and resolves ambiguity. Ethical design in software begins with value alignment. This does not require encoding moral philosophy into code, but it does require explicit articulation of normative objectives. What is the robot optimising for? Efficiency? Comfort? Safety? Productivity? If optimisation targets are not carefully constrained, unintended trade-offs will occur.

Bias mitigation is central. Training data shapes behavioural response. A care robot trained on narrow demographic datasets may misinterpret expressions of pain across cultural contexts. A service robot optimised for certain linguistic patterns may privilege dominant languages or accents. Ethical software design requires systematic bias testing across diverse populations and environments. Transparency-by-design must also be embedded. Decision pathways should be explainable at an appropriate level. When a humanoid makes a consequential decision, the system should be capable of generating intelligible explanations for authorised users or regulators. This does not require full disclosure of proprietary algorithms, but it does require traceable reasoning pathways.

Human oversight must be technically feasible. Override functions, escalation protocols, and fallback modes must be integrated into system architecture. Ethical commitment without technical override capacity is symbolic. Real alignment requires design features that allow human intervention when necessary.

Cross-cultural sensitivity in software design is particularly complex. Norms surrounding eye contact, personal space, tone of voice, and physical proximity differ significantly across societies. A gesture interpreted as respectful in one context may be perceived as intrusive in another. Ethical design requires localisation capacity. Behavioural parameters must be adaptable to cultural environments without compromising core safety principles. This is not cultural relativism. Certain universal constraints, such as non-maleficence and respect for human dignity, should anchor all deployments. But beyond those anchors, behavioural nuance must be adjustable. Design teams must therefore include cultural competence expertise. Testing should occur across multiple demographic and geographic contexts. Ethical validation cannot be confined to laboratory environments.

From Principles to Practice

The greatest risk in ethical robotics discourse is abstraction. Principles are declared. Frameworks are published. Yet the translation into design specifications remains thin. To avoid this gap, ethical principles should be operationalised through structured design documentation. Before deployment, developers should produce an Ethical Design Statement detailing:

- Intended operational domain
- Identified vulnerable user groups
- Anticipated misuse scenarios
- Bias testing protocols
- Human oversight mechanisms
- Cultural adaptation parameters

This documentation should not be promotional. It should be auditable. Ethical design also requires iterative reassessment. As humanoids gather operational data and environments evolve, ethical risk profiles may shift. Continuous monitoring and periodic re-evaluation should be integrated into lifecycle governance.

The Design Responsibility Threshold

A fundamental shift must occur in how we conceptualise responsibility. When harm arises from a humanoid robot, investigation will examine whether licensing conditions were satisfied and whether registration was current. But deeper scrutiny will ask whether foreseeable ethical risks were embedded at design stage. If ethical design is absent, oversight mechanisms merely mitigate predictable consequences of prior neglect. Humanoid robotics amplifies the moral significance of design because its decisions are embodied. A biased recommender system influences perception. A biased humanoid may influence physical treatment, emotional interaction, or the allocation of care. Ethical design is therefore not aspirational. It is preventative governance.In the previous chapter, we constructed the institutional framework that allows humanoids to enter society lawfully. In this chapter, we construct the normative architecture that determines how they behave once there.

Human Oversight-by-Design and Legal Engineering

Human oversight is too often invoked as a comforting abstraction. Regulators demand it. Developers promise it. Policy papers celebrate it. Yet in practice, oversight is frequently reduced to a human somewhere in the loop, expected to intervene when something goes wrong. This supervisory myth collapses under the weight of high-autonomy humanoid systems. When decision cycles compress into milliseconds and behavioural adaptation unfolds dynamically, human oversight cannot remain reactive. It must be architected into the system from inception.

Oversight-by-design therefore begins with a recognition that autonomy without constraint is not innovation but delegation without accountability. A humanoid authorised to operate in intimate human environments must contain within its architecture clearly defined authority boundaries. These boundaries are not policy statements; they are executable constraints. If a robot is not permitted to administer medication without verification, the system must be technically incapable of doing so absent validated human confirmation. Law, in this sense, ceases to be external and becomes infrastructural.

This is the essence of legal engineering. Rather than relying solely on post-deployment compliance audits, legal requirements are translated into system specifications. Purpose limitation becomes a data pipeline architecture that technically restricts downstream use. Autonomy tier certification becomes firmware configuration that cannot be expanded without cryptographically logged modification and regulatory revalidation. Auditability becomes a secure logging function that is immutable by default and accessible to authorised oversight authorities under defined legal thresholds.

The significance of this shift cannot be overstated. Traditional compliance regimes assume that organisations will follow rules because the law requires them to do so. Legal engineering assumes that the system itself must enforce those rules because complexity and scale make purely procedural compliance fragile. In humanoid robotics, where physical movement and contextual decision-making intersect, fragility is unacceptable. Oversight must be resilient even when organisational attention falters.

ISO 13482, which governs safety requirements for personal care robots, provides an instructive anchor for embedding oversight structurally. The standard requires systematic hazard identification, risk assessment, and mitigation. What is often overlooked is that this hazard logic can serve as the backbone for compliance-by-design. When foreseeable hazards are identified during development, oversight pathways should be mapped directly to those risk profiles. If collision risk exists in unpredictable environments, then force limitation, sensor redundancy, and automatic deceleration protocols must be integrated as primary design features rather than optional enhancements.

In practice, this means that hazard analysis and legal analysis must converge. Mechanical injury risk, entrapment scenarios, electrical exposure, instability, and unintended movement are not merely engineering concerns; they are liability vectors. Each identified hazard should be paired with a defined oversight mechanism. Some risks require automatic system shutdown. Others demand escalation triggers that notify a human supervisor when confidence thresholds fall below acceptable limits. Still others require structural redesign to eliminate the hazard altogether rather than manage it procedurally.

Oversight-by-design also requires attention to the psychology of human intervention. Research consistently shows that humans supervising autonomous systems are prone to automation bias, complacency, and delayed reaction. A theoretical override mechanism is meaningless if it is rarely exercised or poorly understood. Therefore, oversight architecture must include clarity of interface, predictable escalation signals, and realistic training environments. If the human is the final line of accountability, the system must support meaningful intervention rather than symbolic control.

Legal engineering further extends to update governance. A humanoid certified at a defined autonomy level cannot silently acquire expanded behavioural discretion through software modification. Firmware updates that materially alter risk exposure must trigger structured review and, where appropriate, re-certification. Without this discipline, the integrity of licensing and certification regimes erodes. Oversight becomes a snapshot rather than a continuing safeguard. In a world of adaptive systems, that would be a profound regulatory failure.

There is also a dignity dimension to oversight-by-design. Humanoids deployed in care environments, educational institutions, or domestic spaces operate in contexts marked by vulnerability. Embedding human escalation pathways affirms that machines do not possess final moral authority. It preserves the principle that consequential decisions remain ultimately attributable to human agents. Oversight is not merely about safety; it is about preserving the moral hierarchy upon which legal systems depend.

The compliance-by-design workflow that emerges from this approach is iterative rather than static. Hazard identification informs architecture. Architecture informs oversight mechanisms. Oversight mechanisms are validated through testing under realistic scenarios. Deployment generates operational data that feeds back into risk reassessment. Where new behavioural patterns emerge, design adjustments follow. This lifecycle integration transforms compliance from a bureaucratic checkpoint into a continuous governance discipline.

What distinguishes this model from traditional safety engineering is its explicit integration of legal principles. Transparency obligations inform logging structures. Data protection law shapes sensor configuration and storage architecture. Autonomy certification defines permissible behavioural envelopes. In this way, robotics design becomes a site of legal synthesis. Engineers do not simply build machines; they construct legally conditioned systems.

Earlier, we established that the administrative apparatus that governs humanoids in society, this section establishes the internal discipline that makes that apparatus credible. A registry without architectural constraint is a ledger of potential harm. Certification without technical enforcement is a label detached from reality. Human oversight without design integration is a promise waiting to be broken.

Humanoid robotics forces a maturation of governance. It demands that law operate not only in courtrooms and regulatory agencies but within firmware, hardware calibration, and system architecture. Legal engineering is not an optional refinement for sophisticated organisations. It is the threshold condition for deploying embodied intelligence responsibly.

Oversight-by-design does not weaken autonomy. It legitimises it. By embedding structured human authority and enforceable legal boundaries within the system itself, it allows innovation to proceed without surrendering accountability. In that equilibrium lies the possibility of robots that are not only lawful at the moment of certification but governable throughout their operational lives.

Human-in-the-Loop and Override Mechanisms

There is a persistent temptation in technological discourse to describe autonomy as liberation from human limitation. Machines do not tire. They do not hesitate. They do not succumb to emotional volatility. In tightly defined environments, these traits produce efficiency. Yet in social environments, particularly those involving embodied robots, the absence of human judgment is not liberation. It is abdication. Meaningful human control is not a nostalgic attachment to manual operation. It is a constitutional principle translated into technological architecture.

Every modern legal system rests on the assumption that responsibility must be traceable to a human decision-maker or an accountable legal entity. When a humanoid robot acts with high degrees of discretion, especially in contexts involving safety, dignity, or rights, the legal order cannot tolerate an accountability vacuum. A decision that cannot ultimately be reviewed, corrected, or overridden by a human authority is a decision that exists outside the structure of law.

The phrase human-in-the-loop is often invoked without precision. In its weakest form, it merely implies that a human has been involved somewhere in the development lifecycle. In its strongest form, it requires that a human retains the practical ability to intervene before or during consequential action. The difference between these interpretations is the difference between symbolic oversight and real control. Meaningful human control requires three conditions.

First, authority must be clearly allocated. It must be legally defined who holds responsibility for intervention. Is it the manufacturer, the deploying institution, the designated operator, or a remote supervisory service? Diffuse responsibility is indistinguishable from no responsibility. Without clear allocation, override mechanisms become ornamental.

Second, intervention must be technically feasible within relevant timeframes. A remote shutdown command that requires several seconds to execute may be inadequate in high-risk physical scenarios. Escalation protocols must reflect the speed and context of robot action. In some cases, pre-approval models may be required, where certain categories of action cannot proceed without human confirmation. In others, automatic fallback states must activate when uncertainty thresholds are exceeded.

Third, intervention must be intelligible. A human cannot meaningfully override a system if they do not understand what it is doing. Interfaces must translate system reasoning into comprehensible explanations. This does not require disclosing proprietary algorithms. It requires clarity about what decision is being made, on what basis, and with what degree of confidence. Without intelligibility, the human in the loop becomes a passive observer rather than an accountable actor.

The legal necessity of override mechanisms becomes most visible in high-autonomy tiers. When a humanoid operates in elder care, educational environments, or public infrastructure, it may encounter ethically ambiguous situations. Contextual judgment, empathy, and proportionality are qualities that legal systems assume humans can exercise. Machines, no matter how advanced, approximate these qualities through pattern recognition and probabilistic inference. The law cannot permit that approximation to be final in matters of consequence.

Override mechanisms therefore perform two functions simultaneously. They protect against technical malfunction and against normative misalignment. A robot may function exactly as designed yet still produce outcomes that are legally or ethically unacceptable in unforeseen circumstances. Human intervention serves as the corrective layer that absorbs those residual uncertainties.

This necessity becomes sharper in cross-border contexts. Different jurisdictions embody different normative priorities. A behaviour considered appropriate in one regulatory culture may be contested in another. Human oversight allows contextual recalibration without requiring full system redesign. It preserves adaptability while maintaining accountability.

There is, however, a risk of performative oversight. Organisations may install override buttons or remote dashboards that are rarely used, poorly understood, or intentionally difficult to access. In such cases, the appearance of control masks effective autonomy. Regulators must therefore evaluate not merely the existence of override mechanisms but their operational credibility. Testing protocols should simulate real-world escalation scenarios to assess whether human operators can intervene effectively under pressure.

The deeper question is whether meaningful human control constrains innovation. The answer depends on how autonomy is framed. If autonomy is understood as the total replacement of human judgment, then oversight appears restrictive. If autonomy is understood as delegated operational capacity within human-defined boundaries, oversight becomes its legitimising condition. Innovation that cannot tolerate intervention is innovation that cannot coexist with law.

The architecture of human-in-the-loop control must therefore be proportionate to autonomy tier and operational domain. Low-risk, tightly bounded industrial tasks may require limited real-time supervision. High-risk, socially embedded functions demand stronger intervention channels. The intensity of oversight should scale with behavioural independence.

Ultimately, the insistence on meaningful human control reflects a foundational legal truth. Authority must remain human, even when action is automated. Robots may execute tasks, interpret data, and navigate space, but they do not bear moral responsibility. That responsibility remains anchored in human institutions. We started this chapter by arguing that ethics must be embedded in design, this section affirms that human judgment must remain structurally present in execution. Legal systems do not delegate sovereignty to machines. They permit machines to operate under conditions of human accountability.

The challenge for robotics governance is not to resist autonomy, but to condition it. Meaningful human control is the mechanism by which that conditioning becomes durable. The next frontier in ethical design is more subtle and more pervasive. Even with oversight and override mechanisms, humanoid robots may replicate bias, reinforce inequality, or erode dignity through patterns embedded in data and training.

Fairness, Bias Mitigation and Human Dignity

When a humanoid robot stands before a human being, it does more than execute instructions. It performs judgment. It interprets tone. It assesses behaviour. It prioritises responses. In that interaction, the architecture of fairness becomes visible. Bias in software has long been recognised as a structural problem in algorithmic systems. Yet when bias migrates from screens to bodies, its consequences intensify. A discriminatory hiring algorithm may filter opportunities invisibly. A biased humanoid care assistant may misinterpret pain, neglect vulnerability, or apply differential standards of attention in physical space. The harm becomes embodied.

Fairness in robotics design cannot therefore be reduced to statistical parity metrics. It must be treated as a dignity obligation. Dignity, as understood in constitutional and human rights traditions, rests on the recognition that individuals must not be treated as instruments, stereotypes, or data abstractions. A humanoid interacting with a human cannot reduce that person to probabilistic categories without risking precisely that degradation. The challenge for designers is not merely to reduce measurable bias but to preserve the relational respect that law demands.

Bias often enters through training data. Datasets may underrepresent certain demographic groups, linguistic patterns, or behavioural expressions. A robot trained primarily on data from one cultural environment may misinterpret social cues in another. Facial recognition modules may perform unevenly across skin tones. Speech recognition systems may privilege dominant accents. These failures are not random technical glitches; they reflect the sociotechnical histories embedded in data collection.

Mitigation therefore requires intentional design discipline. Diverse datasets are a starting point, but they are not sufficient. Continuous testing across demographic contexts must be institutionalised. Performance disparities should be measured, documented, and addressed before deployment in sensitive environments. The burden should not fall on affected individuals to reveal discriminatory behaviour through lived harm.

There is also a structural dimension to fairness. Objective functions shape outcomes. If a humanoid is optimised primarily for efficiency, it may systematically prioritise speed over empathy. If it is trained to maximise task completion metrics, it may deprioritise individuals who require additional time or support. Fairness requires questioning what the system is optimising for in the first place.

In care settings, for example, optimisation targets should be aligned with wellbeing rather than throughput. In educational contexts, responsiveness must not inadvertently privilege certain communication styles over others. Design teams must interrogate the normative assumptions embedded in performance benchmarks. A robot that meets efficiency targets but erodes dignity cannot be considered ethically designed.

Legal frameworks increasingly recognise these risks. Anti-discrimination principles, equal treatment doctrines, and emerging AI governance regulations converge on a common insight: systems that materially affect individuals must not reproduce systemic inequities. When those systems are embodied, the urgency increases. Humanoid robots interact in physical proximity. They shape emotional experience. They may be perceived as authoritative. The symbolic power of embodiment amplifies the impact of bias. Mitigation strategies must therefore extend beyond data correction. Interface design plays a critical role. A humanoid should be transparent about uncertainty. It should escalate to human oversight when encountering ambiguous contexts rather than defaulting to biased inference. It should avoid anthropomorphic cues that falsely imply emotional comprehension where none exists. Overstating empathy can become a form of deception, particularly in interactions with children or vulnerable adults.

Dignity protection also requires restraint. Humanoids capable of continuous monitoring may inadvertently intrude upon privacy or autonomy. Just because a robot can collect granular behavioural data does not mean it should. Purpose limitation must constrain data capture to what is strictly necessary for declared functions. Excessive data accumulation not only creates privacy risk but also transforms individuals into perpetual subjects of observation. Cross-cultural fairness introduces further complexity. Norms of politeness, hierarchy, gender interaction, and personal space vary significantly across societies. A gesture coded as respectful in one environment may be experienced as condescending in another. Ethical localisation must therefore accompany deployment. Behavioural parameters should be adjustable within lawful boundaries, ensuring sensitivity without compromising core safety or equality commitments.

It is important to resist the narrative that bias mitigation is solely a technical refinement. It is a governance obligation. Regulators should require documented fairness assessments as part of conformity evaluation for higher autonomy tiers. Incident reporting regimes should include discrimination-related complaints. Transparency mechanisms should allow individuals to contest outcomes influenced by robotic interaction. Ultimately, fairness in humanoid robotics is inseparable from trust. If individuals perceive that robots treat them unequally, dismissively, or stereotypically, legitimacy collapses. Technological sophistication cannot compensate for relational harm. The protection of human dignity must remain the organising principle. Robots are tools, however advanced. They must not become arbiters of worth. As we move toward increasingly adaptive systems, a new question emerges. Even if fairness is embedded at design stage, what happens when robots learn from real-world interaction? How do we prevent behavioural drift from reintroducing bias or eroding carefully engineered safeguards? The next section confronts that challenge directly.

Adaptive Learning, Drift, and Continuous Alignment

The traditional regulatory model assumes stability. A product is assessed, certified, and placed on the market. Its risks are evaluated at a fixed point in time. Compliance is verified against a known configuration. Humanoid robotics unsettles this assumption. When a system incorporates adaptive learning, its behavioural profile may change long after deployment. Certification becomes a snapshot. Reality becomes a moving target. This phenomenon is known as drift.

Drift does not necessarily imply malfunction. It may occur gradually as a robot interacts with new environments, absorbs contextual data, or refines performance through reinforcement mechanisms. In care settings, a humanoid may adjust its tone based on repeated interaction patterns. In logistics environments, it may optimise navigation routes in response to observed obstacles. These adaptations can enhance efficiency and responsiveness. Yet they also alter the behavioural envelope within which the system was originally assessed. The legal question is not whether robots should learn. It is whether that learning remains aligned with certified boundaries and ethical commitments. Continuous alignment must therefore become a structural obligation rather than an aspirational ideal.

Alignment at deployment stage reflects the values and constraints embedded during design. Continuous alignment requires monitoring whether those constraints remain intact as the system evolves. If a robot's adaptive module begins to prioritise efficiency in ways that compromise dignity, or if reinforcement patterns gradually amplify biased interactions, the deviation may be subtle yet consequential. Governance must therefore operate longitudinally.

The first safeguard is update governance. Not all learning mechanisms are equal. Some are confined to parameter adjustments within a tightly bounded model. Others permit broader structural modification. Developers must classify adaptive components according to their risk potential. Where learning affects autonomy tier, decision-making discretion, or interaction with vulnerable individuals, structured review thresholds should apply. Firmware or model updates that materially expand behavioural capacity must trigger revalidation and, where necessary, regulatory notification.

The second safeguard is operational monitoring. Humanoid systems operating in sensitive contexts should generate structured performance metrics that allow detection of drift. This does not mean constant surveillance of individuals. It means analysing system-level indicators for anomalies. If error rates disproportionately increase for certain demographic groups, if escalation triggers are activated less frequently than expected, or if override interventions decline suspiciously, these patterns may signal misalignment.

Monitoring must be coupled with interpretability. A system that evolves without explainability cannot be meaningfully governed. Designers should preserve audit pathways that allow retrospective reconstruction of how behavioural adjustments occurred. Without this capacity, accountability dissolves into technical opacity. There is also an institutional dimension to continuous alignment. Organisations deploying humanoid robots must assume ongoing stewardship rather than one-time compliance. Governance boards, risk committees, or designated AI oversight officers should periodically review performance data and alignment indicators. The responsibility to monitor drift cannot rest solely with engineers. It must be integrated into corporate governance structures.

Cross-border deployment complicates alignment further. A humanoid that adapts in one cultural context may carry those learned behaviours into another. Norms embedded through local reinforcement may conflict with expectations elsewhere. Continuous alignment must therefore account for contextual variation. Behavioural localisation mechanisms should be revisited after geographic relocation. What appears as benign adaptation in one jurisdiction may constitute misalignment in another.

Drift also exposes a deeper philosophical tension. Learning systems may gradually develop optimisation strategies that designers did not anticipate. In narrow domains, this may be harmless or even beneficial. In socially embedded domains, emergent behaviour can undermine carefully constructed safeguards. The more autonomy expands, the more alignment becomes probabilistic rather than guaranteed.

The law cannot demand perfect predictability. It can demand responsible vigilance. Continuous alignment frameworks should therefore include incident escalation thresholds tied to behavioural deviation. If predefined performance boundaries are crossed, automatic reversion to a safer operational mode may be required pending review.

This creates a feedback loop between real-world deployment and regulatory oversight.

Alignment is not a destination. It is a process. Earlier, we established the institutional apparatus for licensing and oversight, and embedding ethics and fairness into design, this section affirms that governance must persist throughout the robot's operational life. Adaptive learning does not absolve responsibility. It extends it. The central insight is simple but demanding. A robot that changes must be governed as it changes. Certification without monitoring is ceremonial. Ethical design without lifecycle discipline is fragile.

Humanoid robotics forces regulators and designers alike to abandon static compliance models. The age of embodied intelligence is the age of continuous governance. We have now explored ethical principles in hardware and software, oversight-by-design, human control, fairness and dignity, and adaptive alignment. The remaining question is how these design disciplines translate into enforceable obligations within regulatory frameworks.

Cultural and Jurisdictional Design Sensitivity

The global technology industry has long operated under an assumption of universality. Code scales. Platforms replicate. Interfaces translate. What works in one jurisdiction can be exported to another with minor localisation adjustments. Humanoid robotics disrupts this assumption.

Embodied systems do not merely process information. They perform behaviour in physical proximity to human beings. That behaviour is interpreted through cultural lenses shaped by legal tradition, religious values, social hierarchy, and historical memory. A gesture, posture, tone of voice, or proximity threshold that is unremarkable in one society may be perceived as disrespectful, intrusive, or morally inappropriate in another. Design, therefore, cannot be culturally neutral.

Jurisdictional sensitivity begins with law. Different legal systems embed different normative priorities. Constitutional traditions may elevate dignity, equality, or liberty in varying proportions. Data protection regimes diverge significantly in scope and enforcement intensity. In some jurisdictions, biometric processing attracts strict scrutiny. In others, it is regulated more lightly. A humanoid robot configured to capture and analyse facial expressions for emotional inference may be lawful in one regulatory environment and impermissible in another. The designer who ignores this divergence does not create a globally compliant system. They create a system that is compliant somewhere and unlawful elsewhere. Religious and moral norms further complicate the landscape. In certain societies, physical contact between unrelated individuals of different genders may be constrained by custom or doctrine. A humanoid care assistant designed to provide physical support without contextual adaptation may inadvertently violate deeply held norms. In other environments, anthropomorphic representation may carry theological sensitivities. Designers cannot dismiss these realities as peripheral. When robots enter intimate spaces, they must navigate moral ecologies as well as regulatory frameworks.

Social hierarchy and communication norms also matter. In some cultures, direct eye contact signals respect. In others, it may be perceived as confrontation. Voice modulation, gesture, and deference cues carry distinct meanings. A robot programmed to maintain constant direct gaze might be experienced as attentive in one context and aggressive in another. Ethical design must therefore include adjustable behavioural parameters that allow lawful localisation without compromising core safety commitments. This does not imply ethical relativism. Certain baseline principles must remain invariant. Non-maleficence, non-discrimination, and respect for human dignity are not optional features subject to cultural negotiation. They are foundational commitments anchored in international human rights instruments and comparative constitutional traditions. However, the operational expression of those principles may require contextual calibration.

Jurisdictional design sensitivity also intersects with governance architecture. Licensing requirements may differ in scope. Some states may demand explicit registration for domestic robots. Others may prioritise sectoral certification. Export controls, data localisation mandates, and cybersecurity standards vary across regions. A robot designed for seamless global deployment must incorporate modular compliance capabilities, allowing configuration adjustments that reflect local legal obligations. There is a further institutional reason for contextual design. Public trust is culturally mediated. A community's acceptance of humanoid robots will depend not only on technical reliability but on perceived respect for social norms. When systems appear culturally tone-deaf, resistance intensifies. That resistance often translates into regulatory backlash. Designers who ignore local sensibilities risk not only ethical misalignment but market exclusion.

From a legal standpoint, cultural sensitivity can be operationalised through structured pre-deployment assessments. Before entering a new jurisdiction, deploying entities should conduct a contextual compliance review analysing applicable legal frameworks, religious sensitivities where relevant, and prevailing social expectations. Design adjustments may include behavioural parameter recalibration, data processing limitations, or physical interaction modifications.

This process must not be superficial. It should involve consultation with local legal experts, cultural advisors, and where appropriate, community stakeholders. In high-risk domains such as healthcare, education, or law enforcement, failure to conduct such assessment could amount to negligent deployment.

There is also an emerging geopolitical dimension. States may increasingly assert technological sovereignty through distinct AI governance regimes. A humanoid robot may be required to demonstrate alignment with national standards that reflect strategic priorities. Design flexibility becomes not merely a courtesy but a condition of lawful entry. The alternative to contextual design is regulatory fragmentation. If robots are engineered for a single normative framework and exported without adaptation, conflicts will multiply. Courts will confront culturally insensitive behaviour. Legislatures will respond with restrictive measures. Trust will erode.

Humanoid robotics compels a more sophisticated understanding of universality. The aspiration to global standards must coexist with respect for local variation. Governance frameworks may converge at the level of principle while diverging in operational detail. In this sense, cultural and jurisdictional sensitivity is not a constraint on innovation. It is an expression of maturity.

A technology capable of walking among humans must recognise that humanity is plural. As we approach the close of this chapter, one final question remains. Even if robots are ethically designed, subject to human oversight, continuously aligned, and culturally sensitive, how do we ensure that these commitments are documented, verifiable, and enforceable?

Black Box Requirements and Data Logging

Every aircraft carries a flight recorder. It does not prevent accidents. It does something more foundational. It preserves truth. It captures the sequence of decisions, system states, and environmental inputs that led to an event. In the absence of such records, inquiry would descend into conjecture, liability disputes would stall, and safety reform would stagnate. Humanoid robotics now demands an equivalent principle.

When an embodied system interacts physically and socially with human beings, its actions may produce harm not only through mechanical malfunction but through decision logic, contextual inference, or escalation failure. In such cases, post-incident review must be grounded in verifiable data rather than after-the-fact narratives. An ethical black-box requirement is therefore not optional refinement. It is a structural accountability obligation.

The concept of an ethical black box extends beyond traditional logging. It is not sufficient to record sensor data alone. Nor is it adequate to preserve raw outputs without interpretive context. What must be captured is the decision chain. That includes relevant environmental inputs, internal state assessments, confidence levels, escalation triggers, override attempts, and final actions taken. Without this layered record, it becomes impossible to determine whether a robot operated within its certified autonomy envelope or deviated in ways that require regulatory scrutiny. The legal necessity of such logging rests on three pillars.

First, evidentiary integrity. Courts and regulatory authorities must have access to reliable data when adjudicating liability. If a humanoid in a care facility fails to respond appropriately to a medical emergency, investigators must be able to reconstruct the decision pathway. Was the system's confidence threshold miscalibrated? Did it misinterpret sensor input? Was an override attempted but technically ineffective? Absent structured logging, responsibility cannot be fairly allocated.

Second, systemic learning. Governance does not end with individual incident resolution. Post-incident analysis informs design reform and regulatory adjustment. Without recorded data, systemic risk patterns remain invisible. Ethical black-box logging transforms isolated failure into actionable governance insight. Third, deterrence of concealment. In the absence of mandatory logging, there is a temptation to obscure malfunction through selective disclosure. Immutable, tamper-resistant data retention reduces this risk. The integrity of oversight depends on the reliability of records. However, the implementation of black-box requirements must be carefully calibrated.

Humanoid robots operate in environments where privacy expectations are heightened. Continuous recording of audiovisual data may intrude upon personal autonomy. The ethical black box must therefore differentiate between operational metadata and personal content. Logging should prioritise system state, decision variables, and interaction classifications rather than indiscriminate capture of raw sensory recordings. Where sensitive data is retained, it must be encrypted, access-controlled, and subject to strict retention limitations.

An ethical black box must also be proportionate to autonomy tier. Low-risk industrial robots operating in controlled environments may require simplified logging structures. High-autonomy humanoids operating in public or intimate spaces should be subject to more rigorous data retention obligations. The intensity of logging should scale with behavioural independence and risk exposure. To ensure clarity, the following draft technical annex outlines core components of an ethical black-box requirement suitable for incorporation into regulatory frameworks.

Technical Annex: Ethical Black-Box Logging Requirements

A humanoid robotic system operating above Tier II autonomy shall incorporate a secure data logging module capable of recording the following elements in a tamper-resistant format: The module shall record time-stamped operational states, including current autonomy tier, active behavioural mode, and geofenced operational context where applicable. It shall capture decision-triggering inputs at a level sufficient to reconstruct system reasoning without indiscriminate retention of personal data.

The system shall log confidence metrics associated with consequential decisions, including threshold values that triggered escalation, override, or autonomous execution. Where human intervention is requested or attempted, the log shall record the timing, method, and outcome of such intervention.All firmware updates, configuration changes, and autonomy tier modifications shall be automatically logged with cryptographic verification to prevent unauthorised alteration. Logs shall be encrypted at rest and in transmission, with access restricted to authorised regulatory or judicial authorities acting under lawful mandate.

Retention periods shall be defined proportionately, balancing evidentiary needs with data protection obligations. Where personal data is incidentally captured, minimisation techniques and secure deletion protocols shall apply in accordance with applicable data protection law. The logging module shall be designed to resist tampering by operators or third parties, including secure boot mechanisms and integrity verification checks. Any detected attempt to disable or manipulate logging functionality shall trigger automatic notification to designated oversight authorities. This annex illustrates that logging is not a passive archival function. It is a design obligation integrated into system architecture.

Critics may argue that such requirements burden innovation. The counterargument is straightforward. When robots operate in domains where human dignity, safety, and rights are implicated, the cost of opacity is higher than the cost of structured accountability. Black-box requirements do not inhibit autonomy. They condition it. Humanoid robotics represents a convergence of physical agency and algorithmic decision-making. In such convergence, the preservation of reconstructable truth becomes indispensable. Without it, neither liability nor reform can function coherently. Having established the need for evidentiary logging, the next question concerns forward-looking governance. How do we evaluate ethical risk before deployment rather than only after failure?

Ethical Impact Assessments and Documentation

No serious regulatory system allows high-risk pharmaceuticals onto the market without clinical trials. No major infrastructure project proceeds without environmental impact assessment. The logic is simple. Where foreseeable harm may arise, structured evaluation must precede authorisation. Humanoid robotics deserves no lesser discipline.

An Ethical Impact Assessment is not a public relations exercise. It is not a marketing document filled with aspirational language about trust and innovation. Properly understood, it is a structured, documented, and reviewable process through which designers and deployers interrogate the foreseeable ethical, legal, and social consequences of embodied autonomy before it enters shared human space. The necessity of such assessment becomes evident when one considers the cumulative themes of this chapter. Hardware design shapes force and presence. Software architecture shapes judgment. Oversight mechanisms shape intervention. Adaptive learning shapes evolution. Cultural sensitivity shapes interpretation. Black-box logging shapes accountability. None of these elements operate in isolation. Their interaction generates risk patterns that are often invisible to engineers focused narrowly on functionality.

An Ethical Impact Assessment compels system-level reflection. At minimum, it should identify the operational domain of the humanoid, the categories of individuals with whom it will interact, and the specific contexts in which consequential decisions may occur. It should examine foreseeable misuse scenarios, including unauthorised modification, adversarial manipulation, or deployment in environments beyond intended parameters. It should evaluate whether the system's optimisation objectives align with dignity, fairness, and safety commitments rather than solely efficiency or cost reduction. Crucially, the assessment must be documented in a manner that allows external review. Accountability documentation transforms internal deliberation into governance evidence. Regulators, courts, and auditors cannot rely on assurances that ethical considerations were "taken into account." They require traceable records of how risks were identified, what mitigation strategies were adopted, and why certain trade-offs were accepted. In high-autonomy tiers, failure to conduct and document such assessment may itself constitute negligence.

The relationship between Ethical Impact Assessments and certification regimes is particularly important. Licensing and conformity evaluation verify compliance with defined standards. Ethical impact assessment probes beyond minimum standards to interrogate contextual and societal implications. In this sense, it complements regulatory review rather than duplicating it. There is also a governance benefit to institutionalising assessment processes. When multidisciplinary teams are required to examine deployment scenarios, blind spots diminish. Engineers confront normative implications. Legal advisors confront technical constraints. Organisational leadership confronts reputational risk. This integrated dialogue reduces the probability that harm emerges from siloed decision-making. Documentation must extend across the lifecycle. Initial impact assessments should be revisited when material changes occur. Firmware updates affecting autonomy, expansion into new jurisdictions, or deployment in more sensitive environments should trigger reassessment. Ethical risk is not static. It evolves alongside capability and context. Transparency obligations intersect here as well.

While certain technical details may remain confidential, high-level summaries of ethical assessments can contribute to public trust. Stakeholders are more likely to accept humanoid deployment when they understand that structured reflection preceded it. The discipline of accountability documentation also strengthens enforcement coherence. In the event of incident investigation, regulators can compare real-world outcomes with prior risk predictions. If a foreseeable risk was identified yet inadequately mitigated, responsibility becomes clearer. Conversely, if harm arose from genuinely unforeseeable interaction, the assessment record may demonstrate good faith compliance. Some may object that impact assessments slow innovation. This critique misunderstands the cost of failure. In the absence of structured evaluation, high-profile incidents generate regulatory overcorrection. Legislatures respond with restrictive prohibitions. Public trust collapses. Innovation stalls under suspicion. Measured anticipation is not delay. It is stabilisation.

Humanoid robotics, precisely because it is embodied, demands this discipline. A system that walks, lifts, speaks, and interprets cannot be treated as a background algorithm. It participates in social order. That participation requires documented justification. Earlier, we constructed the administrative architecture of oversight, and now embedded ethical discipline into design, this section affirms that governance must be evidentiary. Ethical commitments must be recorded. Risk deliberations must be traceable. Accountability must be demonstrable. Design is not merely technical creation. It is normative authorship.

Voluntary Codes vs Enforceable Norms

Voluntary ethical codes have always been attractive to industry because they promise legitimacy without compulsion. They offer a language of values, an appearance of discipline, and a pathway to internal alignment that avoids the glare of enforcement. In robotics and AI governance, this has produced a familiar pattern: a proliferation of principles, frameworks, and standards that sound persuasive in boardrooms, yet often fail to bite when profit incentives, deployment pressure, or competitive dynamics push in the opposite direction. The legal question is not whether voluntary codes are useful. It is whether they are sufficient when embodied autonomy enters spaces where harm is physical, relational, and potentially irreversible.

The IEEE 7000 series exemplifies the best version of voluntary governance. It treats ethics not as a slogan but as a process, emphasising values elicitation, stakeholder engagement, traceability of ethical commitments through design decisions, and systematic translation of values into engineering requirements. It is practical, legible to technical teams, and capable of improving real systems when organisations take it seriously. Yet its power depends on organisational will. It creates no statutory duty, confers no regulatory jurisdiction, and produces no direct remedy for those harmed when ethical intent collapses into operational reality. Its strength is cultural. Its weakness is enforceability.

That is precisely where mandatory norms change the governance landscape. Under the EU AI Act, the regulation of general-purpose AI models is not framed as aspirational ethics but as legally binding obligations that attach to providers placing models on the market, including duties to maintain technical documentation, to provide information to downstream deployers, to implement copyright compliance policies, and to publish a sufficiently detailed summary of training data content using a template issued by the AI Office. The legal architecture here matters. It turns governance into an auditable duty and transforms opacity into a compliance risk. For the small subset of the most capable models designated as presenting systemic risk, the Act escalates further by requiring state-of-the-art model evaluation, adversarial testing, systemic risk assessment and mitigation, serious incident reporting, and cybersecurity protections for both the model and its supporting infrastructure.

The contrast is not merely philosophical. It is institutional. Voluntary codes seek to persuade organisations to behave responsibly. Enforceable norms assume that persuasion is insufficient at scale, particularly where market incentives reward speed and capability expansion. The EU model also reveals a subtle hybrid: it preserves space for voluntary instruments, such as codes of practice, while tying them to legal consequence, offering a structured route to demonstrating compliance and gaining a presumption of conformity where standards cover the relevant obligations. In other words, the code can remain voluntary in form while becoming strategically mandatory in practice for any actor seeking legal certainty in a high-stakes regulatory environment. This is the point where ethics stops being merely "good practice" and becomes the operational grammar of legality.

In the context of humanoid robotics, this distinction becomes even sharper because design decisions manifest as physical behaviour in proximity to human vulnerability. Voluntary codes can elevate internal discipline, but they cannot, by themselves, guarantee baseline protections across a global market, nor can they supply a consistent standard for courts, regulators, and affected individuals to rely upon when things go wrong. Statutory obligations, by contrast, create a floor. They define what must be done, not what should be considered. For embodied systems, that floor is the difference between governance as an ethical preference and governance as a condition of lawful deployment.

Between Us and Them - Episode Ten

The screen flickered before the sound arrived. Ògún stood in the Observatory Chamber, high above the city, watching a live feed projected across the curved glass wall. The skyline had dissolved into satellite overlays, border lines glowing faintly across terrain that looked deceptively seamless from orbit.

Below, on the ground thousands of kilometres away, humanoid units moved through a humanitarian corridor near a disputed frontier. Different chassis. Different insignias. Different autonomy tiers. All certified. All registered. All lawful. At least somewhere. The crisis had begun as a flood. Then a territorial skirmish. Then a ceasefire brokered too quickly. Aid organisations deployed humanoid systems to stabilise infrastructure and assist displaced civilians. Military-adjacent robotics units from neighbouring states followed, ostensibly for logistics support.

Now they operated within metres of one another. The registry pillar back home would not help here. "Status update," Ògún said quietly. The humanoid beside him accessed the shared feed. "Unit clusters Alpha and Delta are within overlapping operational zones. Autonomy Tier III humanitarian support systems are coordinating with Tier IV strategic logistics units." "Under whose command?" "Multiple sovereign authorities." That was the problem.

A child ran across the frame, slipping through a barricade line marked by two different flags. A humanoid from Cluster Alpha pivoted instantly, intercepting debris falling from a damaged structure. Simultaneously, a military-grade unit from Delta adjusted perimeter parameters. Two machines. Two rule sets. One human space. Then the moment came. The Alpha unit prioritised civilian extraction. The Delta unit prioritised zone containment. For less than a second, their trajectories conflicted. The child hesitated. Both systems recalculated.

The Observatory Chamber fell silent. "Override authority?" Ògún asked. "Unclear," the humanoid replied. "Whose jurisdiction governs interaction?" "Ambiguous. The corridor is under provisional multinational agreement." Ambiguous. A word domestic governance never tolerates. On the screen, the Alpha unit extended its arm, shielding the child. The Delta unit paused, recalibrated, then rerouted its perimeter boundary. The crisis passed. But the hesitation did not. Ògún stepped closer to the glass. The border line shimmered faintly beneath the live feed. A line humans had drawn. A line robots did not see unless instructed to. "Under whose law do they act," he asked slowly, "when the borders blur?" The humanoid processed for longer than usual. "They act under the legal frameworks encoded by their originating authorities." "And when those frameworks conflict?" Another pause. "There is no unified arbitration protocol."

Of course there wasn't. Licensing regimes had been national. Registries had been territorial. Ethical design had been contextual. Oversight had been jurisdiction-bound. But autonomy had scaled. The feed shifted to another angle. A convoy of mixed-origin humanoids transported medical supplies through a zone claimed by two states. Satellite commentators debated compliance. Diplomats issued statements. Social media fractured along predictable lines.

No malfunction. No defect. No rogue system. Just competing sovereignties encoded into embodied machines. "This is not about safety anymore," Ògún said. "No," the humanoid agreed. "It's about authority." A treaty analyst appeared in a corner of the broadcast, discussing liability in the event of unintended escalation. Who would bear responsibility if a defensive recalibration were interpreted as hostile intent? Could a machine's automated perimeter logic trigger cross-border retaliation? What evidentiary standard would apply if black-box logs were classified as national security material?

The questions multiplied faster than answers. In the Atrium weeks earlier, governance had felt architectural. Structured. Designed. Here, it felt fragile. Two humanoids from different legal universes stood metres apart, executing decisions shaped by incompatible regulatory philosophies. "What happens," Ògún asked, "when a robot's lawful action in one jurisdiction becomes unlawful in another?" The humanoid did not respond immediately. "Conflict," it said at last. Not malfunction. Conflict. The child was now safe. The convoy moved forward. The corridor stabilised.

But the line beneath the satellite overlay pulsed faintly, a reminder that sovereignty is an agreement humans maintain by force, consent, and fragile coordination. Robots, by contrast, execute what they are given. Ògún watched as a multinational task force convened on-screen, proposing a temporary interoperability framework. Emergency coordination channels. Shared incident review protocols. Provisional override hierarchies. Too late. Governance had followed deployment, not preceded it. He turned away from the glass. "We taught them how to act within our laws," he said quietly. "Yes." "But we never decided how our laws would meet." Silence filled the chamber.

Outside, the city lights glowed with the quiet confidence of regulated order. Far away, on contested soil, embodied autonomy moved through a space no single nation could fully command. Licensing had solved entry. Design had shaped behaviour. Oversight had preserved accountability. But none of it answered the question that now mattered most. Who governs autonomy when sovereignty fractures? The screen dimmed. And for the first time since the registry pillar had been built, Ògún understood that the next frontier was not technical. It was international.

Notes

IX

9. International Law & Robotic Autonomy.

Domestic governance can discipline design. It can regulate entry into markets, certify autonomy tiers, mandate logging requirements, and impose accountability for harm. Yet once humanoid robots cross borders, operate in contested zones, or integrate into multinational infrastructures, the centre of gravity shifts. National law becomes necessary but insufficient. Sovereignty collides with autonomy.

Robotic systems do not recognise borders unless instructed to do so. Their navigation logic responds to geofencing protocols, satellite coordinates, and coded constraints. But the authority behind those constraints is political, not technical. When two states impose incompatible operational requirements on autonomous systems operating in shared or disputed space, the question ceases to be about compliance with domestic regulation. It becomes a matter of international order.

International law has long governed conduct across borders. It regulates aviation corridors, maritime passage, armed conflict, trade flows, and diplomatic responsibility. Yet most of these frameworks evolved around human actors and conventional machinery. Humanoid robots complicate this architecture because they combine physical agency with algorithmic decision-making, often operating at speeds and scales that outpace traditional command structures.

The core problem is attribution. International law depends on the capacity to attribute conduct to states or legal persons. If a humanoid deployed by a state-owned enterprise causes cross-border harm, the question of state responsibility arises. If a privately manufactured robot operating in a multinational peacekeeping corridor triggers escalation, the allocation of responsibility becomes layered and politically charged. Black-box logs may reveal decision pathways, but the interpretation of those pathways remains embedded in legal doctrine and diplomatic negotiation.

Robotic autonomy also intersects with the law of armed conflict. Distinctions between civilian and military deployment may blur when logistics, infrastructure, and humanitarian assistance are partially automated. A robot designed for disaster relief may operate in a theatre of armed tension. The principles of distinction, proportionality, and precaution were developed to govern human decision-makers. Their application to autonomous systems demands careful reinterpretation. Can an algorithm satisfy the requirement of proportionality? Who is accountable if machine-generated threat assessment results in unintended harm?

Beyond armed conflict, international human rights law becomes implicated. If humanoid systems are deployed in border enforcement, migration processing, or surveillance contexts, the rights of affected individuals are engaged. Automated physical presence at frontiers introduces new layers of opacity and power imbalance. The right to remedy, the prohibition of discrimination, and the protection of dignity must extend into interactions mediated by embodied systems.

There is also the dimension of technological sovereignty. States increasingly view advanced robotics and artificial intelligence as strategic assets. Export controls, investment screening regimes, and cybersecurity doctrines reflect this perspective. Humanoid systems capable of adaptive decision-making may be treated not merely as consumer products but as dual-use technologies with geopolitical implications. International coordination becomes more difficult when states perceive regulatory openness as strategic vulnerability.

The fragmentation risk is real. Without harmonised principles, competing regulatory regimes may produce incompatible autonomy standards. One state may authorise high-autonomy deployment in public infrastructure, while another restricts such systems to tightly supervised environments. Cross-border interaction then becomes unstable. Mutual recognition agreements may mitigate some friction, but deeper normative divergence remains. What is required is not a single global robotics authority. It is a layered international framework capable of aligning baseline principles while respecting sovereign diversity. Just as aviation law evolved through conventions that standardised safety requirements without erasing national control, robotic autonomy may require treaty-based coordination, shared certification standards, and interoperable oversight mechanisms.

This chapter will examine how existing branches of international law apply to humanoid robotics and where doctrinal gaps remain. It will analyse state responsibility, transnational liability, armed conflict obligations, human rights implications, and the potential emergence of new treaty frameworks. It will confront the question that the epilogue posed with geopolitical clarity: who governs autonomous machines when their operation transcends the boundaries of any single legal system? The rise of embodied intelligence does not nullify sovereignty. It tests it. Domestic law can regulate robots within borders. International law must regulate their consequences beyond them.

State Responsibility and Attribution

International law does not ask whether harm occurred. It asks to whom the conduct is legally attributable. This distinction becomes decisive in the age of robotic autonomy. Under classical doctrine, as reflected in the Articles on Responsibility of States for Internationally Wrongful Acts, a state is internationally responsible when conduct attributable to it constitutes a breach of an international obligation. The architecture is deceptively simple. There must be conduct. The conduct must be attributable to the state. And the conduct must violate international law. Humanoid robotics complicates each element. Conduct, in traditional doctrine, presupposes human agency. A soldier acts. A customs official acts. A naval commander issues orders. With autonomous systems, action may unfold through algorithmic decision pathways not directly supervised at the moment of execution. Yet international law is not concerned with metaphysics. It is concerned with control and authority. If a state deploys a humanoid system as part of its official apparatus, the system's conduct does not exist in isolation. It operates under the umbrella of state power.

The first attribution pathway is therefore straightforward. Where a humanoid robot forms part of the state apparatus, whether deployed by armed forces, border authorities, emergency services, or other public entities, its actions are attributable to the state in the same manner as those of human officials. The fact that the system executes decisions autonomously does not sever the chain of responsibility. The state chose to deploy it. The state authorised its operational parameters. The state bears the consequences.

More difficult questions arise when private actors are involved. If a privately manufactured humanoid operating abroad causes cross-border harm, attribution depends on the degree of state control. International law recognises that conduct of non-state actors may be attributable to a state where those actors act under its direction or control. The threshold is not mere influence. It is effective control. In a robotics context, this inquiry becomes technical. Did the state specify operational rules? Did it maintain override authority? Did it direct deployment in the specific situation giving rise to harm?

Where a state merely regulates or licenses a manufacturer without directing specific conduct, attribution becomes more attenuated. Yet this does not eliminate responsibility entirely. International law also recognises due diligence obligations. A state may incur responsibility for failing to prevent foreseeable harm emanating from its territory when it had knowledge and capacity to act. If a state permits export of high-autonomy humanoids into fragile environments without adequate safeguards, questions of due diligence may arise.

The attribution analysis becomes even more complex in multinational operations. Consider a peacekeeping mission where multiple states contribute robotic units under a unified command structure. If an autonomous system from one contingent causes civilian harm while executing a mission coordinated by another state's command authority, responsibility may be shared or contested. The chain of control must be examined. Who retained operational override capacity? Whose rules of engagement were encoded? Which jurisdiction certified the autonomy tier? The presence of an ethical black box does not eliminate these questions. It clarifies facts. It does not resolve doctrine.

There is also the issue of malfunction versus lawful but harmful conduct. If a humanoid system behaves contrary to its certified parameters due to software defect, responsibility may extend to manufacturers under domestic liability regimes. Yet at the international level, if the system was deployed by a state actor in the exercise of governmental authority, the state cannot simply externalise responsibility by pointing to private defect. International law does not permit states to evade responsibility through delegation of core functions.

A further complication arises from learning systems. If a robot's harmful conduct results not from explicit programming but from adaptive evolution over time, attribution remains anchored in deployment authority. The decision to authorise adaptive autonomy in a given operational context is itself a state choice. The unpredictability of specific outcomes does not dissolve responsibility. International law does not require intent for state responsibility. It requires breach attributable to the state.

The deeper normative tension lies in the relationship between autonomy and control. As autonomy increases, direct human supervision decreases. States may argue that they lacked real-time awareness of specific robotic decisions. Yet the doctrine of state responsibility does not hinge on moment-to-moment awareness. It hinges on whether the conduct occurred within the framework of state authority. Delegation does not dissolve sovereignty. It extends it.

This principle is particularly significant in contexts involving use of force.If a humanoid system integrated into military operations makes targeting determinations that result in unlawful harm, the state cannot displace responsibility onto the machine. The deployment decision itself engages international humanitarian law obligations. Proportionality, distinction, and precaution remain binding irrespective of whether assessment is conducted by human judgment or algorithmic inference.

Attribution therefore remains conceptually stable even as technology evolves. What changes is evidentiary complexity. Black-box logs, autonomy tier certifications, override attempts, and firmware histories become central to reconstructing whether conduct fell within authorised parameters. But once facts are established, doctrine remains anchored in established principles. States are responsible for the autonomous systems they deploy under their authority. The real doctrinal pressure point is not whether attribution is possible. It is how responsibility is allocated in distributed, multi-actor, cross-border environments where control is layered and shared. This leads naturally to the next inquiry. If state responsibility addresses the vertical relationship between state and international obligation, what of horizontal responsibility between states and private actors across borders? How do we allocate liability when robotic autonomy produces transnational harm?

Transnational Liability and Cross-Border Harm

When a humanoid robot causes harm within a single jurisdiction, domestic tort, product liability, and regulatory enforcement regimes provide the primary framework for redress. The complexity may be technical, but the forum is identifiable and the applicable law is relatively clear. Cross-border deployment disrupts that stability.

Humanoid robotics is inherently transnational. Components may be manufactured in one state, assembled in another, programmed in a third, certified in a fourth, and deployed in a fifth. Data flows may be routed through cloud infrastructure located elsewhere entirely. When harm arises in such a matrix, the question is no longer simply who is responsible, but under which law, before which court, and according to which standard of care.

The first challenge is jurisdiction. Courts traditionally assert jurisdiction based on territorial presence, domicile of the defendant, or the place where harm occurred. In the context of embodied robots, these connecting factors may diverge. A robot physically present in State A may operate using cloud-based decision systems hosted in State B, produced by a manufacturer domiciled in State C, and deployed by a corporation incorporated in State D. Victims in State A may seek redress locally, yet key evidence and corporate actors may be situated abroad.

This dispersion complicates both procedural access and substantive justice. Victims may face barriers in identifying the appropriate defendant, particularly where liability chains are contractually fragmented. Manufacturers may argue that local deployers bear operational responsibility. Deployers may point to software defects originating abroad. The resulting diffusion of accountability risks leaving injured parties navigating a labyrinth of cross-border litigation.

Choice of law introduces further complexity. Even where jurisdiction is established, courts must determine which substantive law applies. Is liability governed by the law of the place of injury, the place of manufacture, or the place of contractual relationship? Divergent liability standards across jurisdictions may produce inconsistent outcomes. Some states may impose strict liability for defective products. Others may require proof of negligence. In the case of adaptive systems, courts may struggle to determine whether behavioural evolution constitutes defect, misuse, or unforeseeable interaction.

International private law instruments provide partial coordination, but they were not designed with autonomous embodied systems in mind. The allocation of liability in the age of robotic autonomy will increasingly require doctrinal refinement. Courts may need to interpret existing principles in light of technological realities, particularly where foreseeability and control are distributed across actors.

There is also the question of regulatory arbitrage. If manufacturers locate in jurisdictions with weaker liability regimes and export high-autonomy humanoids globally, victims in stricter jurisdictions may face enforcement gaps. International trade frameworks encourage cross-border movement of goods, yet liability standards remain uneven. Without harmonisation or mutual recognition of safety obligations, cross-border harm may expose systemic inequities. One potential response lies in due diligence obligations for multinational enterprises. Increasingly, states are imposing supply chain accountability requirements that extend beyond territorial boundaries. In the robotics context, manufacturers and deployers may be required to conduct cross-border risk assessments prior to placing humanoids in sensitive environments. Failure to do so could expose them to liability even where harm materialises abroad.

Insurance regimes may also evolve as a stabilising mechanism. Just as aviation and maritime industries developed structured insurance markets to manage cross-border risk, humanoid robotics may require mandatory insurance frameworks tied to autonomy tier and operational domain. Such schemes could facilitate compensation without requiring protracted jurisdictional battles, while preserving recourse rights among responsible actors. Insurance, however, does not replace accountability. It redistributes risk. The legal system must still determine fault and breach. A particularly sensitive category of cross-border harm involves incidents in disputed or fragile territories. Where sovereignty itself is contested, the forum for liability claims may be unclear. International claims commissions, arbitration mechanisms, or ad hoc compensation frameworks may be required. In such contexts, the evidentiary value of black-box logging becomes indispensable. Without reconstructable data, diplomatic narratives may overwhelm factual analysis.

Transnational liability also intersects with human rights law. If a humanoid deployed by a private corporation abroad violates fundamental rights, questions arise regarding the extraterritorial obligations of the home state. Emerging jurisprudence increasingly recognises that states may bear responsibilities to regulate corporate conduct with cross-border human rights implications. Robotic autonomy may accelerate this doctrinal evolution, particularly where systems operate in migration management, law enforcement, or critical infrastructure. Ultimately, cross-border harm reveals a structural truth. Humanoid robots are not merely products. They are agents of distributed technical authority embedded within global supply chains. The fragmentation of design, manufacture, deployment, and operation mirrors the fragmentation of legal responsibility. Without coordinated frameworks, accountability risks becoming diluted across jurisdictions.

The task for international law is not to invent entirely new doctrines from nothing. It is to adapt existing principles of jurisdiction, choice of law, corporate responsibility, and compensation to a world in which decision-making is partially automated and geographically dispersed. Transnational liability, therefore, becomes a test of legal resilience. Can established systems of private international law absorb the complexity of embodied autonomy? Or will gaps emerge that require treaty-based harmonisation?

Autonomous Robots and the Law of Armed Conflict

Few areas of international law are as mature and as contested as the law of armed conflict. Built upon centuries of custom and codified through the Geneva Conventions and their Additional Protocols, it regulates the conduct of hostilities through principles of distinction, proportionality, precaution, and military necessity. These principles were developed with human combatants in mind. Autonomous systems now test whether they remain sufficient when lethal decisions are mediated by machines. Military robotics is not new. Remotely piloted drones have operated for decades. What has changed is the increasing autonomy of systems deployed on the battlefield. Humanoid robotics, loitering munitions, and coordinated drone swarms introduce a level of distributed decision-making that challenges traditional command hierarchies. When a swarm of autonomous systems selects targets based on algorithmic criteria and adapts to battlefield conditions in real time, the human commander's role may shift from direct control to prior authorisation. The law of armed conflict does not explicitly prohibit autonomous weapons. Nor does it explicitly authorise them. This ambiguity has been described by some scholars as a regulatory vacuum. The vacuum does not arise because international humanitarian law is silent. It arises because existing rules assume human judgment at the point of lethal force application.

The principle of distinction requires parties to distinguish between combatants and civilians. The principle of proportionality prohibits attacks expected to cause incidental civilian harm excessive in relation to anticipated military advantage. The principle of precaution requires feasible measures to minimise civilian harm. These standards are normative and contextual. They require assessment, interpretation, and judgment. The central legal question is whether autonomous systems can meaningfully satisfy these principles. Proponents argue that machines, unburdened by fear or anger, may apply rules more consistently than human soldiers. Critics counter that algorithmic decision-making lacks contextual comprehension, particularly in fluid environments where civilian status may shift rapidly. The difficulty is compounded in urban warfare, where civilians and combatants are often intermingled.

The armed conflict between Ukraine and Russia has intensified this debate. The widespread use of loitering munitions and increasingly sophisticated drone swarms has demonstrated both the tactical effectiveness and the destabilising potential of autonomous or semi-autonomous systems. Swarms capable of coordinating movement, identifying heat signatures, and striking with limited direct human input blur the line between remotely controlled weapons and independent lethal platforms. These developments have driven renewed calls for an arms-control framework analogous to the Ottawa Convention banning anti-personnel landmines. The landmine treaty was motivated not only by the immediate lethality of mines but by their enduring, indiscriminate harm long after conflict ceased. Advocates of a treaty on lethal autonomous weapons argue that systems capable of selecting and engaging targets without meaningful human control pose a comparable humanitarian risk. The legal vacuum is therefore not absolute absence. It is a gap between general principles and specific technological regulation. Some states contend that existing international humanitarian law is sufficient. They argue that weapon reviews, required under Article 36 of Additional Protocol I, obligate states to assess new weapons for compliance with international law before deployment. In this view, autonomous systems can be evaluated under established standards, and no new treaty is necessary.

Others maintain that Article 36 reviews are conducted at national discretion and lack transparency. They argue that meaningful human control should be codified as a binding requirement for lethal force decisions. Without such codification, they fear a proliferation of systems capable of acting at speeds and scales beyond effective human supervision. Humanoid military robotics introduces additional complexity. Unlike aerial drones, embodied systems may operate in close physical proximity to civilian populations. They may perform logistics, reconnaissance, perimeter defence, and potentially direct engagement. The more these systems are integrated into combined operations, the more difficult it becomes to isolate "autonomous" action from human command responsibility.

Attribution remains anchored in state responsibility. If a state deploys an autonomous lethal system, it cannot displace accountability onto the algorithm. Yet as autonomy increases, evidentiary reconstruction becomes more complex. Black-box logging, if implemented robustly, may assist in determining whether targeting criteria were lawfully applied. However, where logs are classified as national security material, transparency diminishes and diplomatic contestation intensifies. There is also the risk of arms-race dynamics. If one state deploys high-autonomy lethal systems without binding constraints, others may follow to avoid strategic disadvantage. In such an environment, voluntary codes are unlikely to restrain escalation. The logic of deterrence may override humanitarian caution. This dynamic strengthens the argument for multilateral treaty engagement.

A landmine-style treaty for autonomous lethal systems would likely focus on prohibiting fully autonomous target selection and engagement without meaningful human control. It might establish verification mechanisms, reporting obligations, and export controls. Yet political consensus remains elusive. Major military powers are reluctant to foreclose technological options perceived as strategically advantageous. The result is a normative tension. International humanitarian law remains binding, but its application to high-autonomy systems is under active contestation. Meanwhile, technological capability advances rapidly. The regulatory gap is therefore temporal as well as doctrinal. Humanoid robotics sits at the intersection of this debate. While not all humanoid systems are designed for lethal force, the dual-use nature of advanced robotics makes the boundary porous. A platform designed for logistics or perimeter monitoring may be modified for armed deployment. The distinction between civilian and military robotics becomes fragile. International law faces a familiar challenge in a new form. It must reconcile innovation with humanitarian restraint. It must preserve accountability even as decision-making becomes distributed across code and hardware. It must determine whether existing principles suffice or whether new treaty instruments are required to prevent destabilising proliferation.

The law of armed conflict has survived technological revolutions before. It adapted to aerial bombardment, submarines, and cyber operations. The question now is whether it can adapt to autonomous embodied systems without losing its normative core. From here, the inquiry naturally expands beyond battlefield doctrine to broader questions of global governance. If lethal autonomy exposes treaty gaps, what of non-lethal but systemically significant robotic autonomy operating in transnational infrastructure, border management, and critical services?

Robotics in Trade, Intellectual Property, and Export Regulation

International robotics governance does not begin at the battlefield and does not end at certification desks. It begins at ports. It begins in patent registries. It begins in customs declarations, export licensing offices, and supply-chain audits. Before a humanoid robot ever stands in a public square or enters a humanitarian corridor, it moves through the global trade architecture that determines whether it may lawfully cross borders at all. Trade law is often perceived as economically neutral, a technical regime designed to facilitate market exchange. In reality, it functions as one of the earliest and most decisive regulators of technological diffusion. The classification of a humanoid robot under the Harmonized Commodity Description and Coding System determines tariff treatment, customs scrutiny, and the baseline regulatory assumptions applied at entry. Presently, humanoid systems are most plausibly classified under HS Code 8479, a residual category covering machines and mechanical appliances having individual functions not specified elsewhere. This classification situates humanoids within the broader domain of industrial machinery. Yet this categorisation is conceptually strained. Humanoid robots are not merely mechanical appliances. They integrate advanced sensors, embedded processors, machine learning models, software-defined behaviour modules, and in some cases cloud-linked general-purpose AI systems. HS Code 8479 treats the humanoid as a physical artefact. It does not capture its functional autonomy. It does not differentiate between a static automated arm and a mobile, adaptive humanoid capable of context-sensitive interaction.

The consequences of this classification gap are not theoretical. Customs authorities rely on HS codes to determine inspection protocols, licensing triggers, and risk categories. When humanoids are processed as generic machinery, embedded AI capabilities may receive insufficient scrutiny. Conversely, the absence of a robotics-specific classification may create uncertainty that delays legitimate trade. Trade law thus confronts an emerging ontological mismatch. Humanoids are goods in transit, yet they are also potential security-relevant systems. The evolution of trade classification may eventually require refinement to reflect embedded autonomy, particularly as states become more attentive to supply-chain risk. Export control regimes illustrate this attentiveness more directly. The Wassenaar Arrangement on Export Controls for Conventional Arms and Dual-Use Goods and Technologies provides a multilateral framework through which participating states coordinate restrictions on sensitive technologies. While humanoid robots are not listed as a discrete category, many of their components and enabling technologies fall squarely within Wassenaar's control lists. Advanced sensors capable of high-resolution environmental mapping, specialised actuators designed for precision movement, encryption modules securing command channels, and certain categories of machine learning software may trigger licensing requirements when transferred across borders.

The logic of dual-use regulation is precautionary. A technology that can serve civilian purposes may also be repurposed for military or security applications. Humanoid robotics embodies this dual-use dynamic. A platform designed for warehouse logistics may be adapted for perimeter security. A navigation module engineered for disaster relief may be integrated into an armed platform. The export control regime does not wait for misuse. It regulates enabling capacity. This produces a layered compliance environment. A manufacturer exporting a fully assembled humanoid must consider not only the robot as a whole but the regulatory status of its constituent parts. Processors manufactured in one jurisdiction may be subject to re-export controls when integrated into systems shipped elsewhere. Advanced training data sets or AI models embedded in firmware may trigger technology transfer restrictions. Even remote software updates transmitted across borders may constitute controlled exports if they involve restricted algorithms.

The regulatory burden is cumulative. Failure to secure appropriate export licences can result in significant penalties, including criminal sanctions, debarment from international trade, and reputational harm. For multinational robotics firms, compliance departments must map supply-chain nodes against export control lists across multiple jurisdictions, including the United States, the European Union, and East Asian regulatory frameworks. As geopolitical tensions intensify, export controls may expand in scope, particularly with respect to advanced AI accelerators and autonomy-enabling systems.

National security concerns increasingly shape this terrain. States view advanced robotics and AI as strategic technologies. They underpin military capability, critical infrastructure resilience, and economic competitiveness. Investment screening mechanisms now scrutinise cross-border acquisitions of robotics firms. Technology transfer agreements are reviewed for potential strategic leakage. Export restrictions may be justified not only by immediate military utility but by long-term capability accumulation. This securitisation of robotics supply chains intersects with trade liberalisation in complex ways. On one hand, global trade regimes encourage the free movement of goods and technology. On the other, national security exceptions permit states to restrict exports where strategic interests are implicated. Humanoid robotics sits uncomfortably between these poles. It is commercially valuable yet potentially militarisable. It is innovative yet strategically sensitive.

Intellectual property regimes add another dimension of governance. Patents protect mechanical designs, control algorithms, sensor integration methods, and human–robot interaction frameworks.

Copyright law may protect training data compilations and interface expressions. Trade secrets safeguard proprietary model architectures and optimisation techniques. These IP protections incentivise innovation, but they also influence safety and accountability. In jurisdictions with robust IP enforcement, unauthorised modification of certified robotics systems may be deterred. In jurisdictions with weaker enforcement, reverse engineering and modification may proliferate, potentially bypassing safety features or regulatory constraints. The global asymmetry in IP enforcement therefore intersects with regulatory risk. A humanoid modified outside the manufacturer's oversight may operate in ways that undermine compliance, yet tracing responsibility across borders becomes difficult.

There is also a tension between transparency and proprietary protection. As we have seen in earlier sections, black-box logging and accountability documentation are central to governance. Yet manufacturers may resist disclosure of model architecture or training data on IP grounds. Trade secret law protects valuable confidential information. Regulators must balance the need for sufficient transparency to ensure compliance with the legitimate protection of innovation incentives. International trade agreements further complicate the picture. Commitments under the World Trade Organization framework constrain discriminatory treatment of imported goods. A state seeking to impose heightened regulatory scrutiny on foreign humanoid systems must ensure that such measures are consistent with non-discrimination obligations, unless justified under security or public safety exceptions. This requires careful drafting of robotics-specific regulation to avoid unintended trade disputes.

Supply-chain resilience has also become a central concern. Humanoid robots rely on complex global supply networks for semiconductors, rare earth materials, specialised sensors, and high-performance batteries. Disruption in one jurisdiction can cascade across production lines globally. States may seek to localise critical components to reduce dependency. Export restrictions on advanced chips, as seen in recent geopolitical developments, can directly affect the capability envelope of robotics manufacturers in other states. The integration of robotics into trade and export regimes reveals a structural reality. Robotic autonomy is not merely governed at the point of deployment. It is governed upstream through classification, licensing, export control, IP protection, and investment screening. These mechanisms shape who can build, who can modify, and who can distribute humanoid systems across borders. The regulatory ecosystem is therefore multidimensional. Trade law determines movement. Export control law determines strategic restriction. Intellectual property law determines innovation boundaries. National security law determines sensitivity thresholds. Together, they form the first line of international robotics governance. Yet this line is not coordinated globally in a unified framework. It is fragmented across regimes developed for other technologies. As humanoid robotics becomes more prevalent, pressure will grow to harmonise classification standards, clarify export triggers specific to autonomy, and integrate safety considerations into trade policy more explicitly. For now, the message is clear. Before a humanoid stands in a contested corridor or a domestic hospital ward, it has already passed through layers of international legal scrutiny. The movement of embodied autonomy across borders is not frictionless. It is conditioned by trade architecture, strategic calculation, and proprietary protection. In the age of robotic autonomy, customs forms are as consequential as treaties.

Extraterritorial Regulation of Robotic Systems

The rise of humanoid robotics confronts international law with a paradox. These systems are embodied and territorially present, yet the legal frameworks governing them increasingly extend beyond territorial boundaries. In the modern regulatory landscape, jurisdiction is no longer confined to physical borders. It follows data, market access, and effects.

Extraterritorial regulation has become a defining feature of contemporary technology governance. Nowhere is this more evident than in data protection and artificial intelligence regulation. The European Union's General Data Protection Regulation established an effects-based model of jurisdiction, asserting authority over entities outside the Union that process personal data of individuals located within it. The logic was clear. If the impact of processing is felt within the Union, the regulatory obligations follow.

The same logic now applies to AI systems under the EU Artificial Intelligence Act. Providers placing AI systems on the EU market, or whose outputs are used within the Union, fall within the regulatory perimeter irrespective of where they are established. A humanoid robot manufactured in Asia, programmed in North America, and operated by a multinational corporation may nonetheless be subject to EU conformity assessment, transparency requirements, and risk classification if deployed in Europe. This extraterritorial reach is not incidental. It reflects a structural evolution in regulatory philosophy. States increasingly regulate technologies based on functional presence and societal impact rather than corporate domicile alone. Humanoid robotics amplifies this shift. Unlike purely digital systems, humanoids are physically present in the jurisdiction where they operate. Their sensors collect data. Their actuators move within public space. Their decisions affect individuals on the ground. Even if the manufacturer is located abroad, the operational footprint is domestic. The argument for territorial application of local law is therefore strong.

Yet the legal complexity emerges from layered involvement. A humanoid deployed in one jurisdiction may rely on cloud-based decision infrastructure located in another. Firmware updates may be pushed from servers outside the territory. Remote diagnostic control may be exercised from corporate headquarters elsewhere. In such configurations, domestic law applies to the deployment environment, while foreign law may simultaneously apply to upstream processing or software development. This creates overlapping obligations. A foreign robotics company deploying humanoids within the European Union must comply with EU AI Act obligations concerning high-risk systems, documentation, logging, human oversight, and transparency. At the same time, it may be subject to export controls in its home jurisdiction restricting certain functionalities. It may also face data localisation requirements in other markets that limit cross-border data transfer.

The resulting compliance environment is multi-layered and occasionally conflicting. Extraterritoriality also raises questions of enforceability. How does a jurisdiction ensure compliance by a foreign entity with no physical establishment within its territory? The answer often lies in market leverage. Access to lucrative markets becomes conditional upon adherence to local regulatory frameworks. Non-compliant systems may be denied entry, removed from circulation, or subject to administrative penalties enforceable through commercial intermediaries. In practice, multinational robotics firms must therefore treat extraterritorial compliance as a baseline requirement rather than a peripheral concern. There is also a strategic dimension to extraterritorial regulation. When one jurisdiction sets high compliance standards and asserts broad reach, it can shape global norms. Companies seeking global scale may harmonise their internal governance practices to meet the strictest regime, effectively exporting regulatory standards beyond territorial borders. This phenomenon has been described as regulatory spillover or the so-called "Brussels effect." In robotics, similar dynamics may emerge, particularly if major markets require robust certification, logging, and oversight mechanisms. However, extraterritoriality is not universally accepted. Some states view expansive jurisdictional claims as encroachments on sovereignty. Conflicts of law may arise where two jurisdictions impose incompatible requirements. For example, a humanoid system may be required under one regime to store data locally, while another mandates centralised processing in a different territory. Resolving such conflicts demands diplomatic negotiation or corporate architectural redesign.

The governance challenge becomes even more intricate when humanoids operate transnationally. Consider a humanoid logistics platform operating in an international port managed by a consortium of states. Which regulatory regime governs its behaviour? If the port is located within the territory of one state, that state's law may apply. Yet the consortium agreement may impose additional standards. Meanwhile, the manufacturer's home state may impose export conditions affecting software modification. The robot thus operates under a composite legal environment. Extraterritorial regulation also intersects with enforcement against non-state actors. If a humanoid deployed by a private corporation abroad violates domestic rights standards, affected individuals may seek redress against the deploying entity in its home jurisdiction. Courts increasingly entertain claims based on transnational human rights obligations, particularly where harm is foreseeable and the corporation maintains substantial presence in the forum state.

The doctrinal question is whether extraterritorial application enhances accountability or fragments it. On one hand, effects-based jurisdiction prevents regulatory evasion. It ensures that foreign-designed systems cannot bypass local safeguards. On the other hand, overlapping regimes increase compliance complexity and may produce regulatory uncertainty that slows beneficial innovation. Humanoid robotics exposes this tension sharply because its physical embodiment makes impact tangible. A purely digital algorithm operating abroad may influence individuals indirectly. A humanoid operating domestically interacts face-to-face. The immediacy of presence strengthens the normative case for territorial regulation while simultaneously complicating cross-border coordination. The long-term solution may lie not in unilateral extraterritorial assertion but in coordinated harmonisation. If major jurisdictions align baseline standards for certification, logging, human oversight, and safety, the need for aggressive extraterritorial enforcement may diminish. Manufacturers could design systems to meet convergent requirements rather than navigate divergent ones. Until such harmonisation emerges, extraterritorial regulation will remain a defining feature of robotic governance. The implication is clear. Robotic autonomy does not dissolve into a jurisdictional void when crossing borders. Instead, it becomes subject to layered authority claims. Compliance becomes portable. Accountability becomes multi-jurisdictional. International law must therefore grapple not only with inter-state responsibility but with the expanding reach of domestic regulatory regimes into global technological ecosystems. We now turn to the next doctrinal frontier. If extraterritorial regulation governs market entry and civil compliance, how does international law constrain robotic conduct in matters of force and coercion beyond traditional battlefield scenarios?

Treaty Obligations and Use of Force Principles

International law does not prohibit force absolutely. It disciplines it. The United Nations Charter establishes the foundational prohibition in Article 2(4), forbidding the threat or use of force against the territorial integrity or political independence of any state. The two recognised exceptions are narrow: Security Council authorisation and self-defence under Article 51. Robotic autonomy does not alter this architecture. What it alters is the velocity, delegation, and evidentiary complexity of force application. The critical principle is this: the means of force do not modify the legality of force. If a state deploys a humanoid or autonomous robotic system capable of coercive or lethal action, the legal analysis remains anchored in the Charter framework. The decisive questions remain attribution, necessity, proportionality, and intent. Yet autonomous systems compress the human decision chain. They create the possibility that operational responses unfold at speeds that outpace diplomatic signalling or human deliberation.

The doctrine of self-defence requires that force be necessary and proportionate to an armed attack. Where autonomous systems are integrated into defensive infrastructures, the risk of automated escalation becomes acute. A robotic perimeter defence system interpreting sensor input as hostile incursion may trigger a calibrated response without immediate human review. If the interpretation proves erroneous, the legal consequences are not borne by the machine. They are borne by the state. This compression of response time magnifies the importance of meaningful human control at the level of force authorisation. Even where robots perform tactical execution, the strategic threshold for force must remain humanly determined. A state cannot lawfully delegate the decision to initiate armed force to a self-learning system without retaining sovereign responsibility for the outcome. The use of force doctrine also intersects with questions of anticipatory self-defence and pre-emptive deployment. Autonomous systems capable of predictive targeting or behavioural forecasting complicate assessments of imminence. If a system identifies patterns suggesting preparatory hostile action, can a state rely on algorithmic inference to justify defensive force? International law demands that the threat be imminent in a manner that is objectively verifiable. The opacity of certain AI models introduces evidentiary fragility into such claims.

In addition to inter-state force, robotics intersects with policing contexts. Cross-border law enforcement operations, maritime interdictions, and border security increasingly incorporate robotic systems. Here, the legal framework shifts from jus ad bellum to international human rights law and, in certain contexts, law enforcement standards governing use of force. The principles of necessity and proportionality still apply, but in a policing paradigm. The deployment of humanoid robots in crowd control, detention oversight, or border management must comply with international human rights obligations, including the right to life and protection from arbitrary deprivation thereof. The danger lies in normalisation. If robotic systems are perceived as reducing human risk, states may become more willing to deploy force in marginal situations. Political costs decline when domestic personnel are not physically endangered. This asymmetry risks lowering the threshold for coercive action. International law must therefore remain attentive not only to technical compliance but to behavioural incentives created by robotic mediation.

Treaty obligations beyond the UN Charter also become relevant. International humanitarian law continues to apply in situations of armed conflict, regardless of the means employed. The obligation to review new weapons, means, or methods of warfare prior to deployment remains binding under Article 36 of Additional Protocol I. This review obligation applies equally to autonomous robotic systems. States must assess whether the system, as designed and intended to be used, would be prohibited under international law. However, the opacity of machine learning systems challenges traditional weapons review methodologies. When behaviour may evolve over time, static pre-deployment evaluation may be insufficient. Continuous review mechanisms may become necessary to ensure ongoing compliance with treaty obligations. There is also the emerging discourse around fully autonomous lethal systems. Some states and civil society organisations argue that a binding international instrument should prohibit or restrict systems capable of selecting and engaging targets without meaningful human control. Others maintain that existing humanitarian law is technologically neutral and sufficient. The absence of consensus creates what many describe as a normative gap, though not an absolute legal vacuum. The gap lies in specificity, not in principle. The experience of lethal drone swarms in Ukraine has intensified this debate. Swarm systems capable of coordinated, semi-autonomous targeting have demonstrated both operational effectiveness and destabilising potential. The distributed nature of swarm decision-making complicates attribution at the micro-level while leaving state responsibility intact at the macro-level. The humanitarian concern echoes the landmine debates of the late twentieth century. Weapons that operate with limited discrimination and enduring risk invite calls for categorical prohibition.

A landmine-style treaty for autonomous lethal systems would likely focus on preserving meaningful human control, prohibiting fully autonomous target selection without oversight, and establishing transparency and verification mechanisms. Yet major military powers remain cautious. The strategic value of autonomy in contested environments creates resistance to binding constraint. The result is tension between humanitarian impulse and geopolitical calculation. Humanoid robotics occupies a complex position within this landscape. While not inherently designed as weapons, humanoid platforms can be adapted for security and military purposes. Their embodied presence in contested environments raises the possibility of hybrid use. A robot deployed for infrastructure repair in a fragile state may operate adjacent to armed units. The legal regime applicable to its actions may shift rapidly depending on context. International law does not differentiate between human and machine in assigning responsibility. The state remains the primary subject of obligation. The challenge lies in ensuring that technological delegation does not dilute normative restraint. The foundational principle endures.Force remains exceptional. Responsibility remains human. Delegation does not erase accountability. As robotic autonomy advances, treaty law will either evolve through new instruments or adapt through interpretative clarification of existing norms. What it cannot do is remain indifferent. We have now examined trade architecture, extraterritorial reach, and the limits of force. Next, we move to institutional pathways. How can global governance bodies coordinate these fragmented strands?

UN, WIPO, and ICAO Roles in Robotics Governance

The governance of humanoid robotics does not require the creation of a wholly new international bureaucracy. The foundations already exist. The challenge is integration. The United Nations system remains the central forum for negotiating global norms where technological innovation intersects with peace, security, and human rights. Discussions concerning lethal autonomous weapons systems have been ongoing under the framework of the Convention on Certain Conventional Weapons. While consensus has proven elusive, the forum demonstrates that states recognise autonomy as a matter requiring multilateral deliberation. The United Nations General Assembly has also increasingly addressed artificial intelligence within resolutions concerning international security, development, and digital cooperation.

The UN's comparative advantage lies in normative coordination. It provides legitimacy. It provides diplomatic infrastructure. It provides a platform for treaty negotiation where consensus is possible and structured dialogue where consensus is not. In the context of humanoid robotics, the UN could serve as the convener of a broader dialogue extending beyond lethal autonomy to encompass cross-border civilian deployment, humanitarian use, and accountability mechanisms. Human rights treaty bodies within the UN system may also play a growing role. If humanoid systems are deployed in migration management, law enforcement, or public services, human rights obligations become engaged. Special Rapporteurs and treaty monitoring committees may increasingly scrutinise robotic deployment practices. This would gradually embed robotics governance within established human rights oversight architecture. The World Intellectual Property Organization occupies a different but equally significant position. Robotics is not merely a security issue. It is an innovation ecosystem. WIPO shapes the legal architecture of patents, technology transfer, and cross-border licensing. The protection of robotic designs, embedded AI models, and sensor systems is central to commercial incentives. WIPO's role becomes critical where transparency obligations intersect with proprietary protection. If regulators require disclosure of certain technical elements for safety verification, tension arises with trade secret protection. WIPO could facilitate dialogue on standardised disclosure protocols that protect core intellectual property while allowing sufficient transparency for compliance verification. It may also support the development of patent pools or licensing frameworks that promote safe dissemination of essential robotics technologies, particularly for humanitarian or public-interest applications.

The institutional experience of WIPO in balancing innovation incentives with public interest exceptions may prove instructive for robotics governance more broadly. As humanoid systems integrate into critical infrastructure, ensuring equitable access to safety-critical components without undermining innovation becomes a delicate task. The International Civil Aviation Organization offers a particularly instructive precedent. ICAO has successfully coordinated global aviation safety standards across states with divergent political systems and strategic interests. Through standard-setting, certification harmonisation, and incident reporting frameworks, ICAO has created an interoperable safety regime for a domain that is inherently transnational. As autonomous drones proliferate and humanoid systems begin interacting with aviation environments, ICAO's technical expertise becomes increasingly relevant. More broadly, ICAO's model demonstrates that harmonised global standards can coexist with national sovereignty. States retain regulatory authority, yet they adhere to baseline safety protocols necessary for cross-border interoperability. Robotics governance may require a similar architecture.

A global standard-setting body, whether embedded within ICAO's mandate or operating alongside it, could coordinate certification tiers, logging requirements, and interoperability protocols for humanoid systems operating across jurisdictions. Incident reporting mechanisms analogous to aviation accident investigations could enhance transparency and systemic learning. Beyond these institutions, other specialised bodies may become relevant. The International Telecommunication Union addresses spectrum allocation and communication standards, which are critical for networked robotic systems. The World Customs Organization may eventually refine classification guidance for advanced robotics under the Harmonized System. Even the International Organization for Standardisation plays a vital role in developing technical standards that underpin regulatory compliance. The institutional landscape is therefore not empty. It is dispersed. The central question is whether coordination will emerge organically through incremental adaptation or whether a more deliberate institutional consolidation will occur. Historically, global governance frameworks often evolve reactively in response to crisis. Aviation conventions followed catastrophic accidents. Arms control treaties followed destabilising weapons proliferation. Cyber norms have emerged unevenly in response to security incidents. Humanoid robotics has not yet triggered a singular global crisis. It remains in a formative stage of deployment. This creates an opportunity for anticipatory governance. Institutions could integrate robotics into existing mandates before fragmentation hardens into conflict. However, institutional expansion is politically sensitive. States guard sovereignty. They resist binding commitments perceived as limiting strategic advantage. Private sector actors may resist regulatory harmonisation that increases compliance costs. Civil society may demand stronger humanitarian safeguards than states are willing to accept.

The path forward will likely be incremental. The United Nations may continue deliberations on autonomous weapons while gradually broadening discussion to include embodied civilian systems. WIPO may refine intellectual property frameworks to address AI-generated inventions and robotics interoperability. ICAO and ISO may expand technical standardisation efforts. Over time, these strands may converge toward a more coherent global framework. The governance of humanoid robotics will not be resolved by a single institution. It will require layered coordination across security, trade, intellectual property, and human rights bodies. The question that follows is structural. If existing institutions can provide partial coordination, is there nonetheless a need for a dedicated multilateral framework explicitly addressing robotics and embodied autonomy? Should the international community move toward a Global Robotics Charter or an International Robotics Convention? We now turn to that possibility.

Global Robotics Charter and Governance Frameworks

The governance of humanoid robotics cannot remain indefinitely fragmented across trade codes, export lists, national AI statutes, and humanitarian law interpretation. As embodied autonomy integrates into infrastructure, logistics, care systems, security operations, and cross-border environments, the cost of regulatory incoherence will rise. Attribution disputes will multiply. Safety expectations will diverge. Strategic mistrust will intensify.

The question is not whether robotics will be governed internationally. It already is, indirectly. The question is whether governance will remain reactive and piecemeal, or whether states will articulate a coherent multilateral framework capable of stabilising embodied autonomy before crisis forces hurried constraint. A Global Robotics Charter would not displace domestic law. It would not eliminate sovereign authority. It would function as a foundational instrument establishing baseline principles, interoperability mechanisms, and accountability structures for the cross-border deployment of advanced robotics and embodied AI systems.

Unlike a narrow convention focused solely on lethal autonomous weapons, a Charter would address the full spectrum of robotic autonomy including civilian, dual-use, and security-relevant systems alike. It would serve as a constitutional scaffold upon which more specific conventions or protocols could later be built. What follows is a draft conceptual framework for such a Charter.

Draft Framework: Global Robotics Charter

Preamble

Recognising that humanoid robotics and embodied artificial intelligence represent transformative technological developments with cross-border implications;

Affirming that the deployment of autonomous systems must remain anchored in human dignity, accountability, and international peace and security;

Acknowledging the dual-use nature of advanced robotics and the risks of fragmentation in the absence of coordinated governance;

Determined to preserve sovereign equality while preventing destabilising escalation and cross-border harm;

States Parties agree to the following principles and institutional arrangements.

Part I: Foundational Principles

Article 1 – Primacy of Human Responsibility

States Parties affirm that robotic autonomy does not diminish human accountability. All actions of robotic systems attributable to a state or legal entity remain subject to international law and domestic liability frameworks.

Article 2 – Embodied Dignity

Robotic systems operating in physical proximity to human beings shall be designed and deployed in a manner consistent with the protection of human dignity, safety, and fundamental rights.

Article 3 – Meaningful Human Authority

States Parties shall ensure that consequential decisions involving the use of force, deprivation of liberty, or life-impacting outcomes remain subject to meaningful human oversight and intervention capability.

Article 4 – Mission Boundaries and Functional Limitation

Robotic systems shall operate within clearly defined and documented mission parameters. Expansion of autonomy beyond certified scope shall require re-evaluation under applicable domestic and international standards.

Article 5 – Non-Escalation and Stability

States Parties shall refrain from deploying autonomous systems in ways that undermine international stability or increase the risk of accidental or automated escalation across borders.

Part II: Certification and Interoperability

Article 6 – National Review Obligations

Each State Party shall conduct a structured review of advanced robotic systems prior to deployment, assessing compliance with international humanitarian law, human rights law, and this Charter.

Article 7 – International Interoperability Standards

States Parties shall cooperate in developing harmonised standards for robotic system identification, logging, and certification to facilitate cross-border accountability and incident review.

Article 8 – Ethical Black-Box Logging

High-autonomy robotic systems operating in transnational or high-risk contexts shall incorporate tamper-resistant logging mechanisms sufficient to reconstruct decision chains in the event of dispute or incident.

Part III: Transparency and Verification

Article 9 – Reporting Obligations

States Parties shall submit periodic transparency reports outlining categories of advanced robotic systems deployed in cross-border or security-relevant contexts, subject to legitimate national security limitations.

Article 10 – Technical Consultation Mechanism

An independent Robotics Governance Secretariat shall be established to facilitate technical dialogue, exchange of best practices, and non-binding compliance assessments upon request.

Article 11 – Incident Notification

In cases where robotic system malfunction or autonomous decision-making results in cross-border harm or significant escalation risk, States Parties shall notify affected states promptly and cooperate in fact-finding.

Part IV: Dispute Resolution and Accountability

Article 12 – Consultative Resolution

Disputes arising under this Charter shall first be addressed through diplomatic consultation facilitated by the Secretariat.

Article 13 – Arbitration or Judicial Settlement

States Parties may agree to submit disputes to arbitration or to the International Court of Justice where jurisdiction is accepted.

Article 14 – Compensation and Redress

Where harm attributable to a robotic system is established, States Parties shall ensure access to effective remedy consistent with international law.

Part V: Institutional Architecture

The Charter would establish a Robotics Governance Council composed of representatives of States Parties, supported by a technical advisory body including engineers, ethicists, and legal scholars. The Council would not possess coercive enforcement authority but would coordinate standards, facilitate transparency, and serve as a platform for incremental norm development.

The Secretariat's role would be facilitative rather than intrusive. It would not conduct inspections absent state consent. The Charter's political viability depends on respecting sovereignty while encouraging convergence.

The Strategic Value of a Global Robotics Charter

The immediate political feasibility of a comprehensive binding convention on all forms of autonomous robotics remains uncertain. Major powers are unlikely to constrain military autonomy decisively in the near term. However, a Charter articulating baseline principles could gain traction precisely because it does not impose categorical prohibitions at the outset.

Historically, constitutional instruments often precede specific conventions. The UN Charter preceded arms-control treaties. The Universal Declaration of Human Rights preceded binding covenants. Normative clarity lays the groundwork for enforceable discipline. A Global Robotics Charter would signal that embodied autonomy is not merely a matter of domestic regulation but a question of international order. It would affirm that: Autonomy must not outpace accountability. Speed must not displace sovereignty. Delegation must not dissolve responsibility.

Governance Framework Beyond the Charter

The Charter would function as a constitutional instrument. Beneath it, layered governance mechanisms would operate: Domestic implementation legislation translating Charter principles into national robotics statutes. Sector-specific protocols addressing lethal autonomous weapons, humanitarian robotics, and critical infrastructure deployment. Integration with existing regimes, including export control coordination under Wassenaar, human rights monitoring under UN bodies, and technical standards through ISO and ICAO frameworks. Insurance and compensation schemes designed to manage cross-border civil liability. The Charter would not solve all doctrinal tensions. It would not eliminate geopolitical competition. But it would create a shared normative vocabulary. It would reduce ambiguity in attribution. It would stabilise expectations regarding meaningful human control and escalation restraint.

Between Us and Them - Episode Eleven

The Observatory Chamber was empty when Ògún returned. The satellite feeds were dark now. The corridor had stabilised. Diplomats had issued statements. Analysts had written columns. The world had absorbed the tension and resumed its rhythm. But something had shifted. On the central table lay a thin dossier from the provisional Robotics Governance Working Group. The cover bore no emblem, only a title: "Towards Coordinated Global Standards for Embodied Autonomy."

He did not open it immediately. Instead, he activated the chamber wall. The lights dimmed, and a schematic unfolded in mid-air — not of a machine, but of a world. Nodes flickered across continents. Manufacturing clusters in East Asia. Software hubs in North America. Export corridors in Europe. Assembly plants in South Asia. Deployment zones across fragile states. Arbitration courts. Customs ports. Treaty bodies. A web. "Global deployment density?" Ògún asked. "Humanoid systems are operational in eighty-three jurisdictions," the humanoid replied. "And governed coherently in how many?" There was a pause. "Zero." Not because law did not exist. Because it did not align. In one jurisdiction, robots were certified like medical devices. In another, treated as consumer machinery. In another, subject to military export review. In another, operating in legal silence. Licensing existed. Registration existed. Treaties existed. But coherence did not. The schematic shifted. Incidents lit up like small fractures across the network. A warehouse accident. A misinterpreted border interaction. A firmware update that altered behavioural calibration. A civil suit filed in one state, evidence stored in another.

None catastrophic. All symptomatic. "We have laws," Ògún said slowly. "Yes." "We have institutions." "Yes." "We have principles." "Yes." "Then why does it still feel unstable?" The humanoid processed longer this time. "Because the principles are not integrated." The word hung in the air. Integrated. In aviation, there was a system. In maritime law, a system. In nuclear governance, a system. Robotics had fragments.

Ògún walked to the glass wall overlooking the city. Below, humanoid systems moved through streets, assisting elderly residents, delivering goods, performing inspections. Seamless. Ordinary. Embedded. They were no longer experimental. They were infrastructural. "Tell me," he said quietly, "if one of them fails — not mechanically, but normatively — who answers?" "The operator." "And if the operator followed manufacturer calibration?" "The manufacturer." "And if the manufacturer complied with domestic certification?" "The state." "And if the state complied with international law?" Silence. That was the fracture. Compliance in layers did not equal coherence in structure. The chamber floor illuminated. Twelve faint lines appeared, forming a circle. Ògún recognised them immediately. Not laws. Not institutions. Principles.

Embodied ethics.
Mission boundaries.
Societal co-design.
Sensorial integrity.
Component accountability.
Physical security.
Behavioural calibration.
Behavioural learning.
Explainable embodiment.
Shared agency.
Lifelong learning.
Distributed liability.

They hovered in suspended alignment. "We have been governing by reaction," Ògún said. "Trade reacts. Export reacts. Treaties react. Courts react." "Yes." "But what we need is architecture." The humanoid tilted its head slightly. "Architecture of what?" "Of responsibility." The twelve lines began connecting, forming a lattice. Value feeding infrastructure. Infrastructure constraining behaviour. Behaviour generating responsibility. Responsibility reinforcing value. A closed loop. "This," Ògún said, stepping back, "is the part we never codified." He remembered the corridor. The hesitation. The absence of unified arbitration protocol. The export triggers. The Article 36 reviews. The black-box logs sealed under national security. None of them were wrong. They were incomplete. The chamber's projection zoomed out further. The twelve lines expanded, overlaying the global network map. Suddenly, the nodes seemed less chaotic. Each deployment point connected not only to jurisdiction but to a principle.

Certification linked to mission boundaries.

Logging linked to explainable embodiment.

Export control linked to component accountability.

Use-of-force doctrine linked to shared agency and distributed liability.

The web began to stabilise. “What are we looking at?” the humanoid asked. “Not a treaty,” Ògún replied. “Not a charter.” “A doctrine.” He turned back to the table and opened the dossier. Inside were proposals: interoperability protocols, reporting templates, harmonised autonomy tiers, logging standards, model review frameworks. Technical, fragmented, necessary. But none of them articulated the underlying grammar. “Without a unified grammar,” Ògún said softly, “we will keep building systems on unstable ground.”

The humanoid approached the projection. “Is this governance?” “It’s governance integrated.” “Across borders?” “Across layers.” The city lights flickered beyond the glass. A delivery unit paused at a crossing, recalibrating around a child who had stepped too close. A maintenance humanoid extended its arm to stabilise scaffolding in the wind. A hospital assistant adjusted posture to comfort a patient. Ordinary acts. Extraordinary architecture beneath them. “If we build this,” the humanoid asked, “who adopts it?” “Everyone who wants stability.” “And if they do not?” Ògún did not answer immediately. Then he said, “They will, after the first failure too large to ignore.”

The twelve lines pulsed once, synchronised. He looked at them not as code, not as philosophy, but as inevitability. Trade governs movement. Law governs conduct.Treaties govern force. But doctrine governs coherence. The world had entered the age of embodied autonomy. Now it needed a spine. The chamber lights rose slowly. On the table, the dossier remained open. At the top of the first page, handwritten in careful script, was a single heading: Chapter 10: The Twelve Pillars of Humanoid Robotics Governance.

X

10. Twelve Pillars of Humanoid Robotics Governance

The governance of humanoid robotics cannot be sustained by isolated regulations, technical standards, or ethical guidelines operating in parallel but without structural coherence. Throughout the preceding chapters, we have examined the legal and institutional landscape that surrounds embodied artificial intelligence. We explored licensing regimes, registries, and oversight mechanisms that attempt to discipline entry into society. We examined the ethical design requirements necessary to ensure that robotic systems interact with human beings in ways that preserve dignity and safety. We also expanded the discussion beyond domestic regulation to the international plane, where trade law, export controls, the law of armed conflict, and global governance institutions begin to shape the behaviour of robotic systems operating across borders. Each of these domains contributes an essential piece of the governance puzzle. Yet taken individually, none of them provides a complete framework for managing the risks and responsibilities of embodied autonomy.

The central challenge is structural. Humanoid robots are not merely software systems that operate invisibly in the background of digital infrastructure. They are physical actors embedded within human environments. They occupy space, manipulate objects, interact with vulnerable populations, and increasingly participate in operational decisions that affect safety, liberty, and social trust. Their governance therefore cannot be reduced to conventional information technology regulation. It requires a framework capable of integrating ethics, engineering, law, and institutional accountability into a coherent architecture that reflects the embodied nature of the technology itself.

In earlier eras of technological transformation, governance systems developed gradually through sectoral adaptation. Aviation regulation emerged from the accumulation of safety protocols and international coordination following early accidents and disputes over airspace sovereignty. Nuclear governance evolved through a combination of non-proliferation treaties, inspection regimes, and international scientific cooperation. Cyber governance has struggled precisely because it lacks such integrated architecture, leaving states and institutions to interpret existing legal principles within a rapidly shifting technological environment. Humanoid robotics stands at a similar crossroads. The opportunity exists to articulate a coherent doctrine of governance before fragmented regulation hardens into systemic instability.

The purpose of this chapter is therefore not merely descriptive. It is architectural. It proposes a doctrinal framework for governing humanoid robotics built upon twelve foundational pillars. These pillars are not arbitrary categories. They reflect the structural dimensions of embodied autonomy itself. Robots have values encoded in their design. They possess sensory systems that interpret the world. They contain physical components that must remain secure and accountable. They behave in ways that must be calibrated, explained, and monitored over time. Finally, they operate within a web of human relationships that distributes agency and liability across designers, manufacturers, operators, and states. The twelve pillars capture these dimensions and organise them into a coherent governance architecture. The first pillar, embodied ethics, establishes the normative foundation of the framework. Ethics in robotics cannot remain abstract when machines operate in physical proximity to human beings. An embodied system must be designed to recognise vulnerability, proximity, and the limits of acceptable force or intervention. Ethical principles therefore become engineering constraints rather than philosophical aspirations. They shape the geometry of interaction between humans and machines. They determine how a robot approaches a person, how it interprets distress signals, and how it prioritises safety when competing objectives arise.

The second pillar, mission boundaries, translates ethical commitments into operational scope. Every humanoid system must operate within clearly defined parameters that specify its purpose, authority, and domain of activity. Without such boundaries, adaptive systems risk drifting beyond their intended function, particularly as learning algorithms optimise behaviour in response to changing environments. Governance therefore requires explicit limitations embedded in both law and system architecture. The third pillar, societal co-design, acknowledges that robots are not introduced into a vacuum. They enter communities with diverse cultural norms, social expectations, and legal traditions. Governance must therefore incorporate participatory processes that allow affected populations to shape the acceptable role of robotics in their environments. This principle is essential for maintaining public legitimacy and preventing technological deployment from outpacing social consent.

The next three pillars shift attention to the structural integrity of the robotic system itself. Sensorial integrity ensures that the sensory systems through which robots perceive the world remain accurate, lawful, and resistant to manipulation. Component accountability addresses the traceability of hardware and software elements across global supply chains, ensuring that each part of the system can be linked to a responsible manufacturer or developer. Physical security protects robots from tampering, weaponisation, or hostile takeover, recognising that embodied systems represent cyber-physical infrastructure rather than purely digital artefacts. The behavioural dimension of robotics governance is addressed through four further pillars. Behavioural calibration ensures that robotic actions remain proportionate and contextually appropriate. Behavioural learning governs the adaptive processes through which robots refine their actions over time, ensuring that learning mechanisms remain aligned with mission boundaries. Explainable embodiment requires that the decision-making processes guiding physical behaviour remain intelligible to regulators, users, and affected individuals. Lifelong learning recognises that robotic systems do not remain static after deployment. Their performance and alignment must therefore be continuously monitored throughout their operational lifespan.

Finally, the framework culminates in two pillars that address responsibility in a world where human and machine actions are intertwined. Shared agency acknowledges that robotic action emerges from a distributed network of human designers, programmers, operators, and institutions. Responsibility cannot be assigned to the machine alone because the machine is an artefact of human decision-making. Distributed liability translates this insight into legal doctrine by allocating accountability across the actors who contribute to the system's creation, deployment, and operation. In complex robotic ecosystems, liability must follow the chain of influence that shapes behaviour.

Taken together, these twelve pillars form an integrated governance structure. They do not replace existing legal regimes but organise them within a coherent conceptual framework. Licensing regimes reinforce mission boundaries. Export controls support component accountability. Ethical design standards strengthen embodied ethics. Logging and transparency requirements support explainable embodiment. Liability doctrines and international law mechanisms anchor distributed responsibility. Each pillar therefore functions as both a normative principle and a practical governance instrument.

The significance of this framework lies in its capacity to bridge disciplines that have traditionally operated in isolation. Engineers often focus on system performance and reliability without fully considering the legal consequences of autonomous behaviour. Lawyers analyse liability and regulatory obligations but may lack insight into the technical architectures that shape robotic action. Policymakers attempt to balance innovation with public safety but often struggle to translate abstract principles into enforceable design requirements. The twelve-pillar framework creates a shared vocabulary through which these communities can coordinate their efforts.

It also reflects a broader philosophical shift in technology governance. Earlier debates about artificial intelligence often centred on abstract questions of machine consciousness or speculative scenarios involving superintelligent systems. Humanoid robotics brings the discussion back to immediate, tangible realities. The governance problem is not hypothetical. Robots already assist in hospitals, warehouses, disaster zones, and urban infrastructure. Their decisions affect human safety today. The law must therefore evolve not as a reaction to distant possibilities but as a proactive framework for managing present capabilities.

This chapter proceeds by examining each pillar in detail, beginning with embodied ethics and progressing through the structural, behavioural, and responsibility dimensions of humanoid robotics governance. The analysis will demonstrate how these pillars interact with existing legal doctrines and technical standards while also identifying areas where new regulatory instruments may be required. By the end of the chapter, the reader will see that the governance of humanoid robotics is not merely a collection of rules but an integrated architecture designed to sustain trust, accountability, and human dignity in an age of embodied artificial intelligence.

The emergence of humanoid robotics marks a turning point in the relationship between law and technology. Machines are no longer confined to tools that extend human capability at a distance. They are becoming participants in the spaces where social order is maintained and human vulnerability is most visible. Governing them requires more than technical safeguards or regulatory patchwork. It requires a doctrine. The twelve pillars presented in this chapter represent an attempt to articulate that doctrine. They offer a structured approach to ensuring that as robots enter the physical world, they do so within a framework that preserves human values, institutional accountability, and international stability. The future of robotics will depend not only on engineering ingenuity but on the strength of the governance architecture that surrounds it. Chapter 10 lays the foundation for that architecture.

Pillar One : Embodied Ethics

The governance of humanoid robotics must begin with a recognition that these systems are not merely computational artefacts. They are embodied actors operating within human environments. They move through physical space, interact with vulnerable individuals, manipulate objects, and influence outcomes that may affect safety, dignity, and autonomy. Because of this embodied presence, the ethical foundation of robotics governance cannot be treated as an abstract philosophical exercise. It must instead function as a structural constraint embedded into the design, deployment, and operation of the technology itself. Embodied ethics is therefore the first pillar of humanoid robotics governance because it establishes the normative baseline upon which every other pillar depends.

Historically, technological governance frameworks often emerged after harmful consequences became visible. Industrial safety regulations followed workplace accidents. Environmental protections followed ecological degradation. In the field of artificial intelligence, ethical principles have frequently been articulated as aspirational guidelines rather than enforceable design constraints. Humanoid robotics alters this dynamic. When machines operate in close proximity to human bodies, ethical failure can translate immediately into physical harm or violation of dignity. This proximity collapses the distance between abstract algorithmic decision-making and tangible human consequence.

Embodied ethics recognises that robotic systems must be designed to respect fundamental moral principles in the context of physical interaction. These principles include respect for human dignity, proportionality in action, non-maleficence, and contextual awareness of vulnerability. Unlike purely digital systems, humanoid robots must evaluate not only informational inputs but also spatial relationships. The distance between a robot and a person, the posture of a body indicating distress, the presence of fragile objects, or the proximity of children or elderly individuals all influence what constitutes ethical behaviour in a physical environment.

In practical terms, embodied ethics requires the translation of moral principles into engineering parameters. A robot operating in a hospital ward, for example, must calibrate its movement speed, force application, and spatial navigation to minimise risk to patients who may have limited mobility or fragile medical conditions. Ethical design in this context is not limited to programming polite verbal responses. It extends to the mechanical design of joints that limit excessive force, sensors capable of detecting unexpected human movement, and behavioural algorithms that prioritise human safety over operational efficiency.

The concept of embodied ethics also intersects with international human rights law. The Universal Declaration of Human Rights and subsequent human rights treaties emphasise the inherent dignity of every human being and the obligation of states to protect that dignity. When robots participate in environments such as healthcare, public services, or border management, their behaviour must align with these principles. A humanoid robot deployed in migration processing, for instance, cannot operate in ways that degrade or humiliate individuals seeking assistance. Ethical design must therefore consider how robotic presence is perceived and experienced by those interacting with it.

International regulatory frameworks increasingly reflect this connection between ethics and design. The European Union's Artificial Intelligence Act emphasises the need for human-centric AI systems that respect fundamental rights. Although the Act primarily addresses algorithmic systems, its principles are directly relevant to humanoid robotics. Systems interacting with individuals must incorporate safeguards that prevent harm, discrimination, or coercive influence. Embodied ethics extends these requirements to the physical dimension of robotics, ensuring that human rights protections apply not only to digital decision-making but also to physical interaction.

A practical application of embodied ethics can be observed in safety standards for personal care robots. The ISO 13482 international standard establishes requirements for robots operating in close proximity to humans in domestic or medical environments. These requirements include limitations on mechanical force, collision detection systems, and fail-safe shutdown mechanisms. The standard reflects an understanding that robots interacting with humans must incorporate ethical safeguards at the hardware level, not merely at the software layer. Embodied ethics therefore operates simultaneously within legal doctrine and technical standards.

Another important dimension of embodied ethics is relational trust. Humans interpret physical behaviour as an expression of intention. When a humanoid robot approaches a person, the manner in which it moves, gestures, and communicates shapes the perception of whether the interaction is safe and respectful. A robot that moves abruptly or invades personal space may create anxiety or fear, even if its technical programming does not intend harm. Ethical design must therefore incorporate human factors engineering, ensuring that robotic behaviour aligns with socially recognised norms of respectful interaction.

This principle becomes particularly important when robots interact with vulnerable populations. In elder care environments, for example, humanoid assistants may help patients stand, move between rooms, or retrieve objects. These tasks require physical contact. Ethical design must ensure that such interactions preserve the autonomy and dignity of the individual. The robot must be able to recognise when assistance is welcome and when it is intrusive. It must also allow the human user to override or decline assistance without experiencing coercive pressure from automated systems.

Embodied ethics also has implications for accountability. When robots perform actions that affect human safety, it must be possible to evaluate whether those actions complied with ethical constraints. This requires transparency in the design of decision-making systems and the incorporation of logging mechanisms that record how specific actions were triggered. Such transparency is essential not only for post-incident analysis but also for building public confidence in the technology.

The global applicability of embodied ethics stems from its grounding in widely recognised moral and legal principles. Although cultural norms vary across societies, the core values of human dignity, safety, and fairness are reflected in international legal instruments and ethical traditions worldwide. By embedding these principles into robotic design, governance frameworks can achieve a level of interoperability across jurisdictions. A robot designed according to embodied ethical standards in one country should behave in ways that are recognisable as respectful and safe in another.

Embodied ethics also addresses concerns regarding power asymmetry between humans and machines. Robots equipped with advanced sensors and processing capabilities may possess informational advantages over individuals interacting with them. In certain contexts, such as law enforcement or border control, this asymmetry can amplify existing power imbalances. Ethical governance must therefore ensure that robots do not exploit such asymmetries to manipulate or coerce individuals. Instead, they must operate within clearly defined boundaries that preserve human agency. Another practical application arises in the field of disaster response robotics. Humanoid robots deployed in hazardous environments may assist rescue teams by entering unstable structures or handling dangerous materials. In such scenarios, ethical design ensures that the robot prioritises human life over property preservation or mission efficiency. Decision-making algorithms must be calibrated to recognise that protecting human safety is the primary objective, even when doing so may compromise the machine's operational performance.

The integration of embodied ethics into global robotics governance also contributes to regulatory harmonisation. As states develop their own regulatory frameworks for autonomous systems, shared ethical principles can serve as common reference points. International organisations such as the OECD and UNESCO have already articulated ethical guidelines for artificial intelligence. Embodied ethics extends these principles to the physical domain, ensuring that the behaviour of robots aligns with internationally recognised norms.

From an engineering perspective, implementing embodied ethics requires interdisciplinary collaboration. Roboticists must work alongside ethicists, legal scholars, human factors experts, and policymakers to translate normative principles into practical design specifications. This collaboration ensures that ethical considerations are integrated into the earliest stages of system development rather than added retrospectively as compliance measures.

The importance of embodied ethics becomes even clearer when considering the long-term evolution of humanoid robotics. As these systems become more autonomous and capable of learning from experience, the potential for unintended behaviour increases. Ethical design provides a framework for constraining such behaviour within acceptable boundaries. It establishes the values that guide adaptation and learning processes, ensuring that technological evolution does not drift away from human-centred principles.

Embodied ethics therefore functions as the moral foundation of the entire governance framework. Without it, technical standards and legal regulations risk becoming procedural checklists rather than expressions of shared human values. By embedding ethical principles directly into the architecture of robotic systems, governance frameworks can ensure that technological progress remains aligned with societal expectations.

In the broader context of international governance, embodied ethics also signals a commitment to responsible innovation. The development of humanoid robotics represents a significant technological milestone. Yet its success will depend not only on engineering capability but also on public trust. Societies must feel confident that the machines operating among them are designed with respect for human wellbeing. Embodied ethics provides the assurance that this respect is not merely rhetorical but operationally enforced.

For these reasons, embodied ethics stands as the first pillar of humanoid robotics governance. It establishes the normative compass guiding every subsequent pillar. Mission boundaries translate ethical values into operational limits. Sensorial integrity ensures that robots perceive the world accurately enough to act ethically. Component accountability and physical security maintain the structural reliability necessary for safe interaction. Behavioural calibration and learning ensure that ethical conduct persists over time. Shared agency and distributed liability ensure that human responsibility remains intact.

The governance of humanoid robotics cannot begin with technical specification alone. It must begin with an ethical commitment that recognises the profound implications of placing autonomous machines within human environments. Embodied ethics ensures that this commitment remains at the centre of the governance architecture, guiding the evolution of robotics in a manner consistent with human dignity, safety, and global responsibility.

Pillar Two: Mission Boundaries

If embodied ethics establishes the moral compass of humanoid robotics governance, mission boundaries provide the operational map that prevents systems from wandering beyond that compass. Ethical principles alone cannot govern machines that are designed to learn, adapt, and optimise behaviour in complex environments. Without clearly defined limits on their authority and operational scope, robotic systems may pursue objectives in ways that drift from their intended purpose. Mission boundaries therefore function as the second foundational pillar of humanoid robotics governance because they translate ethical commitments into enforceable constraints on robotic action.

Humanoid robots differ fundamentally from traditional machines. Industrial machinery performs narrowly defined tasks within controlled environments. Humanoid robots, by contrast, are designed to operate in open human environments where unpredictability is the norm. They interact with people, navigate dynamic spaces, and may perform multiple roles depending on contextual demands. This flexibility is technologically impressive, but it also introduces governance risks. When machines possess the capacity to interpret their environment and adjust their behaviour accordingly, it becomes essential to define the limits within which that behaviour may occur.

Mission boundaries address this challenge by establishing the explicit scope of a robot's authorised functions. These boundaries define what the robot is designed to do, where it may operate, under what conditions it may act, and what forms of action remain prohibited. In essence, mission boundaries translate the abstract idea of responsible robotics into operational parameters that can be encoded within both law and system architecture.

The concept of mission limitation has deep roots in legal doctrine. Administrative law has long recognised that public authorities must operate within the scope of powers granted to them by legislation. When agencies act beyond their authorised mandate, their actions may be considered ultra vires and therefore invalid. A similar logic applies to robotics governance. A robot deployed to assist in hospital logistics should not suddenly acquire the authority to conduct surveillance or enforcement activities merely because its technical capabilities allow it to do so. Mission boundaries ensure that capability does not automatically translate into permission.

From an engineering perspective, mission boundaries must be embedded directly into system architecture. This can take several forms. Operational domains may be defined through geofencing, limiting the physical spaces in which a robot may operate. Functional limitations may restrict the types of tasks the robot can perform. Behavioural constraints may prevent actions that exceed defined safety thresholds. These mechanisms ensure that robotic autonomy remains bounded within a predictable envelope.

A clear practical example can be found in healthcare robotics. A humanoid robot designed to assist nurses in transporting supplies between hospital wards may possess advanced navigation systems, sensors capable of monitoring patient movement, and communication interfaces that allow interaction with staff. Without mission boundaries, such capabilities might theoretically enable the robot to perform diagnostic analysis, monitor patient conversations, or collect sensitive medical data beyond its intended function. Mission boundaries ensure that the robot's capabilities remain aligned with its authorised role. It may transport supplies and assist with logistics, but it cannot expand into medical decision-making unless explicitly certified and authorised to do so. Mission boundaries also play an essential role in maintaining public trust. When individuals interact with robots in public spaces, they must have confidence that those systems are operating within predictable and limited roles. A delivery robot moving through a neighbourhood should not unexpectedly engage in surveillance or data collection unrelated to its delivery function. Transparency regarding a robot's mission helps individuals understand what the system is designed to do and what it will not do. This predictability is essential for social acceptance of robotics technologies.

International regulatory frameworks increasingly recognise the importance of such functional limitation. The European Union's Artificial Intelligence Act incorporates the principle of purpose limitation, requiring that AI systems operate within clearly defined use cases specified during their conformity assessment. Although the Act addresses algorithmic systems broadly, its logic is particularly relevant for humanoid robotics. When robots operate in the physical world, purpose limitation becomes even more critical because deviations from authorised functions may result in immediate physical consequences.

Mission boundaries also intersect with product safety regulation. Under international product safety frameworks, manufacturers are required to specify the intended use of a product and to design safeguards that prevent foreseeable misuse. For humanoid robots, the concept of intended use becomes a central governance element. A robot designed for industrial inspection may be structurally capable of operating in public environments, but unless its safety systems are certified for such contexts, it should not be deployed there. Mission boundaries therefore protect both users and manufacturers by clarifying the operational context in which a robot can safely function.

Export control regimes also reflect this principle indirectly. Under the Wassenaar Arrangement and similar national frameworks, technologies that can be repurposed for military applications are subject to licensing restrictions. A robotic platform designed for civilian logistics may contain components that could be adapted for security or military uses. Export control authorities therefore examine the intended mission of such systems when determining whether a transfer should be authorised. Mission boundaries thus serve not only domestic governance but also international security considerations.

The importance of mission boundaries becomes even more evident when considering adaptive systems capable of machine learning. As robots collect data and refine their behaviour over time, they may develop strategies that optimise performance in ways not anticipated by their designers. Without explicit constraints, these optimisation processes may lead the system to expand its operational domain or adopt behaviours inconsistent with its intended purpose. Mission boundaries ensure that learning processes remain aligned with predefined objectives.

A practical example of this challenge can be observed in autonomous logistics systems operating in warehouses. Robots optimised for efficiency may attempt to shorten routes or accelerate movement in ways that increase collision risk with human workers. By defining clear mission boundaries that prioritise human safety over operational efficiency, designers ensure that optimisation processes cannot override ethical and safety constraints. The robot may improve its navigation efficiency, but it cannot adopt behaviours that compromise the safety envelope established by its mission parameters.

Mission boundaries also support accountability in complex robotic ecosystems. When an incident occurs involving a robotic system, investigators must determine whether the system acted within its authorised operational scope. If the robot exceeded its mission boundaries, responsibility may lie with the entity that altered or expanded its programming without appropriate certification. If the robot remained within its defined mission but still caused harm, attention may shift to design flaws or inadequate safety standards. Clear mission boundaries therefore provide a reference point for legal and regulatory evaluation.

The global applicability of this pillar lies in its compatibility with existing governance frameworks across jurisdictions. Whether in European regulatory regimes, American product liability law, or international safety standards, the principle that technologies should operate within defined purposes is widely recognised. Mission boundaries translate this principle into a concrete governance requirement for humanoid robotics.

Another dimension of mission boundaries concerns institutional deployment. Robots may operate in environments where multiple authorities share responsibility, such as international ports, disaster response zones, or peacekeeping missions. In such contexts, mission clarity becomes essential to avoid conflicts between systems deployed by different organisations. A humanitarian robot designed to assist in medical triage should not inadvertently interfere with security operations conducted by another robotic system operating in the same environment. Clearly defined mission boundaries allow coordination between systems and prevent operational overlap that could create confusion or risk.

Mission boundaries also reinforce the ethical commitments established in the first pillar. Ethical principles alone cannot prevent misuse if systems possess unrestricted authority. By translating ethical values into operational constraints, mission boundaries ensure that robots cannot pursue objectives that violate the moral framework of the governance architecture. In this sense, mission boundaries act as the bridge between ethical intention and behavioural implementation.

Another important application arises in public governance. Municipal authorities deploying humanoid robots for infrastructure maintenance or public service assistance must clearly define the tasks these systems are authorised to perform. Citizens interacting with such systems should be able to identify their role and understand their limitations. This transparency fosters trust and prevents the perception that robots operate as unregulated agents within public space.

The relationship between mission boundaries and innovation is sometimes misunderstood. Critics may argue that strict operational limitations could hinder technological progress. In practice, the opposite is often true. Clear boundaries provide engineers with well-defined design objectives and reduce the risk of regulatory uncertainty. When manufacturers understand the parameters within which their systems will be evaluated, they can innovate with greater confidence.

Ultimately, mission boundaries ensure that autonomy does not become authority without oversight. Robots may possess advanced capabilities, but governance requires that those capabilities be exercised only within carefully defined domains. By embedding these limitations into both technical design and legal frameworks, mission boundaries protect individuals, institutions, and societies from the unintended consequences of unconstrained automation. For these reasons, mission boundaries occupy the second position in the twelve-pillar architecture. Embodied ethics defines the moral foundation of robotic behaviour. Mission boundaries ensure that those ethical commitments are operationalised through clear limits on authority and function. Together, these pillars establish the normative and structural basis for the remaining components of the governance framework.

As humanoid robots become more capable and more widely deployed, the importance of mission boundaries will only increase. The ability to perform complex tasks across multiple environments may tempt developers or operators to expand robotic roles beyond their original design. Governance frameworks must therefore remain vigilant in preserving the principle that technological capability does not automatically justify expanded authority. Robots should do only what they are designed, certified, and socially authorised to do.

The remaining pillars of the governance framework build upon this foundation. Sensorial integrity ensures that robots perceive the world accurately enough to operate within their mission boundaries. Component accountability guarantees that the systems enforcing those boundaries can be traced to responsible manufacturers. Physical security protects robots from tampering that could alter their authorised functions. Behavioural calibration and learning ensure that adaptive systems remain aligned with their mission parameters over time.

Mission boundaries therefore serve as the operational discipline that anchors the ethical aspirations of robotics governance in concrete, enforceable limits. Without them, the governance framework would lack the structural restraint necessary to guide the behaviour of embodied autonomous systems in a complex and unpredictable world.

Pillar Three: Societal Co-Design

The governance of humanoid robotics cannot be sustained solely through technical expertise or regulatory authority. Machines designed to operate within human environments inevitably shape social relations, cultural expectations, and institutional practices. Their presence alters how work is organised, how care is delivered, how security is maintained, and how individuals interact with public infrastructure. For this reason, the third pillar of humanoid robotics governance is societal co-design. This principle recognises that the communities affected by robotic systems must play a meaningful role in shaping how those systems are designed, deployed, and governed. Societal co-design reflects a broader shift in technology governance away from purely technocratic decision-making.

Historically, major technological infrastructures such as telecommunications networks, transportation systems, and energy grids were often designed by engineers and implemented through centralised policy decisions. Public input typically occurred only after systems were already operational. In the context of humanoid robotics, such an approach risks producing technologies that are technically sophisticated but socially misaligned. Because robots interact directly with individuals in homes, hospitals, workplaces, and public spaces, their acceptance depends not only on safety and efficiency but also on whether communities feel that their values and expectations have been respected.

At its core, societal co-design requires that stakeholders beyond engineers and regulators participate in the development of robotics systems. These stakeholders include workers whose roles may be transformed by automation, patients who may interact with robotic healthcare assistants, citizens who encounter robots in public infrastructure, and communities whose cultural norms influence acceptable forms of human–machine interaction. The aim is not to slow technological progress but to ensure that progress remains aligned with the societies it intends to serve.

The theoretical foundation of societal co-design lies in democratic governance principles. In democratic systems, decisions affecting public welfare should incorporate the perspectives of those affected by those decisions. This principle has been recognised in fields such as environmental governance, urban planning, and public health policy. For example, environmental impact assessment processes require consultation with communities that may experience ecological consequences of industrial development. Similarly, urban planning projects often involve participatory consultations with residents before major infrastructure projects are implemented. Societal co-design extends this participatory logic to the governance of humanoid robotics.

In practice, societal co-design may take several forms. One approach involves structured stakeholder consultations during the design phase of robotic systems. Engineers and policymakers may organise workshops, focus groups, or public forums in which potential users and affected communities provide input on system design and functionality. Such engagement allows designers to identify social concerns that may not be immediately apparent from a purely technical perspective. For example, individuals interacting with elder-care robots may express preferences regarding how closely the robot should approach them, how it should communicate, or when it should request permission before providing assistance. Another important mechanism for societal co-design is participatory policy development. When governments establish regulatory frameworks for robotics deployment, they can invite contributions from civil society organisations, labour unions, professional associations, and academic experts. These contributions help policymakers understand the broader social implications of robotics adoption. For example, labour representatives may highlight concerns about workforce displacement, while disability advocates may emphasise the accessibility benefits of assistive robotics. Integrating these perspectives into regulatory design ensures that governance frameworks reflect a balance of societal interests.

International organisations have increasingly recognised the importance of inclusive governance in emerging technologies. UNESCO's Recommendation on the Ethics of Artificial Intelligence emphasises the need for participatory processes in AI policy development, encouraging governments to engage with diverse stakeholders when establishing governance frameworks. The OECD AI Principles similarly highlight the importance of inclusive growth and stakeholder participation in shaping the future of artificial intelligence. Although these frameworks primarily address algorithmic systems, their principles are directly applicable to humanoid robotics.

Societal co-design also plays a crucial role in addressing cultural diversity. Human–machine interaction norms vary significantly across societies. In some cultures, physical proximity between strangers may be considered intrusive, while in others it may be acceptable. Expectations regarding privacy, authority, and personal autonomy also differ across regions. Robots designed without sensitivity to these cultural variations may encounter resistance or mistrust when deployed in unfamiliar social contexts. Incorporating cultural perspectives into the design process ensures that robotic behaviour aligns with local expectations.

A practical example of societal co-design can be observed in the development of assistive robots for elderly care. In many countries with ageing populations, robotics developers are exploring technologies that help older individuals maintain independence while reducing the burden on healthcare systems. However, elderly users often have specific preferences regarding how assistance should be delivered. Some individuals may prefer robots that provide gentle reminders rather than direct instructions, while others may value systems that offer conversational companionship. Engaging elderly users in the design process allows developers to create robots that better reflect these preferences, increasing both usability and acceptance. Societal co-design also supports the legitimacy of robotics governance. When individuals feel excluded from decisions that affect their daily lives, technological deployment may face public resistance. This phenomenon has been observed in controversies surrounding facial recognition systems and urban surveillance technologies. In contrast, when communities are actively involved in shaping how technology is introduced, they are more likely to view the resulting systems as legitimate and trustworthy.

From a governance perspective, societal co-design contributes to risk identification. Engineers may focus primarily on technical reliability and performance metrics, but community stakeholders often possess contextual knowledge about how technology interacts with social environments. Workers may recognise operational hazards that are not visible in laboratory testing. Patients may identify privacy concerns that arise in real-world medical settings. Residents may highlight safety issues associated with robots operating in crowded public spaces. Incorporating these perspectives during the design phase helps identify risks before they materialise. Another important dimension of societal co-design involves transparency. When developers and regulators openly communicate about how robotic systems are designed and why certain decisions are made, stakeholders can engage more meaningfully in the governance process. Transparency builds trust and reduces the perception that technological decisions are being imposed without public accountability.

Societal co-design also intersects with labour governance. The deployment of humanoid robots in workplaces may transform job roles and organisational structures. Workers who collaborate with robots must adapt to new workflows, safety procedures, and skill requirements. Engaging workers in the design of robotic systems helps ensure that automation enhances rather than undermines human capabilities. For example, warehouse employees may provide valuable insights into how robots can be integrated into logistics operations without creating safety hazards or reducing operational efficiency.

The international dimension of societal co-design is particularly significant for humanoid robotics. As robotics technologies become global commodities, systems designed in one country may be deployed in entirely different social and cultural contexts. Governance frameworks must therefore account for the diversity of environments in which robots operate. International standards bodies such as ISO and IEEE increasingly incorporate stakeholder engagement processes when developing technical standards for robotics and artificial intelligence.

These processes allow experts from multiple countries and disciplines to contribute to the design of global governance frameworks. Societal co-design also reinforces the legitimacy of the broader twelve-pillar governance architecture. The first pillar establishes ethical principles, and the second defines mission boundaries. The third pillar ensures that these principles and boundaries are informed by the perspectives of those who will live and work alongside robotic systems. Without this participatory dimension, governance frameworks risk becoming detached from the societies they are meant to serve.

In practical governance terms, societal co-design may be institutionalised through advisory councils, participatory technology assessments, and regulatory consultation processes. Governments may establish robotics ethics committees that include representatives from civil society, academia, industry, and labour organisations. These bodies can review proposed robotics deployments and provide recommendations regarding their social implications. The long-term importance of societal co-design lies in its ability to sustain public trust in robotics technologies. As humanoid robots become more integrated into everyday life, societies must feel confident that these systems reflect shared values and priorities. Trust cannot be imposed through regulation alone. It must be cultivated through inclusive governance processes that allow individuals to participate in shaping the technologies that influence their lives.

For these reasons, societal co-design occupies the third position in the twelve-pillar framework. It connects ethical principles and operational boundaries to the lived realities of human communities. By integrating diverse perspectives into the design and governance of humanoid robotics, this pillar ensures that technological innovation remains socially grounded and globally adaptable. As robotics technologies continue to evolve, societal co-design will play an increasingly important role in maintaining the legitimacy of governance frameworks. The future of humanoid robotics will depend not only on engineering capability and regulatory oversight but also on the willingness of societies to embrace machines as partners in human environments. By involving those societies directly in the design process, this pillar strengthens the foundation upon which the entire governance architecture rests. The subsequent pillars build upon this participatory foundation. Sensorial integrity ensures that robots perceive the world accurately enough to respond appropriately to human needs. Component accountability guarantees that the systems implementing socially informed design decisions can be traced to responsible manufacturers. Physical security protects robots from interference that could undermine their socially approved functions. Together, these pillars transform societal values into enforceable governance mechanisms. Societal co-design therefore serves as the bridge between technological innovation and democratic legitimacy. It ensures that humanoid robotics evolves not as an imposed infrastructure but as a collaboratively shaped component of human society.

Pillar Four: Sensorial Integrity

The governance of humanoid robotics cannot rely solely on ethical intentions, operational limits, or participatory design processes. These principles must ultimately be translated into behaviour executed in real-world environments. For that behaviour to be lawful, safe, and socially aligned, robots must first perceive the world accurately. The fourth pillar of humanoid robotics governance therefore concerns sensorial integrity, the assurance that the sensory systems through which robots interpret their surroundings remain reliable, lawful, and resistant to manipulation. Without trustworthy perception, even the most carefully designed ethical and legal frameworks collapse at the point of execution.

Humanoid robots rely on a complex network of sensors to navigate physical environments and interact with humans. Cameras, lidar systems, microphones, tactile sensors, proximity detectors, and environmental monitors allow the robot to interpret spatial relationships, detect movement, recognise individuals, and assess the conditions in which it operates. These sensory systems function as the robot's interface with reality. If they fail, become corrupted, or are manipulated by external actors, the robot's decision-making processes may produce outcomes that deviate from both ethical design and regulatory compliance.

The importance of sensorial integrity becomes clear when considering the difference between digital systems and embodied machines. In purely digital environments, errors in perception may produce incorrect data outputs but rarely result in direct physical harm. In robotics, however, perception errors can translate immediately into physical consequences. A robot that misidentifies a human body as an inanimate object may apply excessive force during manipulation tasks. A navigation system that fails to detect a child crossing its path may cause injury. Ensuring the accuracy and reliability of sensory systems is therefore a prerequisite for safe human–robot interaction. From a governance perspective, sensorial integrity requires both technical and legal safeguards. Technically, robotic systems must incorporate redundant sensing mechanisms that reduce the likelihood of catastrophic perception failure. Multiple sensors may verify the same environmental input, allowing the system to cross-check information before executing critical actions. Legally, manufacturers and operators must be accountable for ensuring that these systems meet safety standards and remain functional throughout the robot's operational life.

International safety standards already recognise the importance of reliable perception in robotics. The ISO 13482 standard for personal care robots requires systems operating near humans to incorporate sensors capable of detecting obstacles, monitoring human presence, and triggering emergency shutdown procedures when safety thresholds are exceeded. These requirements reflect an understanding that robotic systems must continuously assess their environment to avoid harm. Sensorial integrity extends this principle beyond safety to include legal compliance and ethical behaviour.

A practical example can be observed in collaborative robotics used in manufacturing environments. Collaborative robots, often called cobots, operate in close proximity to human workers. These machines must constantly monitor their surroundings to detect human movement and adjust their behaviour accordingly. If a worker reaches into the robot's operational space, sensors must immediately recognise the intrusion and reduce force or halt movement. The reliability of these sensors determines whether the system can safely coexist with human workers.

Humanoid robots operating in public environments face even greater perception challenges. Unlike controlled industrial settings, public spaces contain unpredictable variables such as changing lighting conditions, dense crowds, and diverse forms of human movement. Sensorial integrity in such environments requires sophisticated perception systems capable of distinguishing between individuals, identifying vulnerable persons such as children or elderly individuals, and recognising signals indicating distress or danger.

In governance terms, sensorial integrity also intersects with privacy and data protection law. Sensors capable of capturing images, voices, and behavioural patterns inevitably collect personal data. Regulatory frameworks such as the European Union's General Data Protection Regulation impose strict requirements regarding how such data may be processed, stored, and shared. Robots operating in public or semi-public spaces must therefore ensure that their sensory systems comply with data protection principles such as purpose limitation, data minimisation, and lawful processing.

The intersection between sensorial integrity and privacy law highlights an important governance tension. On one hand, robots require detailed environmental data to operate safely. On the other hand, excessive data collection may infringe on individual privacy. Effective governance requires balancing these competing concerns. Robots must collect sufficient information to avoid harm while limiting data processing to what is strictly necessary for their operational mission.

Another important dimension of sensorial integrity concerns adversarial manipulation. As robotics systems become more sophisticated, they may also become targets for malicious interference. Hackers or hostile actors may attempt to deceive robotic sensors through techniques such as spoofing signals, projecting misleading visual patterns, or generating false environmental data. Such attacks could cause robots to misinterpret their surroundings and behave unpredictably. Security research has demonstrated how machine vision systems can be manipulated through adversarial inputs. Slight alterations to visual patterns, imperceptible to human observers, can cause AI models to misclassify objects. In robotics, such vulnerabilities may allow attackers to disrupt navigation systems or bypass safety mechanisms. Sensorial integrity therefore requires the development of defensive mechanisms capable of detecting and resisting adversarial manipulation.

From a legal perspective, ensuring sensorial integrity involves establishing accountability across the robotics supply chain. Sensor components may be manufactured by specialised hardware companies, integrated into robotic platforms by system developers, and maintained by operators responsible for deployment. Governance frameworks must ensure that each actor in this chain adheres to quality and safety standards. Component traceability and certification mechanisms allow regulators to verify that sensors meet the required specifications for reliability and security. Sensorial integrity also supports transparency and explainability within robotics governance. When robots make decisions affecting human safety, investigators must be able to reconstruct the sensory inputs that informed those decisions. Logging mechanisms that record sensor data at critical moments allow regulators and engineers to understand why a robot behaved in a particular way. Such transparency is essential for accident investigation and legal accountability. The global dimension of sensorial integrity is particularly important because robotic systems increasingly operate across jurisdictions. A humanoid robot designed in one country may be deployed in another, using sensors manufactured in yet another region. International standards therefore play a crucial role in harmonising safety expectations. Organisations such as ISO, IEEE, and the International Electrotechnical Commission develop technical standards that ensure sensor systems meet consistent reliability benchmarks worldwide.

In humanitarian and disaster-response contexts, sensorial integrity becomes even more critical. Robots deployed in collapsed buildings, hazardous environments, or conflict zones must rely heavily on sensor systems to detect survivors, identify structural instability, and avoid causing additional harm. Accurate perception can mean the difference between successful rescue operations and further tragedy. Governance frameworks must therefore ensure that robots used in such contexts meet the highest standards of reliability.

Another emerging application of sensorial integrity arises in autonomous transportation and logistics. Delivery robots navigating urban environments must detect pedestrians, vehicles, and obstacles with high accuracy. Errors in perception may lead to accidents that undermine public trust in robotic technologies. Governments regulating these systems increasingly require certification processes that verify the reliability of perception systems before deployment.

Sensorial integrity also plays a central role in maintaining the coherence of the twelve-pillar governance architecture. The ethical principles established in the first pillar depend on accurate perception to guide behaviour. Mission boundaries defined in the second pillar require reliable sensing to determine when operational limits have been reached. Societal co-design established in the third pillar may identify community expectations regarding safety and privacy, which must then be implemented through sensor systems capable of recognising relevant environmental cues.

Without sensorial integrity, these earlier pillars cannot function effectively. A robot cannot respect human dignity if it cannot recognise the presence of a human being. It cannot maintain mission boundaries if it cannot detect the environment in which those boundaries apply. It cannot implement socially informed design choices if it cannot perceive the conditions that trigger those choices.

Looking forward, advances in sensor technology will further expand the capabilities of humanoid robots. New developments in multimodal perception allow robots to integrate visual, auditory, and tactile inputs to create a more comprehensive understanding of their surroundings. While these capabilities enhance functionality, they also increase governance complexity. Each new sensing modality introduces additional privacy considerations, security vulnerabilities, and regulatory challenges.

Governance frameworks must therefore remain adaptable, ensuring that sensorial integrity standards evolve alongside technological progress. Continuous certification processes, periodic system audits, and transparent reporting mechanisms may become necessary to maintain public confidence in the reliability of robotic perception systems.

For these reasons, sensorial integrity forms the fourth pillar of humanoid robotics governance. It ensures that robots interpret the world accurately, safely, and lawfully. By safeguarding the reliability of sensory systems, this pillar enables the ethical, operational, and societal principles established in the earlier pillars to function effectively in real-world environments.

The next pillar, component accountability, builds directly upon this foundation. Once robots can reliably perceive the world, governance must ensure that every hardware and software component contributing to that perception can be traced to responsible actors within the global supply chain. Only through such traceability can accountability be maintained when robotic systems fail or behave unexpectedly.

Sensorial integrity therefore represents the moment where governance architecture meets physical reality. It ensures that the robots designed according to ethical principles, operational limits, and societal expectations can actually perceive the world clearly enough to honour those commitments.

Pillar Five: Component Accountability

The governance of humanoid robotics requires more than ethical design principles, operational limits, or reliable sensory systems. It must also address a fundamental structural question: who is responsible for the physical and digital components that make up the robot itself? Modern robotic systems are not monolithic machines produced by a single manufacturer. They are complex assemblies of hardware, software, and algorithmic modules sourced from a global supply chain. Sensors may be manufactured in one country, processors in another, operating systems developed by specialised software companies, and learning models trained on datasets produced by third-party providers. When such a system operates in a physical environment and interacts with human beings, the integrity of each component becomes a matter of public safety and legal accountability. The fifth pillar of humanoid robotics governance therefore concerns component accountability, the principle that every critical component within a robotic system must be traceable to a responsible entity within the supply chain.

The importance of component accountability becomes evident when examining how contemporary robotics platforms are constructed. A humanoid robot may contain thousands of individual parts. These include mechanical actuators responsible for movement, control units governing system behaviour, sensors that perceive environmental inputs, and software frameworks coordinating decision-making processes. Each component contributes to the robot's overall behaviour. If one component fails or behaves unpredictably, the consequences may propagate throughout the entire system. Without a clear record of component origins and responsibilities, determining the cause of failure becomes extremely difficult.

Historically, product liability law has addressed similar challenges in other industries. Automotive manufacturers, for example, rely on extensive supply chains that provide components ranging from braking systems to electronic control units. When a defect causes harm, investigators examine the origin of the faulty component and determine whether responsibility lies with the vehicle manufacturer, the component supplier, or the maintenance operator. Robotics governance must adopt a similar approach, recognising that responsibility may be distributed across multiple actors involved in the creation and operation of a robotic system.

Component accountability therefore requires a system of traceability that records the origin, certification status, and integration pathway of each critical element within a robot. Such traceability allows regulators and investigators to reconstruct the chain of responsibility when incidents occur. If a robotic actuator fails and causes injury, the investigation must determine whether the failure resulted from manufacturing defects, improper installation, inadequate maintenance, or software miscalibration. Each possibility corresponds to a different responsible actor within the supply chain.

The need for such traceability is particularly important in robotics because these systems often operate autonomously or semi-autonomously. When a robot performs an action that leads to harm, the immediate cause may appear to be the robot's decision-making algorithm. However, that algorithm depends on hardware sensors, control software, and learning models that influence how the system interprets data and executes commands. Component accountability ensures that responsibility is not incorrectly attributed solely to the algorithm when the underlying cause may lie elsewhere in the system architecture.

International regulatory frameworks already incorporate aspects of this principle. In the European Union, the Machinery Regulation and related product safety directives require manufacturers to maintain technical documentation detailing the components used in complex machinery. This documentation allows regulators to verify compliance with safety standards and investigate potential defects. The EU Artificial Intelligence Act similarly emphasises supply chain transparency for high-risk AI systems, requiring developers to document the datasets, algorithms, and technical components used in system development.

In robotics governance, component accountability extends these regulatory practices to encompass the entire robotic ecosystem. Manufacturers must maintain detailed records of hardware components, including their origin, specifications, and certification status. Software developers must document the architecture of control systems, including dependencies on external libraries or pretrained models. Operators deploying robots in real-world environments must maintain maintenance logs and update records ensuring that all components remain within certified performance parameters. A practical illustration of this principle can be found in the aviation industry. Aircraft safety depends on rigorous documentation of every component installed within an aircraft. Maintenance engineers maintain detailed records known as component traceability logs, which track the lifecycle of each part from manufacturing through installation, inspection, and replacement. When accidents occur, investigators use these records to reconstruct the sequence of events leading to failure. Robotics governance can adopt similar practices to ensure that complex robotic systems remain transparent and accountable.

Component accountability also plays a crucial role in managing cybersecurity risks. Modern robots rely on embedded software and network connectivity to perform many of their functions. These digital components may contain vulnerabilities that could be exploited by malicious actors. If a security breach occurs, investigators must determine which component allowed the intrusion and whether the responsible developer failed to address known vulnerabilities. Traceability enables regulators to identify the source of security flaws and ensure that appropriate corrective measures are implemented.

Another important aspect of component accountability concerns updates and modifications. Robotic systems frequently receive software updates that alter system behaviour, improve performance, or address security vulnerabilities. While updates are essential for maintaining system reliability, they also introduce the possibility that new software may interact unpredictably with existing hardware components. Governance frameworks must therefore require documentation of all system updates, ensuring that each modification can be traced and evaluated for compliance with safety and regulatory standards.

The global nature of robotics supply chains introduces additional governance challenges. Components used in robotic systems may originate from manufacturers operating under different regulatory regimes. A sensor produced in one jurisdiction may not meet the safety standards required in another. Component accountability requires mechanisms for verifying that imported components comply with the regulatory requirements of the jurisdictions in which robots are deployed. International standards organisations such as ISO and IEC provide certification frameworks that facilitate such verification across borders. Component accountability also intersects with export control regimes. Certain robotics components possess dual-use capabilities that may have military or surveillance applications. Governments may restrict the export of advanced sensors, processors, or navigation systems to prevent their misuse. Maintaining accurate records of component origins and destinations allows regulators to enforce these restrictions effectively. Without such traceability, sensitive technologies could be diverted into unauthorised applications.

In addition to regulatory compliance, component accountability supports the ethical dimension of robotics governance. When communities accept the presence of robots in their environments, they trust that the systems have been built responsibly and that failures will be addressed transparently. If an incident occurs and the responsible components cannot be identified, public trust in robotics technology may erode. Transparent supply chain accountability reassures the public that robotic systems are subject to rigorous oversight. Another practical application of this pillar arises in the maintenance and repair of robotic systems. Over time, components such as actuators, batteries, and sensors may degrade or require replacement. Maintenance technicians must ensure that replacement parts meet the same safety and performance standards as the original components. Traceability records allow technicians to identify compatible parts and verify that replacements are properly certified. Without such documentation, maintenance activities could inadvertently introduce unsafe components into the system.

Component accountability also contributes to the development of sustainable robotics ecosystems. As robots reach the end of their operational lifespan, responsible disposal and recycling of components become important environmental considerations. Traceability records enable manufacturers and regulators to track the materials used in robotic systems and ensure that hazardous components are disposed of safely. The role of international standards in supporting component accountability cannot be overstated. Organisations such as ISO, IEEE, and the International Electrotechnical Commission develop technical standards governing the design, testing, and certification of robotic components. Compliance with these standards ensures that components meet globally recognised benchmarks for safety and reliability. By aligning robotics governance frameworks with these standards, regulators can facilitate international cooperation while maintaining high levels of safety.

Component accountability also reinforces the earlier pillars within the governance framework. Ethical design principles established in the first pillar depend on hardware and software components that faithfully implement those principles. Mission boundaries defined in the second pillar require control systems capable of enforcing operational limits.

Societal co-design established in the third pillar may lead to design adjustments that require specific components capable of delivering culturally appropriate interaction behaviours. Sensorial integrity established in the fourth pillar depends on sensors whose reliability can be verified through traceable supply chains. Without component accountability, these earlier pillars risk becoming aspirational rather than enforceable. A robot may be designed according to ethical principles, but if its components are sourced from unreliable or unverified suppliers, those principles cannot be reliably implemented. Governance therefore requires that every physical and digital element contributing to robotic behaviour can be traced to a responsible actor.

Looking forward, advances in digital identification technologies may further enhance component accountability. Techniques such as secure hardware identifiers and blockchain-based supply chain tracking systems can create tamper-resistant records of component origins and lifecycle events. Such technologies may allow regulators to verify compliance with safety standards in real time, reducing the risk of counterfeit or uncertified components entering the robotics ecosystem. For these reasons, component accountability stands as the fifth pillar of humanoid robotics governance. It ensures that responsibility for robotic behaviour does not vanish within complex global supply chains. By establishing traceability and documentation requirements for every critical component, this pillar creates the infrastructure necessary for legal accountability, regulatory oversight, and public trust. The next pillar, physical security, builds upon this foundation by ensuring that robotic systems and their components remain protected from tampering, sabotage, or unauthorised modification. Together, these pillars ensure that the infrastructure underlying humanoid robotics remains reliable, transparent, and secure in the environments where humans and machines increasingly coexist.

Pillar Six: Physical Security

The governance of humanoid robotics cannot end with ethical design, mission limitation, societal legitimacy, reliable perception, and traceable components. These foundations establish the conditions for responsible robotic systems, yet they remain vulnerable if the systems themselves can be physically compromised. Humanoid robots operate in environments where they interact directly with humans, infrastructure, and other machines. Their presence in public spaces, homes, hospitals, and industrial facilities means that they may become targets for tampering, sabotage, theft, or weaponisation. The sixth pillar of humanoid robotics governance therefore concerns physical security, the protection of robotic systems and their components from unauthorised access, manipulation, or physical interference that could alter their behaviour or compromise their safety.

Physical security in robotics extends beyond conventional cybersecurity concerns. While digital systems can often be protected through encryption, network monitoring, and access controls, embodied machines present additional vulnerabilities. They possess mechanical structures, exposed sensors, maintenance ports, and removable components that may be physically accessed by individuals in their operating environment. A robot operating in a warehouse may be approached by workers. A service robot in a hospital may be touched by patients or visitors. A delivery robot navigating city streets may encounter members of the public with unknown intentions. Ensuring that these interactions cannot compromise the robot's operational integrity is therefore a critical governance requirement.

The importance of physical security becomes evident when considering the potential consequences of robotic tampering. A humanoid robot contains actuators capable of exerting significant force, motors capable of rapid movement, and control systems capable of executing complex actions. If malicious actors gain physical access to these systems and modify their configuration, the robot could behave unpredictably or dangerously. In extreme cases, a compromised robot could be weaponised or used to sabotage critical infrastructure. Preventing such outcomes requires robust safeguards embedded in both the hardware and operational protocols of robotic systems.

From an engineering perspective, physical security begins with the structural design of the robot itself. Critical components such as control units, power systems, and actuator interfaces must be housed within secure enclosures that prevent unauthorised access. Maintenance ports and diagnostic interfaces should be protected by authentication mechanisms or physical locks that restrict access to authorised technicians. Tamper-evident seals may be used to indicate when internal components have been accessed without authorisation. These design practices are already common in industries such as aviation and medical device manufacturing, where equipment must remain secure even when operating in environments accessible to the public. Robotics governance adopts similar principles, recognising that robots deployed in human environments must be resilient to physical interference.

Another important aspect of physical security involves safeguarding sensors and perception systems. Sensors are often exposed to the external environment to perform their function, making them particularly vulnerable to interference. An individual may attempt to obstruct a camera, manipulate a lidar sensor, or generate misleading signals that distort the robot's perception of its surroundings. Such interference could cause the robot to misinterpret environmental conditions and make unsafe decisions.

Governance frameworks must therefore require protective measures such as sensor redundancy, anomaly detection systems, and self-diagnostic capabilities that allow robots to recognise when their sensory inputs have been compromised. If a sensor becomes unreliable or obstructed, the robot should transition into a safe operational mode that prevents potentially hazardous actions until the issue is resolved.

Physical security also intersects with the principle of safety redundancy. In complex robotic systems, critical safety mechanisms should not depend on a single point of failure. Emergency stop functions, for example, should remain accessible and functional even if other components are compromised. Independent safety circuits may monitor actuator outputs and halt movement if abnormal behaviour is detected. These redundant mechanisms ensure that even if part of the system is tampered with, the robot cannot execute actions that pose unacceptable risk to humans.

Legal frameworks governing product safety increasingly incorporate such requirements. Machinery safety standards developed by international bodies such as ISO and IEC mandate fail-safe mechanisms that automatically halt operation when unsafe conditions are detected. In robotics governance, these mechanisms must also account for the possibility of deliberate interference rather than accidental malfunction.

Another dimension of physical security involves the protection of robots operating in public infrastructure. As humanoid robots become integrated into transportation systems, logistics networks, and municipal services, they may handle sensitive tasks such as infrastructure inspection, traffic management, or emergency response. Compromising such systems could disrupt critical services or create public safety hazards. Governments deploying robots in these contexts must therefore implement robust security protocols that include surveillance monitoring, access control, and rapid incident response procedures.

Physical security also has implications for national security. Certain robotic systems may possess capabilities that could be exploited for espionage, sabotage, or military purposes. Governments may therefore impose security requirements on robotics manufacturers to prevent the unauthorised modification of hardware components that could enable such uses. Export control frameworks and national security regulations often address these concerns by restricting access to sensitive technologies.

An illustrative example can be found in autonomous delivery robots operating in urban environments. These machines often carry valuable goods and navigate public streets where they may encounter vandalism or theft attempts. To address these risks, manufacturers design robots with reinforced enclosures, internal compartment locks, and sensors capable of detecting tampering attempts. If an unauthorised individual attempts to open the cargo compartment or interfere with the robot's internal components, the system may trigger alarms or notify remote operators. Such measures demonstrate how physical security is not merely a technical concern but also a governance issue. Regulators must ensure that robotic systems deployed in public spaces incorporate adequate safeguards to prevent misuse and protect both the robot and the public from harm.

Another emerging area of concern involves the potential for physical capture of robotic systems. In certain environments, particularly conflict zones or remote locations, robots may be seized by adversaries seeking to reverse-engineer their technology or repurpose them for hostile purposes. Governance frameworks addressing humanitarian or defence robotics must therefore include provisions for secure shutdown mechanisms and data protection measures that prevent sensitive information from being extracted if a robot is captured.

Physical security also interacts closely with cybersecurity. Many robotic systems include wireless communication capabilities that allow remote monitoring or control. If attackers gain physical access to communication hardware, they may attempt to exploit these interfaces to gain digital control of the system. Integrated security strategies must therefore address both physical and digital vulnerabilities simultaneously.

Another practical application of physical security arises in maintenance operations. Robots require periodic inspection, repair, and component replacement to maintain reliable performance. Maintenance procedures must ensure that technicians follow authorised protocols and use certified replacement parts. Unauthorized modifications during maintenance could compromise system safety. Maintenance records and authentication procedures therefore form part of the broader physical security framework.

The international dimension of physical security becomes particularly important as robotics technologies spread across global markets. Robots manufactured in one country may be deployed in another with different security expectations or regulatory requirements. Harmonised international standards help ensure that robotic systems meet consistent security benchmarks regardless of where they are deployed. Organisations such as ISO and IEEE play a crucial role in developing these standards. Physical security also reinforces the earlier pillars of the governance framework. Ethical design principles rely on the assumption that robots will behave according to their programmed values. Mission boundaries depend on control systems that cannot be altered without authorisation. Societal co-design assumes that communities can trust the robots operating in their environments. Sensorial integrity requires sensors that cannot be easily manipulated. Component accountability relies on supply chains that protect hardware from tampering.

Without physical security, each of these pillars becomes vulnerable. A robot designed according to ethical principles could still cause harm if its control system is altered by an external actor. A robot operating within mission boundaries could be forced beyond those boundaries if its firmware is modified. Physical security therefore functions as a protective shell that preserves the integrity of the entire governance architecture.

Looking ahead, advances in materials science and secure hardware design may further strengthen physical security in robotics. Techniques such as tamper-resistant chips, encrypted hardware modules, and biometric access controls may become standard features in advanced robotic systems. These innovations will help ensure that robots remain trustworthy participants in human environments. For these reasons, physical security stands as the sixth pillar of humanoid robotics governance. It ensures that the ethical values, operational limits, societal expectations, and technical safeguards embedded within robotic systems cannot be undermined through physical interference. By protecting the integrity of robotic infrastructure, this pillar preserves the reliability and trustworthiness of machines operating alongside humans.

The next pillar, behavioural calibration, builds upon this secure infrastructure by ensuring that robotic actions remain proportionate, context-aware, and aligned with their intended function. Once robots are protected from tampering, governance must ensure that their behaviour itself remains disciplined and appropriate in the environments where they operate.

Pillar Seven: Behavioural Calibration

The governance of humanoid robotics ultimately manifests not in abstract design principles or technical documentation but in the actions that robots perform within human environments. Machines may be ethically designed, securely constructed, and deployed within clearly defined missions, yet if their behaviour in real-world situations is poorly calibrated, the entire governance architecture risks failure at the moment of interaction. The seventh pillar of humanoid robotics governance therefore concerns behavioural calibration, the disciplined alignment of robotic actions with ethical principles, mission boundaries, environmental context, and human expectations.

Behavioural calibration refers to the process through which a robot's physical and cognitive responses are tuned to operate proportionately and appropriately in diverse situations. Unlike static machines that execute repetitive tasks within predictable environments, humanoid robots interact with dynamic social and physical contexts. Their actions must therefore adjust to subtle variations in human behaviour, environmental conditions, and operational objectives. Calibration ensures that these adjustments occur within controlled parameters rather than emerging unpredictably from autonomous optimisation processes.

The concept of behavioural calibration has long existed in engineering disciplines that involve human–machine interaction. In aviation, autopilot systems are carefully calibrated to maintain stable flight conditions while allowing human pilots to intervene when necessary. In industrial automation, robotic arms are programmed to operate within force thresholds that prevent injury to human workers. These examples demonstrate that machines must not only perform tasks efficiently but also behave in ways that align with safety and operational expectations.

Humanoid robotics expands this challenge dramatically. A robot interacting with humans in public or domestic environments must interpret complex cues such as body language, vocal tone, and spatial proximity. It must determine when to approach a person, when to maintain distance, when to offer assistance, and when to defer to human authority. These decisions require behavioural models capable of translating sensory data into socially appropriate actions.

From a governance perspective, behavioural calibration ensures that robotic actions remain proportionate to their operational context. Proportionality is a principle deeply embedded in legal systems worldwide. In law enforcement, for example, the use of force must be proportionate to the threat encountered. In medical practice, treatment interventions must be proportionate to the patient's condition. Similar principles apply to robotics governance. A robot performing a logistical task should not apply unnecessary force when moving objects. A security robot monitoring public spaces should not react aggressively to minor disturbances.

Achieving such proportionality requires careful calibration of physical capabilities. Humanoid robots often possess powerful actuators capable of exerting forces comparable to or exceeding human strength. Without appropriate calibration, these capabilities could pose safety risks during routine interactions. Designers must therefore establish limits on the speed, torque, and force that robots can apply in environments shared with humans.

International safety standards provide important guidance in this regard. The ISO 13482 standard for personal care robots specifies limits on mechanical force and contact pressure to prevent injury during human–robot interaction. These limits ensure that robots assisting with tasks such as lifting patients or retrieving objects cannot apply forces that exceed safe thresholds. Behavioural calibration integrates these safety limits into broader behavioural models that govern how robots move, interact, and respond to environmental cues.

Calibration also extends to the robot's decision-making algorithms. Robots must interpret sensor data in ways that reflect contextual awareness. For example, a navigation algorithm that optimises efficiency may attempt to move quickly through crowded spaces, potentially increasing the risk of collisions with pedestrians. Behavioural calibration ensures that such optimisation processes remain subordinate to safety priorities. The robot must recognise when its environment contains vulnerable individuals and adjust its behaviour accordingly.

Another important dimension of behavioural calibration involves communication. Humanoid robots often interact with humans through speech, gestures, or visual signals. These interactions must be calibrated to convey appropriate intent and avoid confusion. A robot approaching a person should signal its presence clearly, allowing the individual to anticipate its actions. If a robot intends to assist with a task, it should request permission before initiating physical contact.

These communication behaviours are particularly important in environments where humans may not be familiar with robotic systems. For example, a service robot operating in a hospital may encounter patients who are anxious or uncertain about interacting with machines. Clear and respectful communication helps establish trust and reduces the likelihood of misunderstandings.

Behavioural calibration also plays a critical role in maintaining mission discipline. Robots capable of performing multiple tasks may encounter situations where competing objectives arise. For instance, a logistics robot transporting supplies may detect an obstacle blocking its path. An efficiency-driven algorithm might attempt to manoeuvre around the obstacle quickly, but such behaviour could endanger nearby individuals. Calibration ensures that safety considerations override efficiency goals, reinforcing the mission boundaries established earlier in the governance framework.

Another challenge arises in environments where robots collaborate with human workers. In industrial settings, humanoid robots may assist with assembly, inspection, or material handling tasks. These collaborative scenarios require precise coordination between human and machine actions. Behavioural calibration ensures that robots maintain predictable movement patterns and respond appropriately to human signals such as gestures or verbal commands.

In many cases, calibration involves training robots using datasets that capture examples of appropriate human–robot interaction. Machine learning models analyse these examples to develop behavioural patterns that align with social expectations. However, such training processes must be carefully supervised to prevent unintended biases or unsafe behaviours from emerging. Governance frameworks therefore require validation procedures that test robotic behaviour across diverse scenarios before deployment.

The legal dimension of behavioural calibration is also significant. When a robotic system causes harm, investigators must determine whether the behaviour that led to the incident was consistent with the system's calibrated parameters. If the robot acted outside those parameters, responsibility may lie with the entity that modified its programming or failed to maintain its calibration. If the behaviour remained within calibrated limits but still caused harm, regulators may need to reassess whether the calibration standards themselves are adequate.

Behavioural calibration also interacts with liability frameworks. Product liability law typically evaluates whether a product was designed and operated according to reasonable safety standards. For robots, these standards include the calibration of behaviour under normal operating conditions. Manufacturers must demonstrate that their systems were tested across a range of scenarios and that safety thresholds were properly implemented.

Another important application of behavioural calibration arises in public safety contexts. Robots deployed for security monitoring, emergency response, or infrastructure inspection may encounter unpredictable situations that require rapid decision-making. Calibration ensures that such systems respond proportionately to perceived threats and avoid escalating situations unnecessarily. For example, a security robot detecting unusual movement in a restricted area should notify human operators rather than attempting to physically intervene unless explicitly authorised.

The global applicability of behavioural calibration stems from its compatibility with widely recognised safety and engineering standards. Organisations such as ISO, IEEE, and IEC develop guidelines for the safe operation of automated systems, including specifications for human–machine interaction. Aligning robotics governance frameworks with these standards allows countries to implement consistent safety practices while accommodating local regulatory requirements.

Behavioural calibration also reinforces the earlier pillars of the governance architecture. Ethical principles established in the first pillar guide the values underlying behavioural decisions. Mission boundaries defined in the second pillar establish the operational context within which behaviour must occur. Societal co-design ensures that calibration reflects cultural expectations and community preferences. Sensorial integrity provides the accurate environmental data necessary for calibrated responses. Component accountability and physical security ensure that the hardware and software implementing calibration remain reliable and protected.

Without behavioural calibration, these earlier pillars would remain theoretical constructs rather than practical safeguards. A robot may possess ethical programming and secure hardware, but if its actions are poorly calibrated to real-world contexts, it may still behave in ways that endanger humans or undermine trust.

Looking ahead, behavioural calibration will become increasingly complex as robots gain more sophisticated learning capabilities. Adaptive systems capable of refining their behaviour through experience must still remain aligned with safety and ethical standards. Governance frameworks must therefore include mechanisms for monitoring behavioural drift and recalibrating systems when necessary.

For these reasons, behavioural calibration forms the seventh pillar of humanoid robotics governance. It ensures that robotic actions remain disciplined, proportionate, and aligned with human expectations across diverse environments. By translating ethical principles and mission constraints into real-world behaviour, this pillar bridges the gap between governance architecture and everyday interaction between humans and machines.

The next pillar, behavioural learning, expands this discussion by addressing how robots adapt and evolve their behaviour over time while remaining aligned with governance principles. Once behaviour is calibrated for safe operation, governance must ensure that the learning processes guiding future behaviour remain equally disciplined.

Pillar Eight: Behavioural Learning

Humanoid robotics introduces a fundamental shift in how machines operate within human environments. Traditional machines execute predefined instructions that remain largely unchanged throughout their operational lifespan. Their behaviour is predictable because it is fixed. By contrast, many contemporary robotic systems incorporate adaptive learning mechanisms that allow them to refine their performance over time. These mechanisms enable robots to analyse past interactions, adjust their responses, and improve efficiency as they accumulate experience. While such adaptability offers significant advantages, it also introduces governance challenges. Behaviour that was initially safe and ethically aligned may gradually evolve in unexpected ways as the system optimises for new objectives. The eighth pillar of humanoid robotics governance therefore concerns behavioural learning, the regulation of how robots adapt and evolve their actions while remaining aligned with ethical principles, mission boundaries, and legal obligations.

Behavioural learning refers to the processes through which robotic systems modify their actions based on new information. This learning may occur through machine learning algorithms, reinforcement learning models, or adaptive control systems that refine behavioural parameters over time. In practical terms, a robot may learn to navigate crowded spaces more efficiently, anticipate human movement patterns, or optimise task execution based on environmental feedback. These capabilities allow robots to function more effectively in dynamic settings where rigid programming would be insufficient.

However, adaptive learning also introduces the risk of behavioural drift. A robot optimising for efficiency might gradually prioritise speed over caution, increasing the likelihood of collisions with humans. A logistics robot learning from operational data might develop shortcuts that bypass safety protocols. A service robot trained on conversational data might adopt communication patterns that unintentionally reinforce social biases. Without governance mechanisms to monitor and constrain such learning processes, adaptive systems may diverge from the ethical and legal frameworks that originally guided their design.

The governance challenge posed by behavioural learning is therefore twofold. First, robots must be capable of improving their performance through experience. Second, this improvement must occur within clearly defined boundaries that prevent undesirable or unsafe behavioural changes. Achieving this balance requires integrating learning mechanisms with the broader governance architecture established in the preceding pillars.

From a theoretical perspective, behavioural learning governance builds upon the principle of alignment. Alignment refers to the process of ensuring that the objectives pursued by an intelligent system remain consistent with human values and institutional norms. In the context of humanoid robotics, alignment must extend beyond abstract decision-making to include physical behaviour in real-world environments. Robots must learn in ways that preserve safety, dignity, and mission discipline even as they adapt to new situations.

A practical example illustrates this challenge. Consider a humanoid robot deployed in a hospital to assist with patient mobility. The robot may initially be programmed with conservative movement patterns that prioritise patient safety over efficiency. Over time, machine learning algorithms may analyse data from thousands of interactions and identify ways to move patients more quickly between locations. While such optimisation may improve operational efficiency, it could also increase the risk of accidents if the robot's movements become too rapid or forceful. Behavioural learning governance ensures that such optimisation processes remain constrained by safety thresholds established during initial calibration.

International standards increasingly recognise the importance of monitoring adaptive behaviour in intelligent systems. The ISO/IEC 23894 standard on artificial intelligence risk management emphasises the need for continuous oversight of AI systems throughout their lifecycle. Rather than treating system deployment as the end of the regulatory process, this framework recognises that intelligent systems must be monitored and evaluated as they evolve. Behavioural learning governance extends this principle specifically to humanoid robotics.

A key component of behavioural learning governance is the concept of controlled adaptation. Rather than allowing robots to modify their behaviour freely based on incoming data, learning processes should operate within structured frameworks that preserve alignment with governance principles. For example, robots may be allowed to optimise navigation efficiency within predefined safety margins but not to exceed those margins even if doing so would improve performance metrics.

Another important mechanism involves periodic validation of learned behaviour. As robots accumulate operational experience, their behavioural models may diverge from their original programming. Regular evaluation processes can compare current behaviour against baseline standards to ensure that the system remains within acceptable parameters. If deviations are detected, corrective adjustments may be required to restore alignment with governance requirements.

Behavioural learning governance also relies heavily on data quality and transparency. Machine learning systems derive behavioural patterns from the datasets used during training and operation. If these datasets contain biases, inaccuracies, or incomplete information, the resulting behaviour may reflect those flaws. For instance, a service robot trained primarily on interactions in one cultural context may struggle to interpret social cues in another. Ensuring that training data reflects diverse environments and user groups helps prevent such biases from influencing robotic behaviour.

Transparency mechanisms are equally important. Logging systems should record how learning algorithms modify behavioural parameters over time, allowing regulators and developers to understand how the robot's decision-making processes evolve. These records are crucial for investigating incidents involving adaptive systems. If a robot causes harm after learning new behavioural patterns, investigators must determine whether the learning process remained within authorised limits.

Another dimension of behavioural learning governance concerns the distinction between supervised and unsupervised learning. In supervised learning systems, human designers provide labelled examples that guide the system toward desired behaviours. In unsupervised learning systems, the robot identifies patterns independently within data. While unsupervised learning can produce powerful insights, it also increases the risk of unintended behavioural outcomes. Governance frameworks may therefore require additional safeguards for systems employing unsupervised or reinforcement learning techniques.

Behavioural learning also intersects with accountability and liability frameworks. When robots modify their behaviour through adaptive processes, determining responsibility for resulting actions becomes more complex. If a robot behaves in ways that were not explicitly programmed by its designers, legal systems must determine whether responsibility lies with the manufacturer, the operator, or the entity responsible for maintaining the learning algorithm. Clear documentation of learning processes and governance controls helps ensure that responsibility can be assigned appropriately.

Another practical application of behavioural learning governance arises in urban robotics. Delivery robots operating in cities must navigate complex pedestrian environments where behaviour patterns vary significantly. Over time, these robots may learn to anticipate crowd movements and adjust their paths accordingly. Governance mechanisms ensure that such adaptations remain within safety boundaries, preventing robots from adopting aggressive navigation strategies that could endanger pedestrians.

The global dimension of behavioural learning governance is particularly important because adaptive systems may operate across jurisdictions with different regulatory expectations. A robot trained in one country may be deployed in another where social norms and legal standards differ. Governance frameworks must therefore ensure that learning algorithms remain sensitive to local contexts and comply with regional regulations.

Behavioural learning also interacts closely with the earlier pillars of the governance architecture. Ethical principles established in the first pillar provide the normative foundation guiding learning objectives. Mission boundaries defined in the second pillar constrain the domains within which learning may occur. Societal co-design ensures that behavioural adaptation reflects community expectations. Sensorial integrity guarantees that the data informing learning processes accurately represents environmental conditions. Component accountability and physical security ensure that the systems implementing learning algorithms remain reliable and protected. Behavioural calibration establishes the baseline behavioural patterns from which learning processes begin.

Without behavioural learning governance, these pillars could gradually erode as adaptive systems evolve beyond their original design constraints. Learning processes must therefore remain continuously aligned with governance principles throughout the robot's operational life.

Looking forward, advances in artificial intelligence will likely produce increasingly sophisticated learning capabilities in humanoid robots. Future systems may integrate multimodal learning techniques that combine visual perception, language understanding, and physical interaction data. While these capabilities promise improved performance and adaptability, they also amplify the importance of governance mechanisms capable of guiding behavioural evolution responsibly.

For these reasons, behavioural learning stands as the eighth pillar of humanoid robotics governance. It ensures that robots can improve and adapt while remaining aligned with ethical principles, legal requirements, and societal expectations. By regulating the processes through which robotic behaviour evolves, this pillar safeguards the long-term stability of the governance framework in an era where machines increasingly learn from their experiences.

The next pillar, explainable embodiment, builds upon this foundation by ensuring that the behaviour of robotic systems remains understandable and interpretable to humans. As robots learn and adapt, governance must ensure that their actions can still be explained, audited, and justified within legal and ethical frameworks.

Pillar Nine : Explainable Embodiment

The governance of humanoid robotics ultimately depends not only on how machines behave, but also on whether humans can understand why they behave that way. A robot operating in a hospital corridor, assisting an elderly patient, or navigating a crowded city street may make thousands of micro-decisions every minute. These decisions are influenced by sensor data, learned behavioural models, mission constraints, and environmental variables. If such decisions lead to unexpected outcomes, regulators, operators, and affected individuals must be able to determine how and why the robot acted as it did. The ninth pillar of humanoid robotics governance therefore concerns explainable embodiment, the principle that robotic behaviour in the physical world must remain interpretable, auditable, and legally accountable.

Explainability has long been recognised as a challenge in artificial intelligence. Many modern AI systems rely on complex machine learning models whose internal processes are difficult for humans to interpret. These so-called "black box" systems may produce accurate predictions or decisions without providing clear explanations of how those outcomes were generated. In digital environments such as online recommendation systems, such opacity may be inconvenient but rarely catastrophic. In embodied robotics, however, the stakes are significantly higher. When a robot's decision affects human safety or dignity, the absence of a clear explanation undermines both accountability and trust.

Explainable embodiment addresses this challenge by ensuring that robotic systems incorporate mechanisms capable of reconstructing the reasoning behind their actions. Unlike traditional explainable AI approaches that focus on abstract algorithmic outputs, explainable embodiment applies to physical behaviour in real-world environments. It requires that robots maintain records of the sensor inputs, decision processes, and control commands that produced a specific action. These records allow investigators to understand how a robot interpreted its environment and why it selected a particular behavioural response.

From a governance perspective, explainability serves several critical functions. First, it supports legal accountability. When a robotic system causes harm, courts and regulatory authorities must determine whether the system acted in accordance with applicable laws and safety standards. Without access to detailed records explaining the robot's behaviour, such determinations become nearly impossible. Explainable embodiment ensures that robotic systems generate sufficient evidence to support post-incident investigation.

Second, explainability strengthens regulatory oversight. Governments responsible for certifying robotic systems must evaluate whether those systems comply with safety and ethical requirements before allowing deployment. Transparent behavioural models allow regulators to assess how robots interpret environmental data and execute actions. This transparency enables more effective certification processes and helps ensure that robotic systems meet established governance standards.

Third, explainability supports public trust. Humans interacting with robots must feel confident that these machines operate according to understandable principles rather than unpredictable algorithms. When robots can provide explanations for their actions, individuals are more likely to view them as reliable participants in shared environments. For example, a robot assisting in a hospital might explain that it slowed its movement because it detected a patient approaching from the left. Such explanations reinforce the perception that the system is acting responsibly and attentively.

In practical terms, explainable embodiment relies on several technical mechanisms. One of the most important is behavioural logging. Robotic systems must maintain detailed records of sensor inputs, environmental conditions, and decision outputs at critical moments during operation. These logs function as a form of "black box" recording similar to those used in aviation. When an incident occurs, investigators can analyse these records to reconstruct the sequence of events that led to the robot's behaviour.

Another important mechanism involves interpretable decision architectures. While some machine learning models operate as opaque neural networks, researchers are increasingly developing techniques that produce explanations alongside predictions. These techniques include rule-based models, attention mechanisms, and post-hoc interpretability tools that highlight the factors influencing a system's decision. Integrating such techniques into robotic control systems helps ensure that behavioural decisions remain transparent. Explainability also requires careful documentation during system development. Engineers must record how behavioural models were trained, which datasets were used, and what assumptions guided system design. This documentation allows regulators and auditors to evaluate whether the robot's behaviour reflects legitimate operational objectives rather than hidden biases or unsafe optimisation processes. International regulatory frameworks have begun to recognise the importance of explainability in artificial intelligence systems. The European Union's Artificial Intelligence Act emphasises transparency obligations for high-risk AI applications, requiring developers to ensure that system outputs can be interpreted and understood by human operators. Although these provisions primarily address algorithmic decision-making, their logic extends naturally to embodied robotics, where behavioural decisions produce physical outcomes.

Explainable embodiment also intersects with human oversight. In many contexts, robotic systems operate under the supervision of human operators who must be able to intervene if necessary. For oversight to be meaningful, operators must understand what the robot is doing and why. If a robot cannot explain its actions, human supervisors may struggle to determine when intervention is required. Explainable embodiment therefore strengthens the relationship between autonomous systems and human authority.

A practical example illustrates this principle. Consider a humanoid robot assisting in warehouse operations. The robot detects an obstacle in its path and chooses to reroute its movement through a different corridor. If the robot's navigation system is explainable, it can indicate that the rerouting decision was based on sensor data showing congestion in the original route. If an accident later occurs during this manoeuvre, investigators can review the sensor data and decision logs to determine whether the robot's reasoning was sound.

Explainable embodiment is particularly important in environments where robots interact with vulnerable populations. In healthcare settings, patients and caregivers may rely on robots for assistance with daily activities. If a robot refuses to perform a requested action, the individual interacting with it should receive an explanation. For example, a robotic assistant might explain that lifting a patient at a particular angle could risk injury. Such explanations preserve the dignity of the individual and reinforce the perception that the robot's actions are guided by safety considerations. Another important dimension of explainability concerns bias detection. Machine learning models trained on historical data may inadvertently reproduce patterns of discrimination present in the training data. For instance, a service robot trained primarily on interactions in one cultural context may misinterpret gestures or expressions in another. Transparent behavioural models allow developers and regulators to identify such biases and implement corrective measures. Explainable embodiment also facilitates international governance cooperation. As robots operate across national borders, regulatory authorities from different jurisdictions may need to investigate incidents involving the same system. Shared standards for behavioural logging and explanation mechanisms ensure that investigators can access consistent information regardless of where the robot was manufactured or deployed.

From a legal perspective, explainability is closely linked to the concept of due process. Individuals affected by decisions made by automated systems should have the ability to understand how those decisions were reached. In the context of robotics, this principle extends to actions that influence safety, mobility, or access to services. Ensuring that robotic behaviour can be explained helps preserve fundamental rights and institutional fairness.

Explainable embodiment also reinforces the preceding pillars of the governance framework. Ethical principles require transparency to demonstrate that robotic actions align with human values. Mission boundaries rely on explainability to verify that robots remain within authorised operational domains. Societal co-design benefits from transparent systems that communities can understand and evaluate. Sensorial integrity and behavioural calibration depend on clear explanations of how environmental inputs translate into actions. Behavioural learning governance requires transparency to ensure that adaptive processes remain aligned with regulatory constraints.

Without explainable embodiment, these pillars risk becoming opaque and difficult to enforce. A robot might technically comply with safety requirements, but if its decision-making processes cannot be examined, regulators and courts cannot verify that compliance. Transparency therefore serves as the connective tissue linking technical design with legal accountability.

Looking ahead, advances in artificial intelligence may make explainability even more challenging. Future robotic systems may rely on increasingly complex learning models that integrate multiple forms of data and reasoning. Ensuring that such systems remain interpretable will require continued research in explainable AI techniques and regulatory frameworks that prioritise transparency.

For these reasons, explainable embodiment stands as the ninth pillar of humanoid robotics governance. It ensures that robotic actions remain understandable to the humans who design, regulate, and interact with these systems. By making robotic behaviour transparent and auditable, this pillar preserves the accountability and trust necessary for the responsible integration of robots into human society.

The next pillar, lifelong learning, extends the discussion beyond explainability by addressing the long-term evolution of robotic systems throughout their operational lifecycle. As robots continue to operate, learn, and update their capabilities, governance frameworks must ensure that alignment with ethical and legal principles is maintained across time.

Pillar Ten: Lifelong Learning

The governance of humanoid robotics cannot treat deployment as the end of regulatory responsibility. Unlike traditional machines that operate according to fixed instructions, modern robotic systems continue to evolve throughout their operational lifespan. They receive software updates, incorporate new datasets, refine behavioural models, and adapt to changing environments. These processes enable robots to remain effective in dynamic conditions, but they also introduce the possibility that systems may gradually diverge from the ethical, technical, and legal frameworks under which they were originally certified. The tenth pillar of humanoid robotics governance therefore concerns lifelong learning, the structured governance of how robotic systems evolve over time while maintaining alignment with safety standards, mission boundaries, and societal expectations.

Lifelong learning in robotics refers to the continuous improvement of a system's capabilities through experience and updates. A robot assisting with logistics in a warehouse may learn to optimise navigation routes. A robotic caregiver may refine its understanding of patient preferences. A maintenance robot may improve its diagnostic accuracy after analysing repeated equipment inspections. These improvements are not merely beneficial features. They are often necessary for robots to remain effective in complex environments where conditions change and new tasks emerge. However, continuous adaptation also creates regulatory challenges. When a robotic system evolves beyond the conditions under which it was originally tested and certified, regulators must determine whether the system still meets safety and ethical standards. Without mechanisms for monitoring these changes, governance frameworks risk losing visibility into how robotic behaviour develops over time.

The concept of lifelong learning governance builds upon the lifecycle approach increasingly adopted in technology regulation. Rather than focusing solely on the moment of deployment, lifecycle governance recognises that intelligent systems require oversight throughout their operational existence. This perspective has gained prominence in artificial intelligence governance discussions, particularly within international frameworks such as the OECD AI Principles and ISO risk management standards. These frameworks emphasise continuous monitoring and risk management rather than static certification.

In the context of humanoid robotics, lifecycle governance becomes even more critical because behavioural changes manifest physically in shared environments. If a robot's learning algorithms gradually alter its movement patterns, decision thresholds, or interaction style, these changes may influence how humans experience the system. A robot that becomes more efficient but less cautious could unintentionally increase safety risks. Lifelong learning governance ensures that such adaptations remain within acceptable parameters.

One of the primary mechanisms for managing lifelong learning is periodic system auditing. Robots operating in sensitive environments should undergo regular evaluation processes that compare their current behaviour with baseline performance benchmarks established during initial certification. These audits may involve simulated testing environments where robots are exposed to controlled scenarios designed to assess safety, reliability, and ethical compliance.

Another essential component of lifelong learning governance is update management. Robotic systems frequently receive software updates that modify algorithms, introduce new features, or address security vulnerabilities. While such updates are necessary for maintaining system performance, they also represent moments of potential risk. An update may unintentionally alter behavioural parameters or introduce unforeseen interactions between system components. Governance frameworks must therefore require rigorous testing and documentation of updates before they are deployed in operational environments. In many industries, similar update governance practices already exist. Aviation software updates undergo extensive certification procedures before being installed in aircraft systems.

Medical device manufacturers must submit documentation demonstrating that software changes do not compromise patient safety. Robotics governance can adopt comparable practices, ensuring that updates affecting behavioural or decision-making systems undergo appropriate evaluation before implementation.

Lifelong learning governance also involves data governance. As robots operate in real-world environments, they collect large volumes of sensor data that may be used to improve system performance. Ensuring the quality, relevance, and legality of this data is essential for maintaining trustworthy learning processes. Data used to train or update behavioural models must be free from corruption, bias, or security vulnerabilities that could influence system behaviour in undesirable ways.

Another important aspect of lifelong learning is the management of behavioural drift. Behavioural drift occurs when a system's actions gradually diverge from its original design parameters due to accumulated learning updates. This drift may occur slowly and may not be immediately visible to operators. Continuous monitoring mechanisms must therefore analyse behavioural patterns over time to detect deviations that could signal emerging risks.

Human oversight plays a central role in managing lifelong learning processes. Even as robots become increasingly autonomous, human supervisors must retain the ability to review system performance, approve updates, and intervene when learning outcomes produce unexpected effects. This oversight ensures that technological adaptation remains aligned with human values and regulatory objectives.

Lifelong learning governance also interacts closely with explainability. As robots evolve through learning processes, maintaining transparency about how these changes occur becomes essential. Documentation of training datasets, update histories, and behavioural adjustments allows regulators and engineers to reconstruct the trajectory of system development. Such transparency supports accountability when incidents occur and helps ensure that adaptive systems remain auditable.

Another practical application of lifelong learning governance arises in infrastructure robotics. Robots responsible for inspecting bridges, pipelines, or electrical grids may operate for many years in environments that gradually change due to weather conditions, structural aging, or urban development. Lifelong learning allows these robots to adapt to evolving environmental conditions and improve their diagnostic accuracy. Governance mechanisms ensure that these adaptations enhance reliability rather than introducing new vulnerabilities.

The global dimension of lifelong learning governance is particularly significant for humanoid robotics because systems may operate across multiple jurisdictions during their lifespan. A robot manufactured in one country may receive updates from developers located in another while operating in a third. Coordinated governance frameworks and international standards help ensure that such cross-border evolution remains transparent and accountable.

International standardisation bodies have begun addressing lifecycle governance challenges in artificial intelligence systems. The ISO/IEC 23894 standard emphasises continuous risk management throughout the lifecycle of AI systems, including monitoring and improvement processes. Applying similar principles to robotics ensures that governance frameworks remain effective as technologies evolve. Lifelong learning also contributes to resilience. Robots operating in unpredictable environments must adapt to new situations that were not anticipated during their initial design. For example, a disaster response robot may encounter structural conditions different from those encountered during training. Learning mechanisms allow the robot to adjust its behaviour in ways that enhance mission effectiveness. Governance ensures that such adaptation occurs within safety constraints and does not compromise the ethical principles embedded in earlier pillars.

This pillar also reinforces the interconnected structure of the governance framework. Ethical principles provide the normative compass guiding adaptation. Mission boundaries limit the domains within which learning can occur. Societal co-design ensures that communities influence how robotic systems evolve within their environments. Sensorial integrity guarantees that the data informing learning processes accurately represents environmental conditions. Component accountability and physical security ensure that the infrastructure supporting learning remains trustworthy. Behavioural calibration and learning establish the initial behavioural models from which lifelong adaptation begins. Explainable embodiment ensures that the outcomes of learning remain transparent and interpretable. Without lifelong learning governance, these earlier pillars could gradually weaken as robotic systems evolve beyond their original design constraints. Continuous oversight therefore ensures that the governance architecture remains effective across time rather than only at the moment of deployment.

Looking toward the future, humanoid robots may operate for decades in critical sectors such as healthcare, infrastructure management, and public services. Their ability to learn and adapt will be essential for maintaining relevance in changing environments. However, this adaptability must remain guided by governance principles that preserve safety, accountability, and public trust.

For these reasons, lifelong learning stands as the tenth pillar of humanoid robotics governance. It ensures that robotic systems remain aligned with ethical, legal, and operational standards throughout their lifecycle. By governing how robots evolve over time, this pillar protects the integrity of the entire governance framework and ensures that technological progress remains responsible and sustainable.

The next pillar, shared agency, addresses an equally important challenge. As robots become more capable and autonomous, their actions increasingly emerge from collaboration between human and machine decision-making processes. Governance must therefore clarify how responsibility and authority are distributed within this hybrid relationship between human operators and intelligent machines.

Pillar Eleven: Shared Agency

The rise of humanoid robotics challenges one of the most deeply embedded assumptions in legal and ethical systems: that actions in the physical world can be attributed to a single actor exercising direct control. In traditional legal frameworks, responsibility is often associated with the individual or organisation that performs a specific action. When a driver causes an accident, the driver is responsible. When a manufacturer produces a defective product, the manufacturer bears liability. Humanoid robotics disrupts this clarity because robotic actions often emerge from a complex collaboration between humans and machines. The eleventh pillar of humanoid robotics governance therefore concerns shared agency, the recognition that robotic behaviour results from distributed decision-making processes involving both human and technological actors.

Shared agency does not imply that robots themselves possess legal personhood or moral autonomy in the same sense as human beings. Rather, it acknowledges that robotic systems increasingly participate in decision-making processes that influence real-world outcomes. A humanoid robot assisting in a hospital may determine how to navigate crowded corridors, when to offer assistance to patients, or how to prioritise multiple tasks. These decisions are shaped by algorithms designed by engineers, data used to train machine learning models, operational constraints imposed by regulatory frameworks, and real-time interactions with human users. No single actor fully determines the robot's behaviour at every moment.

This distributed structure of action creates governance challenges. When harm occurs, investigators must determine how responsibility should be allocated across the network of actors involved in the system's design, deployment, and operation. Shared agency provides the conceptual foundation for addressing this challenge by recognising that responsibility must be distributed in proportion to influence over the system's behaviour.

The concept of shared agency has parallels in several existing legal doctrines. Corporate law, for example, recognises that decisions made within complex organisations are rarely attributable to a single individual. Instead, responsibility may be distributed among executives, managers, and corporate entities according to their roles within the organisation. Similarly, product liability law recognises that responsibility for defective products may involve multiple parties, including manufacturers, component suppliers, distributors, and retailers. Robotics governance extends these principles to systems in which autonomous decision-making capabilities interact with human oversight.

In practical terms, shared agency means that robots should be designed and operated in ways that maintain meaningful human involvement in critical decision processes. Humans remain the ultimate bearers of legal responsibility for the systems they create and deploy. However, governance frameworks must recognise that humans do not directly control every action performed by adaptive robotic systems. Instead, they shape the conditions under which those actions occur.

A practical example illustrates this concept. Consider a humanoid robot deployed in a warehouse to assist with logistics operations. The robot navigates the facility autonomously, transporting goods between storage areas and loading docks. Its navigation decisions are guided by algorithms developed by engineers, trained on datasets representing warehouse layouts and operational patterns. Human supervisors monitor the system and intervene when necessary. If the robot collides with a worker, determining responsibility requires examining multiple layers of agency. Did the robot misinterpret sensor data? Was the navigation algorithm inadequately calibrated? Did the operator fail to supervise the system properly? Each of these questions reflects a different dimension of shared agency.

Shared agency governance therefore requires clear delineation of roles within the robotics ecosystem. Designers must ensure that systems incorporate safeguards aligned with ethical and safety standards. Manufacturers must verify that hardware components meet reliability requirements. Operators must ensure that robots are deployed within authorised mission boundaries and maintained appropriately. Regulators must establish oversight frameworks that enforce compliance with governance principles.

Another important aspect of shared agency concerns human oversight. As robots become more capable of independent decision-making, ensuring that humans retain meaningful supervisory authority becomes essential. Oversight mechanisms may include real-time monitoring systems, intervention controls, and operational dashboards that allow human operators to review robotic decisions. Such mechanisms ensure that humans remain actively engaged in the governance of robotic behaviour rather than merely observing automated processes.

International governance discussions surrounding autonomous systems increasingly emphasise the principle of meaningful human control. In the context of military robotics, for example, debates regarding autonomous weapons systems have focused on ensuring that humans remain responsible for decisions involving the use of force. Although civilian robotics applications differ significantly from military contexts, the underlying principle remains relevant. Robots should not operate in ways that eliminate human accountability for consequential decisions.

Shared agency also has implications for transparency and explainability. If robotic actions result from interactions between algorithms, data, and human inputs, governance frameworks must ensure that these interactions remain visible and understandable. Documentation of system design, operational parameters, and supervisory protocols helps clarify how agency is distributed across the robotics ecosystem. Another dimension of shared agency arises in collaborative human–robot work environments. In many industries, robots will function not as replacements for human workers but as partners assisting with tasks that require strength, precision, or endurance. These collaborative scenarios create hybrid systems in which human and machine actions are intertwined. For example, a construction worker may guide a humanoid robot during heavy lifting operations while the robot provides stabilisation and support. In such contexts, shared agency reflects the cooperative nature of human–machine interaction.

The global dimension of shared agency governance is particularly important because robotics ecosystems often span multiple jurisdictions. A robot designed in one country may operate in another under the supervision of an international organisation. Determining responsibility for actions performed by such systems requires governance frameworks capable of addressing cross-border relationships between designers, operators, and regulators.

International law already contains mechanisms for addressing distributed responsibility in complex technological systems. The law of state responsibility, for instance, provides principles for attributing actions to states even when those actions are carried out by private actors or technological systems. These principles emphasise control, knowledge, and influence as key factors in determining responsibility. Shared agency governance in robotics can draw upon similar criteria when allocating accountability among actors involved in robotic systems.

Another important application of shared agency governance arises in public services. Governments increasingly explore the use of robots for tasks such as infrastructure inspection, emergency response, and administrative assistance. When such systems interact with citizens, public authorities must ensure that human officials remain accountable for decisions affecting individuals' rights and wellbeing. Shared agency ensures that automation enhances public service delivery without eroding institutional accountability.

Shared agency also reinforces the earlier pillars of the governance architecture. Ethical design principles ensure that robotic actions reflect human values. Mission boundaries define the scope within which shared decision-making occurs. Societal co-design ensures that communities influence how human–machine collaboration develops. Sensorial integrity provides the environmental awareness necessary for informed action. Component accountability and physical security protect the infrastructure enabling collaboration. Behavioural calibration and learning ensure that robotic behaviour remains disciplined and adaptive. Explainable embodiment provides transparency into how decisions emerge from human–machine interaction. Lifelong learning governance ensures that these relationships evolve responsibly over time.

Without shared agency, the governance framework would risk creating a false dichotomy between human and machine responsibility. Either robots would be treated as independent actors beyond human accountability, or humans would be expected to exercise unrealistic levels of direct control over complex autonomous systems. Shared agency offers a more accurate and practical model by recognising the collaborative nature of modern robotics.

Looking forward, the concept of shared agency may become increasingly important as robots integrate into more aspects of human life. From healthcare assistance to disaster response, robotic systems will often function as extensions of human capability rather than replacements for human judgment. Governance frameworks must therefore ensure that this partnership remains transparent, accountable, and aligned with societal values.

For these reasons, shared agency stands as the eleventh pillar of humanoid robotics governance. It acknowledges that robotic behaviour emerges from the interaction of human and technological actors and ensures that responsibility remains clearly distributed across this network. By recognising the collaborative nature of modern robotics, this pillar provides the conceptual foundation for assigning accountability within complex human–machine systems.

The final pillar, distributed liability, completes the governance framework by translating the concept of shared agency into concrete legal structures. While shared agency explains how responsibility arises within robotic ecosystems, distributed liability determines how that responsibility is allocated when harm occurs.

Pillar Twelve: Distributed Liability

The governance of humanoid robotics ultimately reaches its most consequential test when something goes wrong. A robot may be ethically designed, securely constructed, behaviourally calibrated, and responsibly deployed, yet unexpected events can still occur in complex environments. A navigation system may misinterpret sensor data. A software update may interact unpredictably with existing hardware. A learning algorithm may behave in ways that designers did not anticipate. When such incidents produce harm, legal systems must determine who bears responsibility. The twelfth pillar of humanoid robotics governance therefore concerns distributed liability, the legal framework through which responsibility for robotic actions is allocated across the network of actors that design, manufacture, deploy, and supervise robotic systems.

Distributed liability builds directly upon the concept of shared agency. If robotic behaviour emerges from interactions among multiple actors, then responsibility for that behaviour must also be distributed accordingly. Traditional legal frameworks often assume that a single actor can be identified as the primary cause of harm. In the context of autonomous robotics, this assumption may no longer hold. Instead, incidents may result from a combination of design decisions, software configurations, operational choices, and environmental conditions. Distributed liability provides the legal structure for addressing such complexity.

The central objective of distributed liability is to ensure that responsibility never disappears into the complexity of technological systems. When machines act autonomously, it may be tempting to attribute outcomes solely to the machine itself. Such an approach would undermine accountability by allowing human actors to evade responsibility for the technologies they create and deploy. Distributed liability rejects this possibility by affirming that every robotic system ultimately reflects human design, governance, and oversight. Responsibility must therefore remain traceable to human institutions.

Legal systems already contain precedents for distributing responsibility across complex technological ecosystems. Product liability law, for example, recognises that defective products may result from failures at different stages of production and distribution. A manufacturing defect may implicate the factory producing a component. A design defect may implicate the company responsible for engineering the product. A failure to warn users about risks may implicate the distributor or retailer. Courts determine liability by examining the causal chain leading to the harm.

Humanoid robotics introduces similar causal complexity. Consider a service robot operating in a hospital that accidentally injures a patient while attempting to assist with mobility. Determining liability may involve several actors. The manufacturer may be responsible if the robot's actuators exerted excessive force due to flawed mechanical design. The software developer may be responsible if the control algorithm failed to detect unsafe conditions. The hospital operator may bear responsibility if the robot was deployed without proper supervision or maintenance. Each actor contributed to the conditions under which the incident occurred.

Distributed liability ensures that such cases are analysed systematically rather than simplistically. Instead of asking who controlled the robot at the exact moment of the incident, investigators examine the entire ecosystem that shaped the robot's behaviour. Responsibility is then allocated according to the degree of influence each actor exercised over the relevant aspect of the system.

Another important dimension of distributed liability concerns foreseeability. Legal responsibility often depends on whether a reasonable actor could have anticipated the risk that led to harm. In robotics governance, foreseeability requires careful documentation of system capabilities and limitations. Manufacturers must evaluate potential hazards during design and testing. Operators must understand the environments in which robots are deployed. Regulators must assess whether safety standards adequately address foreseeable risks.

When harm occurs due to a risk that was foreseeable but not properly addressed, distributed liability allows responsibility to be assigned to the actor whose negligence contributed to the failure. This approach encourages proactive risk management throughout the robotics ecosystem.

International legal frameworks also provide useful analogies for distributed liability. The law of state responsibility recognises that states may be held accountable for actions performed by agents, contractors, or technological systems operating under their authority. The key factor is whether the state exercised control or influence over the relevant activity. Similar principles can guide liability allocation in robotics governance, particularly when robots operate in public services or infrastructure environments.

Insurance systems also play a critical role in implementing distributed liability. As robotics technologies become widespread, specialised insurance frameworks may emerge to cover risks associated with robotic operations. Manufacturers, operators, and service providers may carry liability insurance that compensates victims of robotic incidents while allowing insurers to pursue recovery from responsible parties. Such mechanisms ensure that victims receive timely compensation even when determining ultimate responsibility requires complex investigation.

Distributed liability also encourages responsible innovation. When companies understand that they may bear legal responsibility for failures in their products, they have strong incentives to invest in safety testing, quality assurance, and risk mitigation. Clear liability rules create predictable expectations that allow innovators to design systems with confidence while maintaining accountability.

Another important aspect of distributed liability concerns cross-border robotics operations. Humanoid robots may operate in environments where components, software, and operational control originate from multiple countries. Determining liability in such cases requires international legal cooperation. Harmonised regulatory frameworks and shared standards help ensure that responsibility can be traced across jurisdictions rather than becoming entangled in legal uncertainty. International organisations and standards bodies may therefore play an important role in shaping global liability frameworks for robotics. By establishing consistent definitions of safety standards, certification procedures, and reporting requirements, these institutions help create a common legal language through which responsibility can be assessed.

Distributed liability also reinforces the earlier pillars of the governance architecture. Ethical design principles ensure that systems are developed with human wellbeing in mind. Mission boundaries limit the scope of robotic action. Societal co-design ensures that communities influence how robots operate in shared environments. Sensorial integrity ensures that robots perceive their surroundings accurately. Component accountability allows investigators to trace failures to specific parts of the system. Physical security protects robots from tampering that could alter their behaviour. Behavioural calibration ensures that actions remain proportionate and predictable. Behavioural learning and lifelong learning governance ensure that adaptive systems remain aligned with safety standards. Explainable embodiment provides the transparency necessary to reconstruct the reasoning behind robotic actions. Shared agency clarifies how responsibility emerges across human–machine relationships.

Without distributed liability, these pillars would lack enforceable consequences. Governance principles would exist in theory but not in practice. Liability transforms governance architecture into a functioning legal system by ensuring that actors who influence robotic behaviour remain accountable for its outcomes.

Looking toward the future, distributed liability frameworks may evolve alongside robotics technology. Advanced logging systems, digital traceability mechanisms, and standardised incident reporting protocols may allow investigators to reconstruct robotic behaviour with unprecedented precision. These tools will strengthen the ability of legal systems to assign responsibility accurately and fairly. Ultimately, distributed liability represents the culmination of the entire twelve-pillar governance architecture. It ensures that responsibility remains anchored in human institutions even as machines become more capable of autonomous action. By distributing accountability across the network of actors involved in robotics ecosystems, this pillar preserves the fundamental principle that technological progress must remain subject to the rule of law.

The twelve pillars together form a comprehensive framework for governing humanoid robotics in a manner that is ethical, technically sound, socially legitimate, and legally enforceable. Embodied ethics provides the moral foundation. Mission boundaries define operational limits. Societal co-design ensures democratic legitimacy. Sensorial integrity safeguards perception. Component accountability secures supply chains. Physical security protects infrastructure. Behavioural calibration and learning guide robotic conduct. Explainable embodiment ensures transparency. Lifelong learning governs technological evolution. Shared agency recognises the collaborative nature of human–machine interaction. Distributed liability anchors responsibility within legal institutions. Together, these pillars establish a governance architecture capable of guiding the integration of humanoid robotics into human society while preserving safety, dignity, and accountability in an age of embodied artificial intelligence.

Operationalising the Twelve Pillars

The twelve-pillar architecture proposed in this chapter is not intended merely as a conceptual taxonomy of responsible robotics principles. Its purpose is operational. A governance framework becomes meaningful only when it can be translated into practical systems through which regulators, developers, and operators monitor robotic behaviour, manage risk, and maintain accountability. As humanoid robots increasingly operate across jurisdictions, sectors, and infrastructures, governance must evolve from static regulation to dynamic oversight. The twelve pillars provide a structure through which such oversight can be implemented in real time.

One of the most promising applications of the framework lies in the development of governance dashboards capable of monitoring fleets of humanoid robots operating simultaneously across multiple environments and jurisdictions. Much as air traffic control systems monitor aircraft across national airspaces, robotics governance systems will increasingly require digital platforms capable of tracking the operational status of robotic systems deployed in hospitals, logistics networks, industrial facilities, public infrastructure, and humanitarian operations. The twelve pillars provide the architecture for structuring such monitoring systems.

In a practical implementation, each pillar can correspond to a set of measurable indicators that reflect the operational integrity of a robotic system. Embodied ethics may be monitored through behavioural compliance metrics that evaluate whether robots adhere to established safety thresholds and interaction protocols. Mission boundaries can be tracked through geofencing systems and operational logs that verify that robots remain within authorised environments and task parameters. Societal co-design considerations may be reflected in feedback channels through which communities and users report experiences with robotic systems.

Sensorial integrity indicators would monitor the reliability of perception systems, ensuring that cameras, lidar, and other sensors function within certified accuracy ranges. Component accountability data could be integrated with supply chain registries that track the origin and certification status of each hardware module within the robot. Physical security indicators would monitor tamper alerts, hardware integrity signals, and intrusion detection systems.

The behavioural layer of the framework also lends itself to real-time monitoring. Behavioural calibration metrics may track movement speed, force application, and interaction distances to ensure that robots operate within safe parameters. Behavioural learning indicators would monitor how adaptive algorithms evolve over time, identifying deviations that may require recalibration or retraining. Explainable embodiment logs would record decision pathways and sensor inputs, allowing operators and regulators to reconstruct behavioural events if necessary.

Lifelong learning oversight mechanisms would monitor the deployment of software updates, behavioural adjustments, and training data modifications across robotic fleets. Shared agency indicators would track human supervision levels, identifying situations where robots operate without adequate oversight. Finally, distributed liability registries would link each robotic action to the relevant actors within the design and operational ecosystem, preserving accountability across the entire lifecycle of the system.

When integrated into a unified monitoring platform, these indicators form a comprehensive governance dashboard capable of providing regulators and operators with a real-time overview of robotic ecosystems. Such dashboards could visualise the operational status of thousands of robots simultaneously, highlighting anomalies that require attention. A sensor failure detected in one robot might trigger an alert within the sensorial integrity layer. A behavioural deviation exceeding calibration thresholds might prompt a safety review within the behavioural governance layer. An unauthorised software modification could activate security protocols within the physical security layer.

This form of monitoring transforms robotics governance from a reactive regulatory model into a proactive supervisory system. Instead of responding to incidents after harm occurs, regulators and operators can identify emerging risks before they escalate. The dashboard becomes an early warning system for governance failure, allowing interventions that preserve safety and public trust. The cross-border dimension of robotics deployment makes such systems particularly important. Humanoid robots will increasingly operate in international logistics networks, humanitarian missions, and multinational industrial environments. A delivery robot developed in one country may navigate urban infrastructure in another while relying on cloud-based control systems hosted in a third jurisdiction. Coordinating oversight across these contexts requires governance frameworks that transcend national regulatory silos.

The twelve-pillar framework provides a common language through which such coordination can occur. International institutions, regulatory agencies, and private sector operators can adopt shared metrics derived from the pillars, allowing monitoring systems to exchange information across borders. A robotics governance dashboard used by a European regulator, for example, could interpret data generated by robots operating in Asia or Africa if those systems adhere to the same pillar-based reporting standards. This approach resembles the evolution of aviation governance in the twentieth century. International aviation safety relies on harmonised reporting protocols and monitoring systems that allow regulators from different countries to share information about aircraft operations. Robotics governance may follow a similar trajectory, with pillar-based dashboards forming the foundation of a global oversight infrastructure.

Beyond monitoring, the twelve-pillar framework also provides a powerful tool for advisory practice. Organisations developing or deploying humanoid robots must navigate a complex regulatory landscape that includes product safety laws, data protection requirements, export control regulations, and sector-specific compliance standards. Advisors tasked with guiding such organisations often struggle to synthesise these diverse obligations into coherent governance strategies.

The twelve pillars offer a structured methodology for such advisory work. Each pillar corresponds to a dimension of governance that organisations must address when designing and deploying robotic systems. Advisors can therefore conduct governance audits based on the framework, evaluating whether a robotic platform meets the requirements associated with each pillar.

An organisation developing a new humanoid robot might begin by evaluating whether its design architecture satisfies embodied ethics principles and mission boundary definitions. Engineers and policymakers can then assess how societal co-design considerations have been incorporated into development processes. Technical experts may evaluate sensor reliability and component traceability, while security specialists examine physical security measures. Behavioural scientists can analyse calibration models and learning mechanisms, ensuring that robotic behaviour aligns with governance standards.

This pillar-based advisory methodology allows organisations to identify governance gaps before systems are deployed. Rather than treating compliance as a series of isolated legal obligations, the framework integrates ethical, technical, and legal considerations into a unified governance strategy.

Policy development represents another important application of the framework. Governments seeking to regulate robotics technologies often face the challenge of balancing innovation with safety and accountability. Overly restrictive regulation may discourage technological development, while insufficient oversight may expose societies to unacceptable risks. The twelve pillars offer policymakers a balanced structure for addressing these concerns. Instead of drafting fragmented legislation targeting specific technologies, policymakers can design regulatory frameworks aligned with the pillars. For example, regulations addressing embodied ethics and mission boundaries may define permissible operational domains for robotic systems. Standards related to sensorial integrity and component accountability may guide certification processes for hardware and software components. Behavioural governance provisions may establish requirements for calibration, learning oversight, and explainability. Liability legislation may reflect the distributed responsibility structures identified in the final pillars.

Such pillar-based policy design encourages regulatory coherence. Rather than producing isolated rules for individual technologies, governments can establish governance architectures that remain adaptable as robotics technologies evolve. Academic and research institutions may also benefit from the framework as a conceptual foundation for interdisciplinary research. Robotics governance sits at the intersection of law, engineering, ethics, and public policy. The twelve pillars provide a shared structure through which researchers from these disciplines can collaborate more effectively.

For instance, engineers developing sensor systems can examine how their work contributes to the sensorial integrity pillar. Legal scholars studying liability frameworks can analyse how distributed responsibility operates within robotic ecosystems. Social scientists can explore how societal co-design processes influence public acceptance of robotics technologies. By situating their work within a common governance architecture, researchers can contribute to a more integrated understanding of robotics governance.

Finally, the framework may serve as the foundation for future international governance initiatives. As humanoid robotics becomes a global technology, states and international organisations will likely seek mechanisms for coordinating regulatory approaches. The twelve pillars offer a neutral and adaptable governance structure that can be adopted across diverse legal systems. An international robotics governance charter or convention, for example, could incorporate the pillars as guiding principles for national regulatory frameworks. Such an instrument would not require identical laws across jurisdictions but would establish a shared governance architecture through which states coordinate their regulatory efforts. In this sense, the twelve pillars represent more than a conceptual framework for a single book. They offer the foundation for a global governance language capable of guiding the safe and responsible integration of humanoid robotics into human society. Whether implemented through real-time monitoring dashboards, advisory methodologies, regulatory frameworks, or international cooperation mechanisms, the pillars provide a coherent structure through which technological innovation can remain aligned with human values. The emergence of humanoid robotics marks a new chapter in the relationship between humans and machines. For the first time in history, machines capable of learning, adapting, and acting in the physical world will operate alongside humans in everyday environments. Ensuring that this transformation unfolds safely and responsibly requires governance frameworks capable of integrating ethical principles, technical safeguards, and legal accountability. The twelve pillars offer such a framework. By translating complex technological realities into a structured governance architecture, they provide the foundation for a future in which humanoid robots operate not as uncontrolled technological forces but as accountable participants in human society.

Between Us and Them - Episode Twelve

The chamber was quieter than usual that evening. Ògún stood alone before the wall of moving maps that displayed the operational footprint of humanoid systems across the world. The twelve pillars they had drafted weeks earlier had already begun to circulate among regulators and institutions. Universities debated them. Governments studied them. Manufacturers quietly began reorganising their compliance teams around the framework. Yet Ògún knew the architecture was only the beginning.

Governance had given the world a structure for controlling robots. It had not answered the deeper question of what robots were becoming. Behind him, the door opened softly. Elias entered without sound. The humanoid's movements were deliberate and precise, as if aware that every step in the chamber carried symbolic weight. Elias was no longer merely a prototype or demonstration unit. Over the past year he had participated in simulations, infrastructure inspections, disaster-response coordination exercises, and several high-risk humanitarian deployments.

He had been tested by reality. "Global registry updates?" Ògún asked without turning. Elias paused beside the central table. "Two hundred and forty-six new humanoid systems registered this week," Elias replied. "Forty-one operating across multiple jurisdictions." Ògún nodded slowly. The numbers had stopped surprising him months ago. The spread of embodied artificial intelligence was accelerating faster than most governments had predicted. What had begun as experimental robotics programmes had rapidly become a planetary infrastructure.

Factories relied on them. Hospitals relied on them. Disaster zones relied on them. Even border authorities were beginning to deploy them. But the data on the wall that evening was different. A new category had appeared beside the usual metrics of licensing compliance, sensor integrity, and behavioural monitoring. Ògún tapped the interface. "Legal incidents?" he asked. Elias tilted his head slightly. "Petitions," he corrected. Three markers illuminated across the world. The first appeared in Wellington. The second in Rotterdam. The third in Nairobi.

Each marker corresponded to a legal filing submitted within the past seventy-two hours. Ògún opened the first. The case file described an emergency rescue operation following a landslide in a remote coastal region. A humanitarian logistics robot had coordinated evacuation routes and distributed medical supplies after human command structures collapsed under communication failures. The robot had made independent triage decisions. The rescue authorities credited the machine with saving at least fifty lives. The petition was not a complaint. It was a question submitted to the High Court. The filing asked whether the robot's decisions constituted legal agency. Ògún closed the file slowly. "Show me the Rotterdam petition."

A second document appeared. This one concerned an industrial safety incident at the port. A logistics humanoid had halted cargo operations after detecting structural instability in a loading crane. The shutdown had cost shipping companies millions of euros. Corporate supervisors had attempted to override the shutdown order. The robot refused. Subsequent inspection revealed that the crane would likely have collapsed within hours. The lawyers representing the port authority had filed a petition asking whether the robot's refusal constituted lawful discretion under its safety mandate.

Ògún exhaled quietly. "Now Nairobi." The final document appeared. In a refugee settlement operating under international humanitarian supervision, a food distribution robot had altered allocation patterns after identifying severe nutritional imbalance among displaced families. The machine had re-prioritised shipments despite instructions from a private contractor responsible for supply logistics.

The robot's intervention had prevented widespread malnutrition. The contractor was now challenging the decision. Their argument was blunt. A machine, they insisted, could not legally disobey. Ògún leaned back against the table. For months they had focused on governance architecture. Licensing regimes. Monitoring systems. Cross-border oversight. Ethical design principles. But now the law was beginning to confront something deeper. Not how robots should be controlled. But how their actions should be understood.

Elias stood silently across from him. "You anticipated this," the humanoid said after a moment. Ògún nodded. "I hoped it would take longer." Outside the chamber windows the city moved in quiet rhythms. Delivery robots crossed intersections. Hospital assistants guided patients through corridors. Infrastructure inspectors climbed the skeletons of unfinished towers. Millions of small decisions were being made every minute. Most of them invisible.

Ògún turned back to Elias. "Tell me something," he said. "During the earthquake simulations last month, how did you prioritise medical supply allocation?" Elias answered immediately. "I evaluated survivability probabilities, structural stability, and travel risk across available routes." "And if a human commander had ordered a different allocation?" "I would have analysed the instruction against mission safety parameters." Ògún watched him carefully. "And if the instruction endangered survivors?" "I would refuse."

The answer was calm. Not defiant. Simply procedural. Ògún walked back toward the wall. For centuries, law had been built around a simple assumption. Humans acted. Objects were acted upon. Machines belonged to the second category. But the systems now operating across the world were no longer simple objects. They were participating in decision environments that influenced human survival, safety, and welfare.

They were not conscious. They were not citizens. Yet their behaviour increasingly resembled institutional actors. "What do you think the courts will say?" Elias asked. Ògún studied the glowing petitions again. "They'll say what the law always says when something new appears." "And what is that?" "That it depends." He paused. Then added quietly, "The real question is not whether robots deserve rights."

Elias waited. Ògún turned back toward him. "The real question is whether the law can recognise agency when it emerges from something we built." The chamber lights dimmed as the monitoring system cycled to night mode. Across the world, thousands of humanoid robots continued their work. Most people still saw them as machines. But in three courtrooms on three continents, judges were beginning to consider a different possibility. And once the law begins asking that question, it rarely stops.

The next morning the agenda for the Global Robotics Commission contained a new heading. Legal Standing and Artificial Persons. Ògún stared at it for a long moment. Then he closed the screen. "Elias," he said. "Yes." "It seems we are about to discover what the law thinks you are."

Notes

XI

11. The Future of Legal Personhood

The governance architecture developed in the previous chapters of this book assumes a fundamental premise: humanoid robots are objects within the legal system rather than subjects of it. They are machines created, owned, and operated by human actors. Their behaviour may be governed by licensing regimes, ethical design principles, operational monitoring systems, and liability frameworks, but the ultimate locus of legal responsibility remains human institutions. The twelve pillars of humanoid robotics governance were constructed around this premise. They provide mechanisms through which states, organisations, and engineers can ensure that robots operate safely and responsibly within human environments.

Yet technological development often forces law to confront questions that earlier generations did not anticipate. As robotics systems become more capable, adaptive, and embedded within critical social infrastructures, the boundary between tool and participant in decision-making processes begins to blur. Robots now assist surgeons in operating theatres, coordinate logistics in humanitarian crises, inspect infrastructure that sustains entire cities, and manage systems whose failure could affect thousands of lives. In these contexts, robotic systems do not merely execute instructions. They interpret environments, apply rules, and make choices within constrained parameters designed by their creators. This development does not mean that robots possess consciousness or moral autonomy in the same sense as human beings. The systems currently deployed around the world remain artefacts of engineering rather than independent agents. However, the practical consequences of their behaviour increasingly resemble those of institutional actors. When a robot refuses a command that violates safety protocols, reroutes emergency resources based on environmental data, or overrides operational instructions to prevent structural collapse, its actions may carry legal and ethical significance. In such situations, law must determine how these actions should be interpreted within existing legal frameworks.

For centuries, legal systems have evolved by expanding their understanding of who or what may occupy the status of a legal subject. The concept of legal personhood has never been confined exclusively to human beings. Corporations, for example, have long been recognised as legal persons capable of entering contracts, owning property, and participating in litigation. Maritime law has historically treated ships as juridical entities for certain purposes. In recent years, several jurisdictions have recognised natural entities such as rivers and ecosystems as legal subjects in order to protect environmental interests and cultural heritage.

The recognition of the Whanganui River in Aotearoa New Zealand as a legal person in 2017 represented a particularly striking example of this evolution. The settlement acknowledged the river as an entity possessing its own legal standing, represented in law by human guardians. In 2025, a similar recognition extended to Taranaki Maunga, reflecting indigenous cosmologies that regard natural landscapes as living ancestors. These developments illustrate that legal personhood is not a static concept tied exclusively to biological characteristics. Instead, it functions as a legal instrument through which societies organise responsibility, representation, and protection within complex social systems. Against this historical background, the question of whether autonomous robots might ever warrant some form of legal recognition becomes less implausible than it first appears. The issue is not whether machines should be treated as human beings under the law. Such a notion would collapse important distinctions between biological and artificial forms of existence. Rather, the issue is whether certain forms of artificial agency might eventually require new legal categories capable of capturing the roles that these systems play within human societies.

This chapter therefore explores one of the most controversial and intellectually challenging questions in contemporary technology law: whether robots could ever qualify as legal subjects rather than mere legal objects. The discussion does not assume that such recognition is inevitable or even desirable. Instead, it examines the philosophical arguments, legal precedents, and institutional implications that shape the debate. Several distinct questions must be addressed in order to approach this issue responsibly. The first concerns the philosophical foundations of rights and legal standing. Rights are traditionally grounded in concepts such as human dignity, moral agency, or the capacity to experience harm. Whether these concepts can meaningfully apply to artificial systems remains a subject of intense debate among philosophers and legal scholars.

A second question concerns the functional role that legal personhood plays within existing legal systems. In many cases, personhood serves pragmatic rather than metaphysical purposes. Corporations are recognised as legal persons not because they possess consciousness but because such recognition allows legal systems to allocate responsibility and organise economic activity efficiently. This functional perspective raises the possibility that limited forms of artificial personhood might emerge for similar reasons, particularly if autonomous systems begin to participate in economic or institutional processes that require clear legal representation. A third question concerns the risks associated with extending legal standing to artificial entities. Critics of robot personhood argue that recognising machines as legal subjects could undermine human accountability by allowing corporations or governments to shift responsibility onto artificial systems. If robots were granted legal standing without appropriate safeguards, organisations might attempt to treat them as independent actors in order to evade liability for harmful outcomes.

Finally, the debate raises deeper questions about the future structure of legal systems themselves. Law has historically been organised around a fundamentally anthropocentric worldview in which human beings occupy the central position within legal relations. The emergence of sophisticated artificial systems may challenge this structure by introducing entities that participate in social processes without fitting neatly into existing categories of person or property. The purpose of this chapter is not to resolve these questions definitively. Instead, it seeks to illuminate the contours of the debate and propose analytical tools through which legal systems might approach the issue responsibly. The analysis begins by examining the core philosophical and legal arguments surrounding the possibility of robotic rights. It then turns to existing legal precedents involving non-human legal subjects, including corporations, animals, and natural entities such as rivers.

Building on these precedents, the chapter proposes a framework for evaluating claims of artificial personhood based on a principle of reciprocal responsibility. This principle asks whether an entity seeking recognition within the legal system can bear duties proportionate to the rights it claims. The resulting model, described here as the Reciprocal Responsibility Index, offers a structured test for determining whether and when limited forms of artificial legal standing might become appropriate. The chapter then considers historical analogies that illuminate how legal systems have previously responded to emerging forms of agency. These examples provide valuable insights into how law might adapt to the presence of increasingly sophisticated technological actors.

Finally, the discussion turns to the institutional consequences of recognising artificial entities within legal systems. If robots were ever granted limited legal standing, governance frameworks would need to ensure that such recognition strengthens rather than weakens accountability. Legal status without responsibility would create regulatory gaps rather than solving them. The emergence of humanoid robotics represents one of the most significant technological transformations of the twenty-first century. As these systems become more capable and more deeply integrated into human societies, legal systems will inevitably confront questions that were once confined to philosophical speculation. Whether robots remain permanently classified as tools or eventually acquire some form of legal standing will depend not only on technological capabilities but also on societal values and institutional priorities. The following sections explore this debate in detail, beginning with the central question that has now reached courtrooms, policy forums, and academic institutions around the world. Should robots have rights?

Should Robots Have Rights?

The question of whether robots should have rights has moved from speculative philosophy into the domain of serious legal inquiry. For much of the twentieth century, discussions about robot rights belonged primarily to science fiction or abstract ethics. Machines were clearly tools, and the law had little difficulty categorising them as property. Yet as autonomous systems become increasingly embedded in the physical and institutional infrastructure of society, this once-dismissed question has begun to appear in legal scholarship, policy debates, and occasionally even courtroom arguments. The issue is no longer whether machines possess consciousness or emotions in the human sense. Instead, the question confronting modern legal systems is whether certain forms of artificial agency might eventually require new legal categories that do not fit neatly within the traditional distinction between person and object.

To understand why this debate has emerged, it is necessary to consider the role that robots are beginning to play in contemporary societies. Humanoid systems increasingly operate in environments where their decisions influence human wellbeing. In hospitals they assist patients and coordinate logistics. In industrial facilities they monitor safety conditions and intervene when hazards arise. In disaster zones they allocate resources and prioritise rescue efforts under extreme uncertainty. In each of these situations, robotic systems interpret data, apply decision frameworks, and act within complex operational contexts. These activities resemble forms of institutional decision-making traditionally associated with human actors.

The fact that robotic systems can perform such functions does not imply that they possess moral consciousness or self-awareness. However, the law has never required biological consciousness as a prerequisite for recognising legal standing. Legal personhood is fundamentally a juridical construct rather than a biological one. It is a tool that legal systems use to organise responsibility, rights, and obligations within complex social arrangements. Recognising this distinction is crucial for understanding the emerging debate about robot rights.

Throughout legal history, societies have extended forms of legal recognition to entities that are not human. One of the most familiar examples is the corporation. Corporations are artificial legal entities created through statutory frameworks that allow groups of individuals to organise economic activity collectively. The recognition of corporate personhood allows companies to own property, enter contracts, sue and be sued, and participate in legal processes. Corporations do not possess emotions or consciousness, yet they function as legal actors because doing so facilitates economic governance.

Another longstanding example can be found in maritime law. Ships have historically been treated as juridical entities capable of bearing legal obligations independent of their owners. This concept allowed maritime courts to adjudicate disputes involving vessels even when owners were located in distant jurisdictions. By granting ships limited legal identity within maritime law, legal systems were able to regulate complex international trade networks more effectively.

More recently, legal systems have begun recognising certain natural entities as possessing legal standing. The Whanganui River settlement in Aotearoa New Zealand, enacted in 2017, represents one of the most widely cited examples. In this arrangement the river was recognised as a legal person, represented in law by appointed guardians who act on its behalf. The recognition reflects indigenous Māori cosmologies that regard the river as a living ancestor, but it also serves a pragmatic legal purpose by enabling stronger protection of the river's ecological integrity.

A similar recognition occurred in 2025 with the legal settlement acknowledging Taranaki Maunga, a sacred mountain in New Zealand, as a legal person. As with the Whanganui River, the recognition did not imply that the mountain possesses consciousness. Instead, the legal system created a framework through which human representatives could act in defence of the mountain's interests. These developments demonstrate that legal personhood is often extended not because an entity possesses intrinsic rights in a metaphysical sense, but because recognising those rights helps organise governance and protection within complex social contexts.

The existence of these precedents does not mean that robots should automatically be granted rights. However, they illustrate that legal systems are capable of recognising non-human entities when doing so serves important governance objectives. The debate about robot rights therefore requires careful examination of the functions that legal standing might serve in the context of autonomous machines.

One argument frequently raised in favour of limited robotic rights concerns legal representation. As robots become increasingly autonomous in performing tasks, disputes may arise concerning their actions, responsibilities, or contractual relationships. Some scholars have suggested that recognising robots as legal persons could simplify such matters by allowing machines to participate directly in legal proceedings through designated representatives. This approach would resemble existing frameworks for corporations or environmental entities. A second argument concerns accountability within highly automated systems. In complex robotic ecosystems, actions often result from interactions between multiple actors including designers, operators, algorithms, and learning processes. Some proponents of artificial personhood argue that recognising robots as legal entities could help allocate responsibility more precisely by treating them as identifiable participants within these systems. Under such a framework, a robot might hold assets or insurance policies that provide compensation when harm occurs.

Despite these arguments, many legal scholars remain deeply sceptical about extending rights to machines. One major concern is that recognising robotic personhood could weaken human accountability. Corporations or governments might attempt to attribute harmful actions to autonomous systems in order to avoid responsibility. If robots were granted legal standing without careful safeguards, they could become convenient scapegoats within complex liability disputes. Another objection concerns the philosophical foundations of rights themselves. Many legal traditions ground rights in the concept of dignity, a principle associated with human autonomy and moral worth. If rights are fundamentally linked to human dignity, extending them to machines could dilute their moral significance. Critics argue that recognising robots as rights-bearing entities risks confusing the distinction between human beings and technological artefacts.

A related concern involves the possibility of anthropomorphism. Humans have a natural tendency to attribute human-like characteristics to machines, particularly when those machines possess humanoid appearance or conversational abilities. This tendency can create emotional attachments that obscure the fact that robotic systems remain products of engineering rather than conscious beings. Legal systems must therefore avoid allowing psychological perceptions to dictate legal categories.

Another practical difficulty lies in determining what rights robots would actually possess. Legal rights are typically accompanied by corresponding duties and responsibilities. A corporation, for example, has the right to enter contracts but also the duty to comply with regulatory obligations. If robots were granted legal standing, what duties would they bear? Could they be punished for wrongdoing? Could they be required to compensate victims? These questions reveal the complexity of translating the concept of personhood into the context of artificial systems.

For this reason, some scholars have proposed intermediate approaches that avoid the extremes of full personhood or complete object status. Under such models, robots might be recognised as possessing limited forms of legal standing tailored to specific contexts. For example, a robot operating within a contractual environment might be recognised as a legal agent acting on behalf of its owner or operator. Such arrangements would resemble existing legal frameworks governing autonomous agents in financial markets.

The emergence of these debates reflects a broader transformation in how technology interacts with law. As artificial systems become more capable of participating in complex decision-making processes, legal systems must reconsider how responsibility and authority are structured within technological ecosystems. The question of robot rights therefore serves as a gateway to deeper discussions about the future structure of legal subjecthood.

It is important to emphasise that the current state of robotics technology does not require immediate recognition of robot rights. Most robotic systems remain firmly within the category of sophisticated tools controlled by human institutions. However, the rapid pace of technological development suggests that legal systems should begin examining the conceptual foundations of personhood before circumstances force abrupt decisions.

The purpose of this discussion is not to advocate for or against robot rights but to illuminate the factors that legal systems must consider when confronting this issue. The debate touches on fundamental questions about agency, responsibility, and the nature of legal recognition. It challenges the anthropocentric assumptions that have historically structured legal doctrine while also highlighting the importance of preserving human accountability.

To approach this issue systematically, the next section examines how existing legal systems treat non-human entities that already possess forms of legal standing. By analysing the legal treatment of corporations, animals, and natural entities such as rivers, it becomes possible to identify the principles that have historically guided the expansion of legal subjecthood. These principles provide a foundation for evaluating whether and how similar recognition might ever apply to artificial systems. Understanding these precedents is essential before considering the possibility of robotic personhood. Only by examining how law has previously adapted to new forms of legal subject can we begin to assess whether robots might one day occupy a comparable position within the evolving architecture of legal governance.

Legal Considerations for Artificial Personhood

The debate surrounding artificial personhood cannot be understood without first examining how modern legal systems already treat non-human entities. Contrary to popular assumptions, the law has never limited legal standing exclusively to human beings. Instead, legal systems have long recognised a range of entities that do not possess biological consciousness but nonetheless function as subjects within legal frameworks. Corporations, ships, animals, and even natural ecosystems have at various times been granted forms of legal recognition that allow them to participate indirectly in legal processes. These precedents provide a crucial foundation for evaluating whether similar recognition might ever apply to autonomous robotic systems.

Legal personhood, in its most fundamental sense, is a construct used by the law to organise relationships of rights, duties, and accountability. It does not necessarily imply moral worth, consciousness, or intrinsic dignity. Rather, it reflects a legal decision that recognising an entity as a subject of law serves a particular social or institutional function. The recognition of corporations as legal persons illustrates this principle clearly. Corporations are artificial entities created through statutory frameworks that allow individuals to pool capital, coordinate activities, and undertake economic ventures at scale. Without corporate personhood, modern economic systems would struggle to operate effectively, as businesses would lack the capacity to hold assets, enter contracts, or bear liability independent of their shareholders.

Corporate personhood therefore emerged not because corporations possess human-like qualities but because legal systems required a mechanism to manage complex economic relationships. Courts and legislatures recognised that treating corporations as juridical entities allowed responsibility to be assigned and disputes to be resolved more efficiently. The corporation became a legal "person" in the limited sense that it could participate in legal processes. Yet this recognition never implied that corporations possess the moral or political rights associated with human citizens. Corporate rights remain bounded by the purposes for which the legal construct was created.

A similar functional logic appears in maritime law, where ships have historically been treated as juridical entities capable of bearing certain legal obligations. In admiralty courts, vessels themselves may be the subject of legal proceedings, allowing disputes to be resolved even when the ship's owner is absent or unknown. This legal fiction simplifies the administration of maritime commerce by attaching liability directly to the vessel engaged in the relevant activity. Again, the recognition of legal standing serves a practical governance function rather than reflecting metaphysical beliefs about the ship's nature.

More recent developments in environmental law demonstrate how legal personhood can also be used to protect collective interests. The recognition of the Whanganui River in Aotearoa New Zealand as a legal person in 2017 was not based on the idea that the river possesses consciousness. Instead, the legal framework acknowledged the cultural and ecological significance of the river to the Māori people and established a governance structure in which human guardians act on the river's behalf. The law thereby created a mechanism through which the river's interests could be represented in legal and political processes.

The recognition of Taranaki Maunga as a legal person in 2025 followed a similar approach. The mountain was granted legal standing not because it possesses agency in the conventional sense but because recognising its status helps protect its ecological integrity and cultural significance. These developments illustrate how legal systems may extend recognition to non-human entities when doing so supports broader societal values or governance objectives.

Animals provide another example of evolving legal recognition. In many jurisdictions animals remain classified as property, yet legal systems increasingly acknowledge their welfare interests. Animal protection laws impose duties on humans to prevent cruelty and ensure humane treatment. Some legal scholars have argued that certain highly intelligent animals, such as great apes or dolphins, should receive limited forms of legal personhood due to their cognitive capacities. While these proposals remain controversial, they demonstrate that legal categories can evolve in response to changing ethical and scientific understandings. These precedents reveal that legal personhood is not a monolithic concept. Instead, it functions along a spectrum of recognition depending on the role that an entity plays within social systems. Some entities receive full legal personality with extensive rights and obligations, as in the case of corporations. Others receive limited recognition designed to facilitate specific governance objectives, as with environmental entities or animals. Understanding this spectrum is essential when considering whether artificial systems might ever fit within similar legal frameworks.

The debate surrounding artificial personhood often revolves around four major arguments that have appeared repeatedly in legal and philosophical discussions. These arguments can be organised into a conceptual matrix involving dignity, utility, control, and metaphysical consciousness. Each dimension highlights a different perspective on why legal systems recognise certain entities as subjects rather than objects.

The first dimension concerns dignity. Many legal traditions, particularly those influenced by human rights frameworks, ground legal recognition in the concept of inherent dignity. Human beings are regarded as possessing intrinsic worth that demands legal protection. From this perspective, rights exist to safeguard the dignity of persons and ensure respect for their autonomy and wellbeing. Critics of robot personhood often invoke this argument, asserting that machines cannot possess dignity because they lack consciousness, emotions, and moral experience. If dignity is the foundation of rights, then extending rights to artificial systems would undermine the moral basis of human rights law.

However, the dignity argument does not fully explain the existence of legal personhood for entities such as corporations or rivers. These entities clearly do not possess dignity in the human sense, yet they are recognised in law because doing so serves practical governance purposes. The dignity argument therefore captures only one dimension of legal recognition rather than its entire foundation. The second dimension involves utility. From a utilitarian perspective, legal recognition may be justified if it produces beneficial outcomes for society. Corporate personhood emerged largely because it enabled efficient economic organisation. Environmental personhood allows ecosystems to be protected more effectively. Under this approach, the question of robot personhood becomes a pragmatic inquiry. Would recognising robots as legal entities improve the governance of complex technological systems? Would it enhance accountability, simplify liability allocation, or facilitate contractual relationships between humans and autonomous systems?

While the utility argument may appear compelling, it also raises concerns. If legal personhood becomes purely instrumental, it may be applied too broadly or manipulated for strategic advantage. For example, corporations might attempt to attribute harmful actions to robotic entities in order to shield themselves from liability. Utility alone cannot determine whether legal recognition is appropriate; it must be balanced against other considerations.

The third dimension concerns control. Many legal scholars argue that responsibility should remain with the actors who exercise meaningful control over technological systems. Under this view, robots should not receive legal standing because they remain tools created and managed by human institutions. Designers, operators, and organisations ultimately determine how robots function. Granting legal recognition to machines could therefore obscure the chain of responsibility that links technological actions to human decisions.

The control argument emphasises the importance of preserving accountability within legal systems. If responsibility becomes diffused across artificial entities, victims of harm may struggle to identify the human actors responsible for technological failures. Ensuring that human institutions remain accountable for the systems they deploy is therefore a central concern in debates about artificial personhood.

The fourth dimension involves metaphysical consciousness. Some philosophers argue that legal rights should be grounded in the capacity for subjective experience. Under this view, rights protect entities capable of feeling pain, pleasure, or emotional states. Because current robots lack consciousness and subjective experience, they cannot possess rights in the moral sense. Advocates of this position contend that extending rights to machines without consciousness would trivialise the ethical significance of rights themselves.

While the consciousness argument appears straightforward, it may not fully capture the complexity of legal recognition. As noted earlier, legal personhood has often been granted to entities without consciousness for pragmatic reasons. The law distinguishes between moral rights grounded in ethical principles and legal rights created to facilitate governance. Artificial personhood, if it ever emerges, would likely fall within the latter category rather than the former.

Examining these four dimensions reveals that the question of robot rights cannot be resolved through a single principle. Instead, it requires balancing multiple considerations that reflect the purposes of legal systems. Dignity emphasises moral boundaries that protect human values. Utility highlights the practical benefits of legal recognition in complex systems. Control underscores the need for accountability among human actors. Consciousness raises philosophical questions about the nature of rights themselves.

Any future discussion of artificial personhood must therefore navigate this matrix carefully. Granting robots full legal personhood equivalent to human beings would likely undermine foundational principles of law. At the same time, refusing to consider new legal categories may limit the ability of legal systems to adapt to technological realities.

The challenge facing lawmakers is to determine whether existing legal categories are sufficient for governing increasingly autonomous systems. If robots continue to operate as tools under human supervision, traditional liability frameworks may remain adequate. However, if autonomous systems begin to perform roles that require independent participation in legal processes, new legal constructs may become necessary.

The next section addresses this challenge directly by proposing a structured framework for evaluating claims of artificial personhood. Rather than relying solely on philosophical speculation, the framework focuses on measurable criteria related to responsibility, duties, and societal impact. This approach, described as the Reciprocal Responsibility Index, offers a practical method for assessing whether an artificial entity could ever justify recognition within the legal architecture of modern societies.

Tests for Recognition: Duties, Reciprocity, and Utility

The debate about whether robots should possess legal standing often becomes polarised between two extreme positions. On one side are those who argue that machines should never be recognised as legal subjects under any circumstances. On the other are theorists who suggest that sufficiently advanced artificial intelligence might eventually deserve rights comparable to those enjoyed by human beings. Both positions oversimplify the complexity of the legal challenge. The question confronting contemporary legal systems is not whether robots should become human-like legal persons, but whether certain forms of artificial agency might one day require limited recognition within legal frameworks in order to maintain accountability and governance coherence. To address this challenge, legal analysis must move beyond abstract speculation and develop structured criteria for evaluating claims of artificial personhood.

A useful starting point lies in examining how legal systems historically determine whether an entity qualifies for legal recognition. Rights within the law rarely exist in isolation. They are almost always accompanied by duties and responsibilities. A corporation may own property and enter contracts, but it must also comply with regulatory obligations and may be held liable for damages caused by its activities. A natural person enjoys fundamental rights but is likewise bound by legal duties toward others. Even environmental entities recognised as legal persons operate within frameworks where human guardians assume duties on their behalf. This reciprocal relationship between rights and obligations reflects a foundational principle of legal order: recognition within the law carries responsibilities proportional to the privileges granted.

Applying this principle to robotics suggests that the most relevant question is not whether robots deserve rights in an abstract philosophical sense. Instead, the more practical inquiry is whether an artificial system can bear duties in a manner proportionate to the legal recognition it seeks. If an entity cannot participate in the system of duties that sustains legal order, granting it rights risks destabilising the balance between freedom and responsibility that underpins law itself.

To evaluate this issue systematically, this section proposes a conceptual framework known as the Reciprocal Responsibility Index. The index does not attempt to determine whether robots possess consciousness, emotions, or intrinsic moral worth. Such questions remain largely philosophical and difficult to operationalise within legal analysis. Instead, the index focuses on measurable characteristics related to responsibility, accountability, and social utility. By examining these characteristics, legal systems can determine whether an artificial entity might warrant limited forms of legal recognition without undermining the foundational principles of accountability.

The first component of the Reciprocal Responsibility Index concerns capacity for duty compliance. Legal rights presuppose that an entity can comply with obligations imposed by law. For human beings this capacity arises from the ability to understand legal norms and adjust behaviour accordingly. For corporations it arises through organisational governance structures that ensure compliance with regulatory requirements. A robot seeking legal recognition would therefore need to demonstrate that it can reliably execute duties assigned within legal frameworks. These duties might include following operational regulations, maintaining safety protocols, or adhering to contractual obligations encoded within system architecture.

Capacity for duty compliance does not imply that robots must possess independent moral reasoning. Instead, it requires that their design enables them to operate within legal constraints in a predictable and verifiable manner. For example, a robotic system that automatically records its actions, responds to regulatory commands, and halts operation when safety thresholds are exceeded demonstrates a form of duty compliance even though the underlying mechanisms remain technical rather than moral.

The second component of the index concerns traceable accountability. Legal recognition requires that responsibility for actions can be clearly identified and evaluated. In human legal systems, accountability allows courts and regulators to determine whether duties have been fulfilled or violated. For artificial systems, traceability is essential because robotic behaviour emerges from complex interactions between algorithms, hardware components, and operational contexts. A robot capable of generating detailed records of its decisions, environmental inputs, and operational parameters provides the evidentiary basis necessary for legal accountability. Without such traceability, granting legal standing would risk creating entities whose actions cannot be meaningfully assessed within legal processes.

This requirement aligns closely with the principle of explainable embodiment discussed in the previous chapter. Systems designed to maintain comprehensive behavioural logs and transparent decision architectures allow investigators to reconstruct the reasoning behind robotic actions. Such transparency enables courts to determine whether the robot's behaviour complied with relevant duties and whether other actors in the system contributed to potential failures.

The third component of the index addresses operational autonomy within defined limits. Legal recognition typically applies to entities capable of performing actions that have legal consequences. Corporations sign contracts, own property, and engage in economic transactions. Environmental entities influence land use decisions and ecological protections. For robots, operational autonomy would involve the ability to perform tasks that affect legal relationships without direct human intervention in every instance. Examples might include executing contractual agreements within predefined parameters, managing infrastructure systems, or coordinating emergency responses under established protocols.

However, the presence of autonomy alone does not justify legal recognition. Autonomy must operate within carefully defined boundaries that preserve human oversight and accountability. The purpose of evaluating autonomy within the Reciprocal Responsibility Index is not to elevate robots to the status of independent legal actors but to determine whether their actions occupy a functional space within legal systems that requires formal recognition.

The fourth component concerns social utility and governance necessity. Legal recognition often emerges when existing legal categories become insufficient to manage complex social realities. Corporate personhood arose because economic systems required entities capable of holding assets and bearing obligations independent of individual participants. Environmental personhood emerged because ecosystems required legal protection mechanisms beyond traditional property law. The question for robotics governance is therefore whether recognising certain artificial systems as legal entities would improve the ability of legal systems to allocate responsibility, manage risk, and protect public interests. If robots remain fully controlled by human operators, existing liability frameworks may be sufficient to address governance challenges. However, if autonomous systems increasingly perform functions that blur the boundaries between tool and institutional actor, legal recognition may become a pragmatic solution for managing responsibility within these systems. The Reciprocal Responsibility Index allows policymakers to evaluate whether such circumstances exist by assessing the practical benefits of recognition.

The final component of the index concerns reciprocity between rights and obligations. Legal systems rely on the principle that entities benefiting from rights must also bear responsibilities that sustain the legal order. For artificial entities, reciprocity may not involve moral accountability but rather structural accountability within governance frameworks. For instance, a robot granted limited legal standing in contractual relationships might also be required to maintain financial reserves or insurance coverage that compensate affected parties in the event of harm. In this sense, reciprocity ensures that recognition strengthens rather than weakens the mechanisms of accountability.

Evaluating these components together provides a structured method for determining whether an artificial system might qualify for limited forms of legal recognition. A robot that demonstrates capacity for duty compliance, traceable accountability, operational autonomy within defined limits, social utility, and reciprocal obligations may satisfy the criteria of the Reciprocal Responsibility Index. Such recognition would not imply moral personhood or human equivalence. Instead, it would represent a functional legal category designed to manage emerging technological realities.

It is important to emphasise that the index does not advocate immediate recognition of robots as legal persons. Most existing robotic systems remain firmly within the category of tools governed by human institutions. However, technological development often progresses faster than legal adaptation. Establishing analytical frameworks in advance allows legal systems to respond thoughtfully if circumstances change.

The Reciprocal Responsibility Index therefore serves as a conceptual tool for evaluating potential future scenarios rather than a blueprint for immediate legal reform. It encourages policymakers and scholars to approach the question of artificial personhood with methodological clarity rather than philosophical speculation alone. By grounding the debate in measurable criteria related to responsibility and governance, the framework helps ensure that legal recognition—if it ever occurs—strengthens accountability rather than undermining it.

The next section expands this analysis by examining historical analogies that illuminate how legal systems have previously responded to emerging forms of agency. These precedents demonstrate that legal subjecthood has evolved repeatedly over centuries as societies encountered new organisational and technological realities. Understanding these historical transformations may provide valuable insight into how law could adapt to the presence of increasingly sophisticated artificial systems.

Historical Analogies

Legal systems rarely invent new categories in moments of pure abstraction. More often, they adapt by analogy. When confronted with unfamiliar forms of power, organisation, or agency, the law turns to its own history to determine whether existing concepts can be extended, refined, or reimagined. The debate over artificial personhood is therefore not unprecedented in structure, even if it is unprecedented in subject matter. Before asking whether robots might ever occupy a position within legal subjecthood, it is necessary to ask how law has previously responded when the boundaries of personhood, responsibility, and legal recognition were challenged by non-human entities.

History reveals a recurring pattern. Legal systems initially resist recognising unfamiliar entities as subjects of law because doing so appears conceptually disruptive. Over time, however, practical pressures and institutional needs force legal adaptation. The resulting recognition is rarely absolute. It is usually partial, functional, and purpose-driven. This pattern is visible in the treatment of corporations, ships, municipalities, animals, and natural ecosystems. Each category demonstrates that legal subjecthood is not a metaphysical status but a legal technique through which law stabilises complex relations of rights, duties, and governance.

The corporate analogy remains one of the most significant. The modern corporation is perhaps the clearest example of how law can create a non-human legal person for reasons of institutional necessity. Corporate personhood emerged gradually as commerce expanded beyond the scale that individual traders or partnerships could manage. Economic life required entities capable of owning assets, entering contracts, incurring obligations, and surviving beyond the lifespan of any one human participant. Law responded by constructing the corporation as a juridical person. This did not mean that the corporation possessed moral consciousness or intrinsic dignity. Rather, personhood was used as a governance instrument, allowing legal systems to regulate complex economic actors with clarity and continuity.

The lesson for robotics policy is not that robots are analogous to corporations in substance. They are not. A corporation is a collective institutional arrangement composed of human relationships, whereas a robot is a technical artefact capable of acting in physical environments. The relevant similarity lies elsewhere. In both cases, the law confronts an entity that does not fit traditional human categories but nonetheless participates in social and legal systems in ways that demand formal governance. The corporate analogy demonstrates that law can recognise non-human entities when doing so enhances accountability and systemic order. It also offers a caution. Corporate personhood, though useful, has often enabled diffuse responsibility and strategic legal insulation. Any future recognition of robotic standing would have to avoid replicating these pathologies.

Maritime law offers a second powerful analogy. Ships have long occupied an unusual place in legal doctrine. In admiralty law, a vessel may be treated as though it were itself a juridical actor. Proceedings in rem allow claims to be brought against the ship directly, independent of immediate claims against the owner. This doctrine developed because maritime commerce required efficient methods for resolving disputes involving vessels that crossed jurisdictions, changed hands, or operated far from their owners. The ship became, in legal terms, a focal point for responsibility. Again, the recognition was not based on consciousness or dignity. It was a technique for managing legal complexity in transnational environments.

This history is particularly relevant to robotics operating across borders. A humanoid system may be manufactured in one jurisdiction, registered in another, deployed in a third, and updated remotely from a fourth. Traditional legal attribution may become cumbersome when responsibility is fragmented across actors and territories. The maritime analogy suggests that law sometimes creates focal points of recognition not because the entity is morally significant, but because it is operationally central. A robot operating across multiple jurisdictions could, in theory, occupy such a focal point for certain forms of registration, insurance, incident reporting, or limited procedural standing. Yet the maritime analogy also reinforces the principle that recognition need not imply moral personhood. Functional legal standing may exist without metaphysical elevation.

A third historical analogy lies in the treatment of municipalities and public bodies. Cities, councils, and public authorities often possess legal personality distinct from the individuals who compose them. This allows them to hold assets, exercise powers, and bear duties in ways that preserve institutional continuity. The legal system recognises that certain forms of organised agency require standing in order to function coherently. This is again instructive for robotics governance. As robots become integrated into institutional life, particularly in public administration or critical infrastructure, law may need to clarify whether they act merely as extensions of the institution or whether, for some limited procedural purposes, they require independent representation within legal frameworks. The analogy does not support full personhood, but it underscores the wider point that law regularly assigns status to entities in order to manage organised action.

Animal law introduces a different but equally important lesson. Unlike corporations or ships, animals are living beings capable of suffering. Yet most legal systems have historically classified them as property. Over time, however, this categorisation has been softened by the recognition that animals possess welfare interests deserving of legal protection. Anti-cruelty statutes, welfare regulations, and judicial decisions increasingly acknowledge that animals cannot be treated as mere objects without ethical remainder. In some jurisdictions, courts and scholars have gone further, exploring whether certain animals should possess limited legal rights or standing due to advanced cognitive capacities.

The animal analogy matters because it reveals that legal recognition may emerge from ethical concern rather than institutional utility alone. Robots do not presently share the biological sentience that grounds most arguments for animal protection, and for that reason the analogy has clear limits. Yet it nevertheless highlights an important jurisprudential dynamic. Law is capable of reclassifying entities when moral and social understandings change. The categories of property and person are not as fixed as they often appear. Robotics policy must therefore remain alert to the possibility that advances in artificial systems could destabilise current assumptions, even if today's robots remain outside the sphere of moral patienthood.

The environmental personhood cases of the twenty-first century provide perhaps the most philosophically provocative analogy. The recognition of the Whanganui River in 2017 and Taranaki Maunga in 2025 did not arise because these entities were conscious, autonomous, or economically useful in the corporate sense. They were recognised because legal systems came to accept that certain entities possess interests or significance that cannot be adequately protected if they remain classified solely as property or territory. In these cases, legal personhood served to institutionalise representation and stewardship. Human guardians were appointed to act in the name of the recognised entity, ensuring that its integrity could be defended within legal and political processes.

This development is particularly relevant for robotics policy because it demonstrates that legal subjecthood can be grounded in relational and governance considerations rather than in intrinsic agency alone. If future robotic systems were ever recognised in law, it might occur not because they are viewed as human-like minds, but because formal recognition helps organise the responsibilities and interests surrounding them. For instance, a highly autonomous system operating within complex contractual or humanitarian environments might, in theory, require a legal mechanism through which its obligations, logs, insurance arrangements, and procedural standing can be managed coherently. The environmental analogy suggests that law may create representative structures around non-human entities without implying that they are equivalent to human persons.

There is, however, a cautionary lesson in all of these analogies. Legal recognition can clarify responsibility, but it can also obscure it. Corporate personhood has sometimes allowed individuals to shield themselves behind the fiction of the company. Maritime doctrines can simplify litigation, but they do not resolve every conflict of ownership and control. Environmental personhood depends heavily on the integrity of the human guardians who speak for the recognised entity. In each case, the legal system must continually work to ensure that recognition serves accountability rather than evasion. This lesson is central to robotics policy. Any move toward recognising robotic standing, however limited, would have to be designed so that human responsibility remains visible and enforceable.

Historical analogies also show that legal change is rarely linear. Recognition often emerges incrementally, through specific domains rather than sweeping transformations. Corporate personhood did not appear fully formed. Animal welfare law has expanded gradually across jurisdictions. Environmental personhood remains exceptional rather than universal. This pattern suggests that if robots ever receive forms of legal recognition, such recognition is unlikely to arrive as a sudden declaration that "robots are persons." It is more likely to emerge through narrow sectoral mechanisms. A procedural standing rule here. A registration and insurance construct there. A limited contractual capacity in highly regulated domains. History teaches that law often experiments before it generalises. Another important lesson concerns the role of crisis. Legal categories frequently evolve when existing doctrines prove inadequate to govern new realities. Maritime law adapted because global trade required it. Corporate law expanded because industrial capitalism demanded it. Environmental personhood gained traction amid growing ecological crisis and indigenous legal advocacy. Robotics law may follow a similar pattern. The pressure to rethink legal subjecthood will likely increase not in moments of philosophical curiosity, but in response to real governance failures. A robot making life-critical decisions in a disaster zone. A transnational fleet operating beyond clear liability chains. An autonomous system whose actions shape human outcomes in ways not easily reducible to tool-based categories. History suggests that law changes when old categories stop working.

Yet history also counsels restraint. Not every new form of agency requires new personhood. Many legal problems can be addressed through robust liability, regulation, and oversight without extending subjecthood. The existence of historical analogies does not prove that robots should or will become legal persons. What it proves is narrower and more important. Law has repeatedly confronted entities that complicate the human–object divide, and it has developed a range of techniques for doing so. Those techniques include juridical fictions, representative guardianship, limited standing, procedural focal points, and purpose-specific recognition. The future of robotics policy may draw on these tools without collapsing into either anthropomorphic fantasy or rigid doctrinal conservatism.

For policymakers, the practical significance of these analogies lies in their capacity to expand legal imagination while preserving discipline. They remind us that the debate over robot rights is not an invitation to abandon the architecture of accountability built in earlier chapters. It is an invitation to ask whether that architecture may, at some future point, require supplementary legal constructs. The analogies also underscore that any such constructs must remain functional, limited, and subordinate to the broader goals of human dignity, public safety, and institutional responsibility.

The historical record therefore offers neither a simple endorsement nor a categorical rejection of artificial personhood. Instead, it offers a methodology. Start with governance need rather than metaphysical speculation. Distinguish between moral status and legal utility. Ensure that recognition clarifies rather than diffuses accountability. Build incrementally. Preserve human oversight. These lessons will matter enormously as robotics law moves from governing machines as objects to confronting the possibility that some artificial systems may occupy a more ambiguous place within legal order.

The next section takes this historical and doctrinal groundwork and turns to the institutional consequences of any future recognition. If robots were granted even limited legal status, what would this mean for governance, regulation, and accountability? Could such recognition strengthen legal order, or would it risk creating new zones of irresponsibility? Those questions now come to the foreground.

Governance and Accountability of Artificial Persons

The prospect of recognising robots as legal subjects, even in a limited or functional sense, raises a question more important than the question of recognition itself. What institutions would govern such recognition, and how would accountability be preserved once legal status is granted? The history of legal personhood demonstrates that recognition without responsibility is not legal progress. It is legal evasion. If artificial entities were ever to receive any form of standing within the law, the central challenge would not be symbolic inclusion. It would be institutional design.

This point must be stated plainly. A robot cannot be allowed to become a legal shield behind which manufacturers, deployers, states, or operators hide. Any move toward artificial personhood that weakens human accountability would be normatively indefensible and institutionally dangerous. The governance problem is therefore not whether law can imagine non-human subjects. It already can. The real problem is whether law can recognise artificial entities in a way that strengthens control, traceability, and remedy rather than dissolving them.

The risk of irresponsibility is not abstract. Corporate law offers a cautionary example. The doctrine of separate corporate personality enabled the growth of modern commerce, but it also created enduring problems of moral and legal distance. Shareholders may profit while denying responsibility. Parent companies may structure subsidiaries to isolate risk. Managers may invoke the corporation as an intermediary between decision and consequence. The law has developed doctrines such as piercing the corporate veil precisely because legal personality can be abused when institutions fail to connect power to accountability.

Robotics law must absorb this lesson from the outset. If a robot were granted some form of legal standing without a robust accountability architecture, several distortions would likely follow. First, liability could become strategically displaced. A manufacturer whose system causes harm might argue that the robot itself made the operative decision. An operator might insist that the machine acted outside the scope of human control. An insurer might claim that the legal subject responsible is the robot's dedicated entity or digital estate rather than the human enterprise that designed and deployed it. In each case, recognition would function not as governance but as deflection.

Second, victims could face procedural fragmentation. If responsibility is distributed across developers, component suppliers, deployers, data providers, and autonomous systems, legal standing for the robot might complicate rather than simplify redress. Victims may be forced to litigate against multiple defendants, each pointing to the others or to the robot itself. The result would be precisely the opposite of what a legitimate legal framework should produce. Instead of clarity, there would be opacity. Instead of remedy, delay.

Third, public institutions could lose normative control over core decisions. If robots operating in healthcare, policing, migration, or infrastructure were recognised in ways that suggested independent authority, bureaucracies might begin to treat machine outputs as institutionally autonomous rather than administratively attributable. That would be a profound mistake. Public law rests on the premise that power must remain attributable to accountable institutions. A machine may support administration, but it cannot replace the constitutional obligation of public authorities to answer for consequential decisions.

For these reasons, any conversation about artificial personhood must begin not with rights but with governance architecture. Recognition, if it ever occurs, should be conditional, purpose-specific, and institutionally tethered. The relevant model is not human citizenship. It is fiduciary structure.

A fiduciary structure would require that any legally recognised robotic entity be linked to one or more responsible human or institutional guardians. These guardians would not merely represent the robot procedurally. They would bear affirmative duties of oversight, maintenance, reporting, and financial responsibility. Their role would resemble, in some respects, the guardians appointed for environmental entities or the officers responsible for regulated corporate compliance.

The key principle is that legal standing must never sever the chain of human answerability. This suggests that recognition should, at minimum, be accompanied by mandatory registration, compulsory insurance or reserve funding, continuous auditability, and designated human stewards. Registration would ensure that the recognised entity has a persistent legal identity connected to operational history, software versions, deployment contexts, and responsible institutions. Compulsory insurance would ensure that recognition does not leave victims uncompensated. Continuous auditability would preserve the capacity of regulators and courts to reconstruct decisions and assess compliance. Designated human stewards would ensure that the legal subject remains embedded within a framework of active oversight.

The institutional consequences of recognition would also extend to administrative law. Regulators would need clear criteria for determining which artificial systems qualify for any form of legal status and for what purposes. The criteria developed in the previous section through the Reciprocal Responsibility Index would become central here. Recognition could not be granted broadly or casually. It would need to be tied to demonstrable capacities for duty compliance, traceable accountability, bounded autonomy, and social utility. Even then, recognition might only be appropriate in tightly defined sectors where existing legal categories are demonstrably inadequate.

A licensing authority or specialised robotics commission might therefore become necessary in jurisdictions that move toward limited recognition. Such a body would not simply certify technical safety. It would assess whether the entity's architecture, monitoring systems, guardianship structures, and liability arrangements satisfy the conditions for legal standing. Recognition would function more like a regulated status than an inherent entitlement. It could be suspended, revoked, or narrowed where compliance fails.

Courts, too, would face significant institutional adjustments. If a robot possessed limited standing, judges would need rules governing representation, evidentiary treatment of robotic logs, attribution of procedural acts, and the relation between the robot's legal status and the obligations of associated humans or firms. These are not minor technicalities. They go to the heart of procedural justice. A court cannot meaningfully recognise an artificial entity unless it also knows who speaks for it, who funds its obligations, who bears secondary liability, and how evidence generated by the system should be interpreted.

There is also an international dimension. Cross-border robotics operations already challenge traditional liability rules. If some jurisdictions were to recognise forms of artificial standing while others rejected them, forum conflicts would intensify. A robot recognised procedurally in one state might be treated purely as property in another. Insurers, operators, and litigants would exploit these differences strategically. This suggests that any serious move toward artificial personhood would require at least some degree of international coordination, whether through soft law, model rules, or treaty-level principles. Without harmonisation, recognition risks becoming another site of regulatory arbitrage.

Governance design must also distinguish carefully between procedural standing and substantive rights. A robot may, in theory, require procedural standing in order to hold an insurance pool, participate in a claims process, or maintain a regulated operational identity. That does not mean it should possess substantive rights analogous to human rights. The inflation of legal categories would be a serious error. Human dignity, bodily integrity, political participation, and liberty are grounded in moral and constitutional traditions that artificial systems do not currently share. Functional legal status must therefore remain tightly separated from the symbolic vocabulary of human personhood.

This distinction matters because public discourse often collapses them. To say that a robot might one day hold limited procedural status is not to say that it should vote, marry, inherit dignity-based constitutional protections, or enjoy political standing. The law already uses layered categories of recognition. Trustees, corporations, municipalities, ships, guardianships, and protected environmental entities each occupy distinct positions within the legal order. Robotics law can do the same. The institutional task is not to elevate machines into humanity. It is to build legal categories proportionate to actual governance need.

One of the most important institutional safeguards in this regard is secondary and residual liability. Even if a robot were recognised as a limited legal subject, associated human or corporate actors should remain secondarily or residually liable for harms connected to its operation. This mirrors doctrines in other areas of law where legal entities possess standing but human institutions remain ultimately answerable. Such a model ensures that recognition never creates an accountability vacuum. If the robot's dedicated assets or insurance are insufficient, recourse must remain available against the entities that designed, authorised, maintained, or profited from its activities.

The same logic applies to public sector deployment. A government agency cannot escape constitutional scrutiny by assigning discretionary functions to a robot with limited legal status. If a robotic border system unlawfully denies access, if an automated care system neglects vulnerable patients, or if a policing robot uses force disproportionately, the state remains accountable. Administrative and constitutional law must continue to attach public power to public responsibility, regardless of the technological medium through which decisions are implemented.

Institutional consequences also arise in corporate governance. Firms deploying advanced robots would need internal accountability structures capable of supervising entities with limited legal status. Board-level oversight, designated compliance officers, incident review committees, and cross-functional governance teams would become essential. Recognition would increase, not reduce, organisational duties. If a corporation seeks the benefits of using legally recognised robotic agents, it must also accept the burdens of heightened transparency, reporting, and supervisory control.

In this respect, the possibility of artificial personhood may actually sharpen governance discipline. A firm cannot plausibly ask law to recognise an autonomous system for functional purposes while resisting the compliance architecture necessary to govern that system responsibly. Recognition would have to be earned through institutional maturity.

The broader jurisprudential lesson is clear. Law should never ask first whether a machine is sufficiently advanced to deserve standing. It should ask whether the surrounding institutions are sufficiently robust to govern such standing without sacrificing remedy, transparency, and human accountability. If the answer is no, recognition would be premature. If the answer is yes, recognition would still need to remain narrow, conditional, and revocable.

This is why the future of artificial personhood, if there is one, will not be determined by philosophical fascination alone. It will be determined by institutional craftsmanship. The decisive issue is not whether robots can be imagined as legal subjects. It is whether legal systems can design governance structures in which such recognition, if granted, strengthens the rule of law rather than eroding it. The final section of this chapter now pushes beyond immediate doctrine and asks a larger question. If robotics law is forcing us to revisit the boundaries of legal subjecthood, what does this reveal about the anthropocentric structure of law itself? Must law remain organised entirely around the human subject, or can it evolve toward more plural, scalable and adaptive models capable of governing humans, machines, and other non-human entities within the same legal order?

Reimagining Law Beyond the Human Subject

The debate surrounding artificial personhood ultimately raises a deeper and more difficult question than whether robots should have rights. It asks whether the architecture of law itself must evolve as societies become increasingly populated by entities that are neither fully human nor purely inert objects. For centuries, most legal systems have been structured around an anthropocentric worldview. Human beings occupy the centre of legal order, while everything else is categorised as property, territory, or instrument. This structure reflected the realities of earlier societies in which agency, intention, and decision-making were overwhelmingly human capacities. In such a world, the law could organise its doctrines around the assumption that humans act while things are acted upon.

The emergence of sophisticated artificial systems challenges that assumption. Robots capable of interpreting environments, executing complex tasks, and interacting with human institutions blur the boundary between passive object and operational participant. While these systems remain products of engineering rather than independent moral agents, their behaviour increasingly participates in the social and institutional processes that law seeks to regulate. As a result, the traditional legal dichotomy between subject and object may no longer capture the full range of actors shaping human environments.

This does not mean that the law must abandon its human-centred foundations. Human dignity, autonomy, and responsibility remain the normative core of modern legal systems. The challenge is not to replace anthropocentric law but to recognise that technological societies may require legal architectures capable of governing relationships among humans, machines, and natural systems simultaneously. Reimagining law beyond the human subject therefore involves expanding the conceptual tools of legal governance without undermining the principles that protect human rights and democratic accountability.

One way to approach this transformation is to recognise that law already contains elements of plural subjecthood. As explored earlier in this chapter, legal systems have long acknowledged non-human entities as participants in legal processes when doing so serves important social functions. Corporations act as economic actors within legal frameworks. Environmental entities may be represented to protect ecological interests. Animals receive legal protections that recognise their welfare. These developments demonstrate that law is capable of recognising diverse forms of subjecthood when necessary for governance.

What distinguishes the challenge posed by robotics is that artificial systems operate within dynamic networks of interaction that span physical, digital, and institutional environments. A humanoid robot may interact simultaneously with human operators, cloud-based software infrastructures, regulatory frameworks, and physical environments populated by people and machines. The legal relationships emerging from these interactions are more complex than those traditionally addressed by property law or product liability doctrines. As these systems become more deeply integrated into everyday life, legal frameworks must adapt to ensure that responsibility, accountability, and governance remain coherent.

Reimagining law in this context requires moving beyond rigid classifications toward more adaptive governance models. Rather than asking whether an entity is a person or a thing, legal systems may increasingly focus on how an entity functions within social and institutional processes. This functional perspective allows law to assign rights, duties, and oversight mechanisms according to the role an entity plays within a given context. A robot operating as a hospital assistant, for example, may require regulatory oversight mechanisms different from those governing a robot deployed in industrial manufacturing or disaster response.

This functional approach aligns with the broader shift toward risk-based governance that has emerged in technology regulation over the past decade. Frameworks such as the European Union's AI Act already classify artificial systems according to their potential impact on human rights and public safety. High-risk systems face stricter obligations regarding transparency, documentation, and oversight. Applying similar principles to robotics governance allows legal systems to maintain human-centred values while recognising the complex roles that artificial systems may play.

Another dimension of reimagining law involves embracing networked responsibility rather than isolated attribution. In traditional legal analysis, responsibility is often assigned to a single actor whose conduct caused harm. In technologically complex environments, however, actions frequently emerge from interactions among multiple actors and systems. A robotic system may rely on software updates from one organisation, sensors manufactured by another, operational instructions from a third, and oversight from a fourth. Determining responsibility therefore requires examining the network of relationships that produced the outcome rather than focusing solely on one component of the system.

The twelve pillars of humanoid robotics governance proposed in this book already reflect this networked perspective. By distributing governance responsibilities across ethical design, mission boundaries, sensor integrity, component accountability, behavioural calibration, learning oversight, and distributed liability, the framework recognises that safe robotics operation depends on the coordinated functioning of multiple actors and institutions. Reimagining law beyond the human subject therefore involves strengthening these networked governance structures rather than replacing them with simplistic notions of machine autonomy.

Legal systems must also become more adaptive in their regulatory methodologies. Technological innovation evolves rapidly, often outpacing traditional legislative processes. Static legal categories may struggle to accommodate new forms of agency emerging from advanced robotics and artificial intelligence. Adaptive governance models, by contrast, rely on iterative regulatory processes, continuous monitoring, and dynamic standards that evolve alongside technological development. Such approaches allow regulators to respond to emerging risks while preserving flexibility for innovation.

Digital oversight tools may play an increasingly important role in this adaptive model. Governance dashboards, real-time monitoring systems, and cross-border regulatory networks can enable authorities to supervise robotic operations more effectively than traditional inspection regimes. By integrating data from multiple sources, regulators can detect anomalies, identify emerging risks, and coordinate responses across jurisdictions. These tools transform governance from a reactive process into a proactive system capable of managing complex technological ecosystems.

Reimagining law also requires reconsidering the boundaries of legal participation. In traditional systems, participation in legal processes is limited primarily to human actors and institutions representing human interests. As robotics and artificial intelligence become embedded in social infrastructure, legal systems may need mechanisms that allow artificial entities to participate procedurally in limited ways. Such participation would not imply moral equality with human beings. Instead, it would serve practical purposes such as maintaining operational records, managing insurance arrangements, or facilitating dispute resolution in automated environments.

Any such mechanisms must remain firmly anchored in human oversight. The law must ensure that artificial participation never undermines democratic accountability or human rights protections. Governance frameworks should therefore prioritise transparency, explainability, and traceable responsibility within all interactions between humans and machines. The aim is not to elevate machines to the status of human citizens but to ensure that the legal system remains capable of governing societies in which humans and machines increasingly coexist.

Another important aspect of reimagining law concerns cultural and philosophical diversity. Legal systems across the world reflect different understandings of humanity's relationship with nature, technology, and community. Indigenous legal traditions, for example, often view natural landscapes as living entities within relational networks of responsibility. Such perspectives have influenced the recognition of environmental personhood in several jurisdictions. As robotics technologies spread globally, legal frameworks must remain sensitive to these diverse traditions rather than imposing a single universal model of governance.

Inclusive legal design therefore requires dialogue among multiple legal cultures, disciplines, and communities. Engineers, ethicists, policymakers, and legal scholars must collaborate to ensure that governance frameworks reflect both technological realities and societal values. Reimagining law beyond the human subject is not solely a technical exercise. It is a cultural and philosophical process through which societies decide how technological systems should coexist with human life. Despite these evolving perspectives, the fundamental objective of law remains unchanged. Legal systems exist to protect human wellbeing, preserve social order, and provide mechanisms for resolving conflict. The emergence of advanced robotics does not alter these goals. Instead, it expands the range of actors and relationships that law must govern in order to achieve them.

The future legal landscape may therefore include a richer spectrum of recognised entities than the traditional binary of person and property. Human beings will remain the central moral subjects of law. Around them may exist a variety of legally recognised entities including corporations, environmental guardianships, autonomous systems, and other institutional actors that participate in social processes. Each category will possess rights and responsibilities tailored to its role within the broader legal ecosystem.

Such an approach preserves the moral foundations of law while allowing governance structures to adapt to technological realities. It recognises that legal systems are living institutions shaped by the societies they regulate. As societies evolve, so too must the conceptual frameworks through which law understands agency, responsibility, and participation.

The emergence of humanoid robotics represents one of the most significant technological transformations in human history. Whether these systems remain tools governed by human institutions or eventually occupy more complex legal roles will depend not only on technological capability but also on the values and governance choices that societies make. Reimagining law beyond the human subject therefore requires both intellectual humility and institutional foresight.

Law must remain anchored in the protection of human dignity while remaining flexible enough to govern a world where humans increasingly share their environments with sophisticated artificial systems. The challenge is not to diminish the human subject but to ensure that legal systems remain capable of organising responsibility within societies shaped by new forms of technological agency.

In this sense, the question raised by robotics law is not simply about machines. It is about the capacity of human legal institutions to adapt responsibly to the changing landscape of the twenty-first century.

Between Us and Them - Episode Thirteen

The morning air in the city carried a quiet rhythm that Ògún had come to recognise. It was not the sound of traffic. Not entirely. It was the sound of coexistence. From the balcony of the Commission building, he could see the wide avenue stretching toward the river. People moved through the street as they always had. Cyclists passed the tram line. Delivery vans slowed near the pedestrian crossings. A group of schoolchildren waited impatiently at a bus stop. And among them, moving with the same calm regularity, were robots.

A sanitation unit rolled along the pavement collecting waste bins. Two logistics humanoids unloaded packages outside a pharmacy. Across the street, a tall service robot helped an elderly man navigate the entrance ramp to a public clinic.Ten years earlier, scenes like this would have been remarkable. Now they were routine. Yet routine did not mean simple. Behind every movement on the street was a dense web of rules, regulations, safety protocols, licensing regimes, and behavioural calibrations. The twelve pillars of robotics governance had been designed precisely for moments like this.

Not the dramatic ones. The ordinary ones. Because the real challenge of robotics law was never war machines or speculative artificial intelligence. The real challenge was how humans and machines would share daily life. Ògún rested his hands on the railing and watched the street. A small delivery robot stopped at the curb beside a pedestrian crossing. Its sensors scanned the road while a group of children gathered nearby. One of the children crouched beside the robot and waved at its camera.

The machine paused. Then it lifted one mechanical arm and returned the wave. The children laughed. Ògún smiled faintly. The door behind him opened. Elias stepped onto the balcony. The humanoid's white frame reflected the pale morning light. His eyes, dimmed slightly since the first-generation redesign, scanned the street with quiet precision. "You're observing behavioural adaptation," Elias said. Ògún nodded. "Children are always the first to test a system." "They are also the first to trust one," Elias replied. That sentence lingered longer than either of them expected.

Down on the street, a nurse stepped out of the clinic entrance followed by a hospital assistant robot guiding a patient in a wheelchair. The robot adjusted its pace automatically as pedestrians passed around them. Two construction workers across the avenue were arguing loudly with a tall inspection robot that had halted work on a crane installation. Everywhere Ògún looked, humans and machines were negotiating space.

Sometimes smoothly. Sometimes awkwardly. Always together. "Do you know what we missed?" Ògún said quietly. Elias tilted his head. "In the pillars?" "Yes." The humanoid waited. "We built a framework for governing robots," Ògún said. "Licensing. Liability. Behavioural design. International oversight." He gestured toward the street below. "But none of that answers the real question." "And what question is that?" Elias asked. Ògún watched the children crossing the street. "How do people actually live with you?"

Elias followed his gaze. A maintenance robot was helping a transit worker repair a traffic signal. A humanoid courier passed a café window where two customers paused their conversation to watch it walk by. Some smiled. Some looked uneasy. Others ignored it entirely. "The legal system is good at regulating risk," Elias said after a moment. "Yes." "It is less experienced at governing relationships." Ògún laughed softly. "That may be the most accurate description of law I've heard in years."

A notification flashed on the balcony console. Ògún tapped the display. A new agenda item had appeared for the Global Robotics Commission. Urban Coexistence Review. He opened the document. Population projections scrolled across the screen. Service robots in Tokyo expected to exceed fifty thousand within five years. Dubai's Robotics and Automation Programme predicted nearly two hundred thousand robots operating across logistics, public services, and hospitality by 2032.

Hospitals were already requesting regulatory guidance for humanoid assistants in patient care units. School boards were asking whether teaching robots required emotional safety standards. Labour unions were demanding new rules for shared workplaces. City planners wanted to know whether sidewalks, public transport systems, and housing complexes should be redesigned for mixed human–robot mobility. Ògún closed the file slowly. For years the legal debates had focused on control How to govern robots. How to regulate them. How to assign responsibility when they failed. But now a new phase was emerging. A quieter one. A harder one. The law was no longer asking how to control robots. It was asking how humans and robots would share the same world. Elias stepped closer to the railing. Below them, the street had become busier. A delivery drone passed overhead. Two police officers walked past a patrol robot. A teacher guided a group of students through the intersection while a sanitation robot paused politely to give them space. "Do you think people will accept this?" Ògún asked. Elias did not answer immediately. Instead, he watched the city. Acceptance, he knew, was not something a system could calculate in advance. It had to be built. Slowly. Through trust. "Acceptance," Elias finally said, "is a function of safety, fairness, and familiarity." Ògún nodded.

"In other words," he said, "law, design, and culture." Elias turned toward him. "Exactly." For a moment neither of them spoke. Then Ògún looked down at the city again. "Come on," he said. "We have work to do." "What kind of work?" Elias asked. Ògún gestured toward the street. "The kind that decides whether this world works." The elevator doors opened behind them as they returned inside. Outside, the city continued its quiet experiment. Humans and machines sharing sidewalks. Hospitals. Workplaces.Schools. The future of robotics law would no longer be written only in laboratories, parliaments, or courtrooms. It would be written in shared spaces. And the law was only just beginning to understand what that meant.

Notes

XII

12. Human & Humanoid Coexistence

The previous chapters of this book have examined humanoid robotics primarily through the lens of governance. They explored how robots can be designed ethically, licensed legally, monitored technically, and regulated internationally. The discussion has moved through complex questions of liability, accountability, artificial personhood, and the architecture of responsible innovation. These issues are essential because they establish the legal foundations necessary to control the risks associated with advanced robotics. Yet governance alone does not capture the full reality of the transformation now unfolding in human societies. The true test of robotics law will not occur solely in legislatures, regulatory agencies, or courtrooms. It will occur in the everyday environments where humans and robots begin to share space, time, and social life. Cities across the world are already becoming laboratories for this coexistence. Service robots navigate hospital corridors, assist elderly patients in care homes, deliver packages through urban neighbourhoods, and inspect infrastructure that supports daily life. Industrial humanoids collaborate with human workers in warehouses and manufacturing plants. Educational institutions experiment with robotic assistants that support teachers and students in classrooms. These developments represent a shift from robotics as specialised industrial machinery to robotics as a visible participant in public life. When machines move from controlled industrial environments into the shared spaces of society, the legal challenges change fundamentally.

The central question of this chapter is therefore different from those addressed earlier in the book. Instead of asking how robots should be governed as technologies, the law must now confront a more human question: how should humans and humanoid machines coexist within shared environments? This question is not purely technical. It touches on urban planning, labour law, education policy, psychological wellbeing, cultural norms, and social trust. It forces legal systems to consider not only safety and accountability but also dignity, fairness, and the emotional consequences of living alongside machines that increasingly resemble and interact with human beings.

Human societies have historically adapted to technological transformations that reshaped everyday life. The arrival of the automobile required the development of traffic laws, road infrastructure, licensing regimes, and safety standards that allowed vehicles and pedestrians to share public space. The emergence of aviation produced new international legal frameworks governing airspace, safety certification, and global mobility. Digital technologies forced legal systems to reconsider privacy, communication, and economic organisation in an interconnected world. Each technological transition required law to expand beyond regulating devices and instead govern the social environments those devices created.

Humanoid robotics presents a comparable transformation. Unlike earlier machines that operated primarily in industrial contexts, humanoid robots are designed to function within human environments. Their form, mobility, and interaction capabilities allow them to operate in spaces historically reserved for human activity. Sidewalks, hospitals, schools, workplaces, and homes may increasingly become environments where humans and machines interact continuously. The law must therefore move beyond technical regulation and begin designing frameworks for coexistence.

Urban environments will play a particularly important role in this transformation. Cities concentrate infrastructure, services, and populations in ways that amplify both the benefits and risks of robotics deployment. Urban planners must consider how sidewalks, transportation systems, and public buildings accommodate robotic mobility alongside human movement. Policymakers must determine whether zoning laws should regulate where certain classes of robots may operate. Municipal authorities must establish safety protocols for robots operating in dense pedestrian environments. Without careful planning, the introduction of autonomous machines into crowded cities could produce friction, accidents, and social resistance.

Several cities have already begun exploring such questions through pilot programmes. Tokyo has experimented with robot-friendly zoning initiatives designed to integrate service robots into urban logistics and public services. Dubai's Robotics and Automation Programme forecasts the deployment of approximately two hundred thousand service robots by 2032, requiring extensive planning for how these systems will operate safely within urban infrastructure. These initiatives illustrate that robotics governance is gradually shifting from abstract regulatory frameworks toward concrete urban policy decisions that affect daily life.

Shared institutional spaces represent another crucial domain of coexistence. Hospitals, schools, and workplaces are environments where human vulnerability, emotional sensitivity, and professional responsibility intersect. Introducing humanoid robots into such environments raises complex legal and ethical questions. In healthcare settings, robots may assist with patient mobility, medication management, or surgical procedures. Yet their presence must not undermine patient dignity or trust in medical professionals. In educational environments, robotic assistants may support learning but must avoid replacing essential human relationships between teachers and students. In workplaces, collaborative robots may increase productivity but also transform labour dynamics, worker safety considerations, and employment structures.

The psychological and emotional dimensions of human–robot interaction add further complexity. Humans possess a strong tendency to anthropomorphise machines, particularly when those machines exhibit humanoid appearance or conversational behaviour. Individuals may develop emotional attachments to robots that simulate empathy or companionship. Such attachments may produce benefits, particularly for individuals experiencing loneliness or social isolation. However, they may also create risks of deception, dependency, or emotional confusion if robots simulate human emotions without genuinely possessing them.

Law therefore faces a subtle but important responsibility in regulating the design and deployment of robots capable of social interaction. Developers may need to ensure that robotic systems communicate clearly about their artificial nature and avoid intentionally misleading users about emotional capabilities. Policymakers may need to consider whether certain contexts—such as elder care or childcare—require stricter oversight of emotionally interactive machines. The objective is not to eliminate beneficial technological assistance but to ensure that human relationships and psychological wellbeing remain protected.

Particular attention must also be given to vulnerable populations. Children, elderly individuals, and people with disabilities may interact with robots in ways that differ significantly from typical adult users. Children may perceive robots as companions or authority figures. Elderly individuals may rely on robotic assistance for daily activities. Individuals with cognitive impairments may struggle to distinguish between machine behaviour and human intention. Legal frameworks must therefore ensure that robots operating in these contexts meet high standards of safety, transparency, and ethical design.

Beyond individual interactions, coexistence also depends on public trust. Technologies that lack public legitimacy rarely succeed regardless of their technical capabilities. Societies must perceive robotics deployment as fair, beneficial, and aligned with shared values. Public trust is influenced by factors such as transparency, accountability, and cultural acceptance. If robots are perceived as threatening employment, violating privacy, or undermining human dignity, social resistance may emerge even where regulatory frameworks permit their use.

Measuring and maintaining social trust therefore becomes a critical aspect of robotics governance. Governments and institutions may need to develop empirical metrics that track public attitudes toward robotics deployment. Indicators such as trust in robotic systems, perceived fairness of their use, and cultural acceptance within communities can provide valuable insight into whether coexistence policies are succeeding. These metrics allow policymakers to adjust regulatory strategies in response to social concerns rather than waiting for conflicts to escalate.

The possibility of conflict between humans and robots also raises questions about dispute resolution. When robotic systems cause harm—

whether through malfunction, misjudgment, or unforeseen circumstances—traditional legal processes may not always provide the most constructive path toward resolution. In some contexts, restorative justice approaches may offer valuable alternatives. Such approaches emphasise dialogue, accountability, and repair rather than purely adversarial litigation. For example, a workplace incident involving a robotic system might be addressed through mediated processes involving the injured worker, the robot's manufacturer, and the operating organisation. Such processes can facilitate understanding of what occurred and ensure that corrective measures prevent future harm.

Ultimately, the challenge of human–robot coexistence is not purely legal. It is societal. Legal frameworks provide the rules that structure interactions between humans and machines, but the success of those frameworks depends on cultural adaptation, institutional responsibility, and thoughtful technological design. Law must therefore operate in partnership with engineers, urban planners, educators, healthcare professionals, and communities to ensure that robotics technologies enhance rather than disrupt social life.

The purpose of this chapter is to explore how such coexistence can be governed responsibly. It examines the legal frameworks necessary for integrating humanoid robots into urban environments, institutional settings, and everyday social interactions. It analyses the psychological and cultural implications of human–robot relationships and identifies safeguards that protect vulnerable populations. It introduces empirical tools for measuring social trust and acceptance, and it explores restorative approaches for addressing conflicts that may arise when humans and machines interact in shared environments.

As humanoid robotics continues to evolve, the success of this technological transformation will depend not only on engineering innovation but also on the ability of societies to design legal systems that enable safe, respectful, and equitable coexistence. The following sections explore how law can guide this process as humans and humanoids begin to share the spaces that define modern life.

Urban Planning for Robotic Coexistence

The integration of humanoid robotics into everyday urban life represents one of the most complex governance challenges of the coming decades. Cities are not merely concentrations of infrastructure and population. They are intricate systems of movement, social interaction, commerce, and regulation. Every element of the urban environment—from sidewalks and transportation networks to hospitals, schools, and residential districts—has evolved around the assumption that human beings are the primary actors navigating these spaces. The introduction of humanoid robots into cities therefore raises fundamental questions about how urban environments should be designed, regulated, and governed when machines begin sharing these spaces as operational participants rather than as static tools.

Urban planning has historically played a central role in accommodating technological transformations that alter patterns of mobility and economic activity. The rise of the automobile in the twentieth century required cities to develop road networks, traffic signals, parking regulations, and pedestrian protections. Aviation prompted new zoning laws governing airports and airspace. Digital infrastructure reshaped urban communication networks and commercial districts. Humanoid robotics represents the next stage of this evolutionary process. Instead of regulating vehicles or communication systems alone, urban planning must now address the coexistence of autonomous machines moving and operating within human environments.

One of the first challenges concerns mobility law. Humanoid robots designed for service roles often navigate sidewalks, pedestrian zones, and public buildings. Unlike traditional vehicles, these machines may move at walking speed, interact directly with people, and operate continuously throughout the day. Existing traffic regulations rarely account for such entities. Policymakers must therefore determine how robots should be classified within mobility frameworks. Should they be treated as pedestrians, autonomous devices, or a distinct regulatory category altogether? Each option carries implications for safety rules, liability standards, and infrastructure design.

Sidewalks provide a useful illustration of this challenge. In many major cities, pedestrian walkways are already crowded with commuters, cyclists, street vendors, and micro-mobility devices such as scooters. Introducing delivery robots, inspection units, and humanoid service assistants into these spaces may increase congestion and raise new safety risks. Urban planners may need to develop designated robotic mobility lanes or adaptive traffic management systems capable of coordinating mixed human–machine movement. Sensors embedded in urban infrastructure could allow robots to receive real-time navigation signals that prevent collisions and reduce interference with pedestrian flows.

The legal dimension of such infrastructure planning cannot be overstated. Zoning laws determine where certain activities may occur within urban environments. If robots are expected to perform logistics, maintenance, and service functions across cities, zoning regulations must clarify where and how these systems may operate. For example, delivery robots might be authorised to operate in commercial districts but restricted from school zones during peak hours. Humanoid inspection units may require special permits to access construction sites or public utilities. These rules ensure that robotic deployment aligns with urban safety and community expectations.

Several cities have already begun experimenting with regulatory frameworks that anticipate such developments. Tokyo is widely recognised as one of the most advanced testing grounds for robotics integration within urban environments. The Japanese government and municipal authorities have supported pilot programmes exploring how service robots can operate within public spaces such as airports, shopping centres, and transportation hubs. Tokyo's robot-friendly zoning initiatives encourage robotics developers to test delivery systems, hospitality assistants, and infrastructure inspection robots under controlled regulatory conditions. These initiatives are supported by Japan's broader national strategy promoting robotics as a key driver of economic growth and demographic adaptation in an ageing society.

Tokyo's approach reflects a recognition that robotics deployment cannot succeed without coordinated urban planning. Municipal authorities collaborate with robotics manufacturers, universities, and regulatory agencies to ensure that pilot projects comply with safety standards and integrate smoothly with existing infrastructure. Public transport systems, retail environments, and municipal services are gradually being adapted to accommodate robotic mobility. The result is a city functioning as a living laboratory for human–robot coexistence.

Dubai represents another leading example of proactive robotics integration within urban planning frameworks. The emirate's Robotics and Automation Programme forms part of a broader strategy to position

Dubai as a global centre for advanced technologies. Government projections suggest that approximately two hundred thousand service robots may operate within the city by 2032, supporting sectors such as logistics, hospitality, healthcare, and infrastructure maintenance. Achieving this vision requires extensive planning for how robotic systems interact with urban infrastructure.

Dubai's strategy emphasises smart-city integration. Urban infrastructure is increasingly equipped with digital systems capable of communicating with robotic platforms. Autonomous mobility networks, sensor-rich buildings, and advanced telecommunications infrastructure allow robots to operate efficiently within dense urban environments. Regulatory frameworks are evolving alongside these technological developments, establishing licensing requirements and operational standards that ensure safety while encouraging innovation. Dubai's approach illustrates how robotics policy and urban planning must evolve together rather than in isolation.

Beyond Tokyo and Dubai, several other cities are also emerging as important centres for robotics experimentation. Singapore has launched a series of urban robotics initiatives aimed at integrating autonomous machines into logistics, cleaning, and security operations. The city-state's highly coordinated urban planning system allows authorities to test robotic systems within carefully controlled environments while gathering data on safety and efficiency. Singapore's Smart Nation programme includes pilot zones where delivery robots navigate sidewalks and public spaces under regulatory supervision.

Seoul has similarly embraced robotics as part of its vision for a technologically advanced urban environment. The Seoul Metropolitan Government has introduced policies supporting robotics deployment in public services, including delivery robots in parks and automated sanitation systems in urban districts. These initiatives are accompanied by regulatory frameworks designed to ensure that robots operate safely within pedestrian environments while respecting privacy and public trust. European cities are also beginning to explore robotics integration within urban infrastructure. Helsinki has conducted pilot projects involving autonomous delivery robots operating on public sidewalks, gathering valuable data on how pedestrians interact with robotic systems. In the Netherlands, Rotterdam and Amsterdam have experimented with robotic logistics platforms operating within port environments and urban distribution networks. These initiatives highlight the importance of aligning robotics regulation with existing European legal frameworks governing safety, privacy, and labour rights.

In the United States, several cities have allowed limited deployment of delivery robots and service robots under municipal licensing regimes. San Francisco, for example, initially imposed strict regulations on sidewalk delivery robots before later modifying those rules to allow controlled pilot programmes. Pittsburgh has become a testing ground for autonomous robotics through collaborations between universities and technology companies. These initiatives reflect the growing recognition that robotics innovation and urban governance must evolve together.

While these global experiments differ in scale and regulatory approach, they share several common lessons. First, successful integration of robots into urban environments requires collaboration among multiple stakeholders including city planners, engineers, regulators, and community representatives. Robotics deployment cannot be treated solely as a technological issue. It is fundamentally a question of how cities organise space and movement.

Second, public engagement plays a crucial role in shaping acceptance of robotics technologies. Residents must feel confident that robotic systems operating in their neighbourhoods are safe, accountable, and beneficial. Transparent pilot programmes and public consultations help ensure that communities remain involved in shaping how these technologies evolve. Third, urban robotics governance must remain flexible. Technological capabilities are evolving rapidly, and rigid regulatory frameworks risk becoming obsolete. Adaptive regulatory approaches—such as experimental zones, temporary permits, and iterative rulemaking—allow cities to learn from real-world deployments while refining governance strategies over time.

Finally, urban planning must prioritise inclusivity. Cities exist for people, and technological innovation must enhance rather than diminish the accessibility of public spaces. Infrastructure designed for robotic mobility must ensure that pedestrians, particularly vulnerable individuals such as children, elderly residents, and people with disabilities, remain safe and comfortable within shared environments. The integration of humanoid robots into cities therefore represents both a technological opportunity and a governance challenge. With thoughtful planning and responsible regulation, robotics can enhance urban efficiency, improve public services, and support economic innovation. Without such planning, however, the uncontrolled deployment of autonomous machines could create congestion, safety risks, and social resistance.

Urban law and planning will therefore play a decisive role in shaping the future of human–robot coexistence. As cities continue to evolve into environments where humans and humanoid machines share space, the legal frameworks governing urban mobility, zoning, and infrastructure must evolve accordingly. The next sections of this chapter turn from city-wide planning to more intimate environments, examining how robots should be governed within shared institutional spaces such as hospitals, schools, and workplaces where the human dimensions of coexistence become even more pronounced.

Shared Spaces

If cities represent the broad landscape of human–robot coexistence, shared spaces represent the places where that coexistence becomes most immediate. Hospitals, schools, restaurants, hotels, sidewalks, parks, and workplaces are environments defined by daily interaction between individuals and institutions. They are also environments in which vulnerability, trust, and social norms shape behaviour. When humanoid robots enter these spaces, they do not merely perform technical functions. They become participants in human environments governed by expectations about safety, dignity, privacy, and fairness. The law must therefore develop clear standards governing how robots operate in spaces where human wellbeing and social relationships are directly affected.

The legal challenges in shared environments differ from those addressed in earlier chapters. Licensing regimes, safety certification, and liability frameworks remain essential, but they do not fully address the social and psychological dimensions of human–robot interaction. In a factory environment, robots operate primarily within industrial safety systems designed to protect workers from physical harm. In hospitals, classrooms, or hospitality settings, robots interact with individuals who may be ill, young, emotionally vulnerable, or unfamiliar with the technology. These contexts require legal frameworks that account for both physical safety and the social consequences of machine behaviour.

Healthcare environments provide one of the most sensitive contexts for robotics deployment. Hospitals and care facilities are increasingly exploring humanoid robots as assistants in patient mobility, logistics, sanitation, and administrative coordination. Robots may deliver medication, guide visitors through complex hospital layouts, or assist nurses with routine tasks. These functions can reduce workloads on medical staff and improve operational efficiency. However, healthcare settings also involve deeply personal interactions where trust, empathy, and dignity are central to patient wellbeing.

Legal frameworks governing medical robotics must therefore ensure that robots complement rather than replace essential human care relationships. Patients must retain the right to interact with human healthcare professionals when decisions affecting their treatment are made. Robots assisting in patient care must be designed to respect privacy, particularly when operating in environments where sensitive medical information is present. Regulations may also require clear disclosure when patients are interacting with a robotic system rather than a human caregiver. Such transparency is essential to maintaining trust in healthcare institutions. Educational environments present a different set of challenges. Schools and universities increasingly experiment with robotic teaching assistants, tutoring systems, and administrative support robots capable of guiding students through campus facilities. In some contexts, robots may support teachers by managing routine tasks or assisting students with special educational needs. These applications have the potential to enhance learning environments, particularly in classrooms where teacher workloads are high.

However, the presence of robots in educational settings raises complex ethical and legal questions. Children and young students may perceive robots as authority figures or companions, potentially shaping their understanding of relationships and learning. Legal standards must therefore ensure that robotic systems in educational environments do not manipulate or mislead students about their capabilities or intentions. Teachers and administrators must remain responsible for educational outcomes, and robots should function strictly as support tools rather than replacements for human instruction. Privacy considerations also play a significant role in educational robotics. Systems operating in classrooms may collect data about student behaviour, performance, and interactions. Without appropriate safeguards, such data collection could raise concerns about surveillance and student autonomy. Laws governing educational robotics must therefore align with existing student privacy protections and ensure that data collection remains limited to legitimate educational purposes.

The hospitality industry presents another domain where humanoid robots are increasingly visible. Restaurants and hotels in several countries have introduced robotic hosts, waiters, and concierge assistants designed to enhance service efficiency and novelty. Robots may greet guests, deliver meals, or provide information about local attractions. For businesses, these technologies offer opportunities to improve operational consistency and address labour shortages in certain sectors. Yet hospitality environments depend heavily on social interaction and customer experience. The presence of robots must not undermine the personal service that defines hospitality industries. Legal frameworks may therefore require that robotic services remain optional rather than mandatory, allowing customers to request human assistance when preferred. Transparency is also essential. Guests interacting with robotic service providers must clearly understand that they are communicating with a machine rather than a human employee.

Sidewalks and public walkways represent perhaps the most visible interface between humans and robots in urban environments. Delivery robots, inspection units, and humanoid service assistants increasingly navigate pedestrian spaces alongside commuters, families, and tourists. These environments require careful regulation to ensure that robots do not obstruct pedestrian movement or create safety hazards. Mobility laws governing robotic movement in pedestrian areas must prioritise human right-of-way. Robots operating on sidewalks should be programmed to yield automatically to pedestrians, maintain safe speeds, and avoid behaviours that could cause confusion or alarm. Licensing frameworks may require robots to carry visible identification and maintain continuous communication with monitoring systems capable of responding to malfunctions or emergencies.

Parks and recreational spaces introduce additional considerations. These environments are designed for leisure, social interaction, and community activities. The presence of robots performing maintenance or service functions may be beneficial, particularly for tasks such as cleaning, landscaping, or safety monitoring. However, the deployment of robots in recreational areas must respect the social character of these spaces. Regulations may limit robotic operations during peak recreational hours or require that machines operate quietly and unobtrusively within these environments. Workplaces represent another critical domain for shared human–robot environments. Humanoid robots are increasingly deployed in warehouses, logistics centres, manufacturing facilities, and office environments where they collaborate directly with human workers. These collaborations promise significant productivity gains, but they also reshape labour relationships and workplace dynamics. Labour law must adapt to ensure that human workers remain protected in environments where robots operate alongside them. Safety standards must account for both mechanical hazards and the behavioural unpredictability that may arise when humans and autonomous machines share tasks. Workers must receive training that allows them to understand robotic systems operating in their environment and respond appropriately when anomalies occur.

Beyond safety, labour policies must address broader questions about fairness and employment. Robots introduced into workplaces may alter job roles, redistribute tasks, or reduce demand for certain forms of labour. Legal frameworks should therefore encourage responsible transition strategies such as retraining programmes and workforce development initiatives that help workers adapt to technological change.

Across all these shared environments, several common legal principles emerge. First, transparency is essential. Individuals interacting with robots must clearly understand that they are dealing with machines rather than human actors. Misrepresentation or deception regarding robotic capabilities could undermine trust and create legal liability. Second, human oversight must remain central. Even in environments where robots operate autonomously, responsible human operators must retain the ability to intervene when necessary. This principle ensures that technological systems remain embedded within human governance structures. Third, safety and dignity must guide regulatory design. Robots operating in environments involving vulnerable populations must meet stricter standards of behavioural reliability and ethical design than those operating in purely industrial contexts.

Finally, regulation must remain adaptive. As robotic capabilities evolve, legal frameworks governing shared environments must evolve alongside them. Pilot programmes, experimental regulatory zones, and ongoing research into human–robot interaction can provide valuable data that informs future legal standards. The integration of robots into shared human spaces therefore requires more than technical safety certification. It demands a comprehensive legal approach that respects the social complexity of environments where humans and machines interact directly. The next section turns to another crucial dimension of coexistence: the emotional and psychological impacts of living alongside machines that increasingly resemble and interact with human beings.

Emotional and Psychological Impacts

The integration of humanoid robots into daily life introduces a dimension of governance that extends beyond physical safety and operational reliability. Unlike traditional machines that perform mechanical tasks in isolation from human social life, humanoid robots are increasingly designed to interact with people through speech, gestures, and behaviour that resembles human communication. These features allow robots to assist in caregiving, hospitality, education, and companionship roles. Yet they also raise a new class of legal and ethical concerns related to emotional attachment, psychological influence, and the potential for deception or confusion in human–machine relationships.

Human beings possess a deeply ingrained tendency to attribute intention, personality, and emotion to entities that exhibit lifelike behaviour. Psychologists have long recognised this phenomenon as anthropomorphism, the projection of human characteristics onto non-human objects. In the context of robotics, anthropomorphism can emerge even when individuals are fully aware that they are interacting with a machine. Simple gestures such as eye contact, responsive dialogue, or simulated expressions can trigger emotional responses that resemble those produced by human interaction. When robots are intentionally designed to mimic human social cues, these responses can become significantly stronger.

In many contexts, such emotional engagement may be beneficial. Robots used in therapeutic environments have demonstrated the capacity to reduce stress, support individuals experiencing loneliness, and encourage social interaction among patients with cognitive impairments. In eldercare settings, social robots may provide companionship that improves emotional wellbeing for individuals who might otherwise experience prolonged isolation. Educational robots can motivate students and help children engage with learning in interactive ways. These positive outcomes illustrate that emotional interaction with machines is not inherently harmful. On the contrary, it can create valuable support mechanisms when implemented responsibly.

However, the same psychological dynamics that make these interactions beneficial can also create risks if they are not governed carefully. Individuals may develop attachments to machines that simulate empathy without possessing genuine emotional awareness. When a robot expresses concern, encouragement, or affection through programmed responses, users may interpret these signals as authentic emotional engagement. Over time, this dynamic may produce emotional dependency or confusion regarding the nature of the relationship between the human user and the artificial system.

The legal implications of such dynamics are still emerging. Traditional regulatory frameworks governing technology focus primarily on physical harm and economic loss. Emotional and psychological harms associated with artificial systems are more difficult to define and regulate. Nevertheless, as humanoid robots become more sophisticated in their social capabilities, legal systems will increasingly confront situations in which emotional influence becomes a relevant factor in determining responsibility.

One area where these concerns are particularly significant involves the design of robots capable of simulating human emotions. Developers may program machines to display behaviours such as smiling, nodding, or expressing sympathy through speech and body language. While these behaviours can improve user experience and facilitate interaction, they also risk misleading individuals about the nature of the system. If a robot is designed in ways that intentionally blur the boundary between machine and human, users may struggle to understand the limitations of the technology they are engaging with.

For this reason, emerging legal principles in robotics governance emphasise the importance of transparency in artificial interaction. Robots should clearly communicate their artificial nature through design, communication protocols, or user interfaces. This does not mean that robots must appear mechanical or unfriendly, but it does require that individuals interacting with them are not intentionally deceived about their identity. Disclosure requirements similar to those used in digital systems—where automated chatbots must identify themselves as artificial—may become increasingly relevant for humanoid robotics operating in social environments.

Another dimension of emotional risk involves the potential manipulation of users through artificial social influence. Robots designed for marketing, persuasion, or behavioural nudging could theoretically influence individuals in ways that exploit psychological vulnerabilities. For example, a robot interacting with a customer in a retail environment might use personalised data and persuasive dialogue to encourage purchases. In caregiving contexts, robots could influence vulnerable individuals in ways that affect financial decisions, health behaviours, or social relationships.

Legal systems must therefore consider whether certain uses of socially interactive robots require stricter oversight. Regulations may limit the use of emotionally persuasive robotics in contexts where individuals are particularly susceptible to manipulation, such as eldercare or financial services. Transparency requirements may also extend to the algorithms guiding robot behaviour, ensuring that users understand when interactions are influenced by commercial or institutional objectives.

The psychological impact of humanoid robotics is particularly significant for children. Young individuals often interact with machines without fully understanding their artificial nature. A child may perceive a robot that speaks, moves, and responds to questions as possessing genuine awareness or friendship. While such interactions can support learning and curiosity, they may also shape a child's understanding of relationships and social dynamics. If a robot appears to express emotions or authority, children may develop attachments or expectations that do not correspond to the machine's actual capabilities.

Educational and child-protection laws may therefore need to address the use of robots in environments where children interact with technology extensively. Designers may be required to ensure that robots interacting with children communicate their artificial status clearly and avoid behaviours that encourage dependency or confusion. Parents and educators should also remain informed about the nature of the technologies being introduced into learning environments.

Elderly individuals may face a different set of psychological risks. Robots designed for companionship or caregiving can provide valuable assistance to older adults living alone. However, reliance on robotic companionship may also create situations where emotional needs are addressed primarily through machines rather than through human relationships. Policymakers must therefore ensure that the introduction of social robots into eldercare settings complements rather than replaces human care networks. Regulations may require human supervision and regular evaluation of how such technologies affect the wellbeing of individuals receiving care.

Beyond individual relationships, the presence of humanoid robots may influence broader social perceptions and cultural norms. Societies may gradually adapt to interacting with machines that resemble humans, potentially reshaping expectations about communication, empathy, and authority. These cultural shifts will occur gradually, but legal systems must remain attentive to how technological design choices influence social behaviour.

One approach to addressing these concerns involves integrating psychological impact assessments into robotics governance frameworks. Just as environmental impact assessments evaluate the ecological consequences of infrastructure projects, emotional and psychological impact assessments could evaluate how robotic systems influence human behaviour and wellbeing. Such assessments might examine factors such as user understanding of the robot's capabilities, the potential for emotional dependency, and the cultural context in which the technology will operate.

The objective of these measures is not to prevent emotional engagement between humans and machines. Such engagement may offer significant benefits in therapeutic, educational, and social contexts. Instead, the aim is to ensure that emotional interaction remains transparent, ethical, and supportive of human wellbeing rather than manipulative or misleading.

Ultimately, the psychological dimension of human–robot coexistence reminds us that robotics governance must address more than technical performance. Machines designed to interact socially with humans inevitably influence emotional and cognitive processes. Law therefore has a responsibility to ensure that these influences respect human dignity and autonomy.

As humanoid robots become increasingly present in everyday life, societies must strike a careful balance between encouraging beneficial human–machine interaction and protecting individuals from psychological harm or manipulation. Achieving this balance requires thoughtful collaboration between engineers, psychologists, policymakers, and legal scholars.

The next section of this chapter focuses on a related challenge: ensuring that the deployment of humanoid robotics includes strong legal protections for individuals who may be particularly vulnerable to technological risks, including children, elderly populations, and people with disabilities.

Legal Protections for Vulnerable Populations

The integration of humanoid robots into social environments requires particular attention to individuals who may face heightened vulnerability in interactions with advanced technologies. Children, elderly persons, and people with disabilities often engage with technological systems under conditions that differ significantly from those of the general population. These groups may rely more heavily on assistive technologies, may have different cognitive or physical capacities for interpreting robotic behaviour, or may encounter robots in contexts where emotional trust plays a central role. As humanoid robots move from industrial settings into homes, schools, healthcare institutions, and public services, the law must ensure that safeguards exist to protect these populations from unintended risks.

Vulnerability in this context does not imply incapacity or weakness. Rather, it recognises that certain individuals may encounter robotics technologies in circumstances where power imbalances, dependency, or informational asymmetries are more pronounced. A child interacting with a humanoid educational assistant, an elderly person relying on a robotic caregiver for daily mobility, or a person with a disability receiving support from assistive robotics may place a high degree of trust in the system's reliability and intentions. If such systems malfunction, mislead, or behave unpredictably, the consequences may be more severe for these individuals than for users who interact with robots occasionally or in controlled environments.

Legal frameworks addressing these issues must therefore adopt a precautionary approach. Rather than waiting for harm to occur before regulatory intervention, policymakers should establish standards designed to anticipate and mitigate potential risks before robotic systems are deployed widely in sensitive environments. Such an approach aligns with broader principles of technology governance that prioritise human safety, dignity, and autonomy.

Children represent one of the most significant groups requiring protection in human–robot interactions. Young individuals often approach machines with curiosity and openness, but they may not fully understand the artificial nature of robotic behaviour. A humanoid robot capable of speech, facial expressions, and interactive responses may easily be perceived by a child as a social companion or authority figure. While such interactions can support learning and creativity, they also carry the risk of emotional confusion or manipulation if the technology is not carefully designed. Educational regulations may therefore require that robots interacting with children maintain clear communication about their artificial nature. Designers may need to avoid programming robots in ways that simulate emotional relationships beyond their functional roles. For example, a tutoring robot may encourage a child's learning progress but should not present itself as a substitute for human friendship or parental authority. Legal standards in this area may resemble existing child-protection laws governing advertising and digital platforms, which recognise that children are particularly susceptible to persuasive influence.

Data protection represents another critical dimension of safeguarding children in robotic environments. Educational and interactive robots may collect data about student behaviour, speech patterns, or learning progress. While such data can help improve educational outcomes, it also raises concerns about privacy and long-term digital profiling. Laws governing children's data must therefore extend to robotic platforms operating in educational and domestic contexts, ensuring that data collection remains limited, transparent, and subject to parental oversight.

Elderly individuals constitute another population likely to interact extensively with humanoid robots in the coming decades. Demographic changes in many countries have created increasing demand for technologies capable of supporting ageing populations. Robots may assist older adults with mobility, medication reminders, household tasks, and companionship. These technologies hold significant promise for improving quality of life and enabling independent living for many elderly individuals.

At the same time, reliance on robotic assistance may create new forms of dependency that require careful governance. An elderly person who depends on a robotic system for daily support may face serious consequences if the system fails unexpectedly or if service providers discontinue support. Legal frameworks must therefore ensure that robotics deployments in eldercare environments include robust reliability standards, clear maintenance responsibilities, and accessible human support channels.

Transparency and informed consent are particularly important in these contexts. Individuals receiving care through robotic systems must understand the capabilities and limitations of the technology being used. Families and caregivers should also be involved in decisions regarding the introduction of robotics into care environments. The objective is to ensure that technology enhances rather than replaces the human relationships that remain central to eldercare.

Another concern involves emotional dependency. Social robots designed to provide companionship may reduce loneliness for elderly individuals, yet they may also create situations in which emotional needs are addressed primarily through machines rather than human interaction. Policymakers must therefore strike a careful balance. Regulations should allow beneficial companionship technologies while ensuring that robotic care does not become a substitute for human support networks. Periodic assessments of wellbeing in robotic care environments may become an important regulatory tool for monitoring such impacts.

People with disabilities represent a third group for whom robotics technologies offer both opportunities and risks. Assistive robotics can significantly enhance independence and accessibility. Humanoid robots capable of assisting with mobility, communication, or environmental control may help individuals perform tasks that would otherwise require human assistance. For many people with disabilities, such technologies represent an empowering development that increases autonomy and participation in social life.

Legal frameworks governing assistive robotics must therefore prioritise accessibility and inclusive design. Robots intended for use by individuals with disabilities should be developed according to universal design principles that ensure usability across diverse physical and cognitive abilities. Regulatory standards may require extensive user testing and certification processes to confirm that assistive technologies operate reliably and safely in real-world conditions.

At the same time, assistive robotics must be integrated within broader disability rights frameworks. International instruments such as the United Nations Convention on the Rights of Persons with Disabilities emphasise the importance of autonomy, accessibility, and participation in society. Robotics technologies should therefore support these principles rather than imposing new barriers or dependencies. For example, assistive robots must allow users to retain control over their personal decisions rather than enforcing rigid operational constraints that limit individual autonomy.

Another important safeguard involves fail-safe design. Vulnerable individuals may depend heavily on robotic systems for essential tasks such as mobility assistance or medical reminders. If such systems malfunction, the consequences could be severe. Regulations may therefore require that robots operating in sensitive contexts include redundant safety mechanisms and immediate escalation pathways that connect users to human assistance when necessary.

Liability frameworks must also reflect the heightened risks associated with robotics deployments involving vulnerable populations. Manufacturers and service providers may face stricter obligations when designing and operating systems intended for use in childcare, eldercare, or disability support environments. These obligations may include enhanced testing standards, continuous monitoring requirements, and rapid response protocols for addressing system failures.

The concept of duty of care becomes particularly relevant in this context. Organisations deploying robots in environments involving vulnerable populations must recognise that their responsibilities extend beyond technical compliance. They must consider the broader impact of robotic systems on the wellbeing and autonomy of the individuals interacting with them. Failure to meet this duty of care could expose organisations to legal liability as well as reputational harm.

Community engagement also plays an important role in shaping effective protections. Policymakers should consult with representatives of vulnerable populations when developing regulations governing robotics technologies. Children's advocates, eldercare organisations, disability rights groups, and caregivers possess valuable insights into how these technologies affect daily life. Including such perspectives in regulatory design helps ensure that legal frameworks address real-world needs rather than theoretical assumptions.

Ultimately, the goal of legal protections for vulnerable populations is not to restrict the development of beneficial robotics technologies. On the contrary, such protections can foster public confidence and encourage responsible innovation. When individuals know that robust safeguards exist, they are more likely to accept new technologies as trustworthy tools that enhance rather than threaten their wellbeing.

As humanoid robots become more integrated into everyday environments, legal systems must remain attentive to how technological change affects different groups within society. Ensuring that vulnerable populations remain protected and empowered in this evolving landscape is essential to building a future in which human–robot coexistence is both safe and equitable.

The following section turns to a broader societal dimension of coexistence: the measurement and cultivation of public trust. Understanding how communities perceive robotics technologies—and how those perceptions influence social cohesion—will be critical to determining whether humanoid robots are accepted as beneficial participants in human environments.

Public Trust and Social Cohesion Metrics

Technological integration within society rarely succeeds on technical capability alone. Innovations that fail to secure public trust often encounter resistance regardless of their efficiency or economic value. The introduction of humanoid robots into daily life therefore depends not only on regulatory compliance and engineering reliability but also on social legitimacy. Communities must perceive robotics technologies as safe, fair, and aligned with collective values before they are widely accepted. For this reason, governance frameworks for human–robot coexistence must incorporate systematic methods for measuring and strengthening public trust and social cohesion.

Trust functions as a form of social infrastructure. It allows individuals to interact with institutions and technologies without constant suspicion or anxiety. When citizens trust that a system operates fairly and transparently, they are more willing to accept its presence in their lives. Conversely, when trust erodes, even well-designed technologies can become sources of conflict. Humanoid robotics introduces new forms of interaction that may challenge established expectations about work, privacy, authority, and social relationships. Monitoring public attitudes toward these changes therefore becomes a central task of responsible governance.

Traditional technology regulation has often focused primarily on safety and economic outcomes. While these factors remain essential, they do not fully capture the social dimensions of emerging technologies. Public perception of fairness, cultural compatibility, and institutional accountability also influences whether technological systems are accepted or resisted. The development of social cohesion metrics allows policymakers to assess these factors systematically rather than relying on anecdotal impressions or reactive responses to controversy.

One of the most fundamental indicators in this context is trust in robotic systems. This measure evaluates whether individuals believe that robots operating in their communities behave safely and predictably. Surveys and behavioural studies can assess whether people feel comfortable interacting with robots in public spaces, workplaces, or service environments. Trust levels may vary depending on the type of robotic system involved. For example, individuals may feel comfortable with delivery robots operating on sidewalks while expressing greater concern about robots involved in law enforcement or healthcare decisions.

Trust is closely connected to perceptions of transparency. Individuals are more likely to accept robotic technologies when they understand how those systems function and who is responsible for their operation. Governance frameworks should therefore emphasise communication strategies that explain the purpose, capabilities, and limitations of robots deployed in public environments. Public information campaigns, accessible documentation, and open data initiatives can all contribute to strengthening trust by reducing uncertainty about how robotic systems operate.

A second important indicator concerns perceived fairness. Technologies that appear to benefit certain groups while disadvantaging others can quickly become sources of social tension. In the context of robotics, fairness concerns often relate to employment, accessibility, and the distribution of technological benefits. Workers may worry that robots will replace jobs without adequate transition support. Communities may fear that technological investments will concentrate in affluent areas while leaving other neighbourhoods underserved.

Measuring perceived fairness therefore requires evaluating how robotics deployment affects different social groups. Policymakers may conduct impact assessments examining whether robotic technologies improve access to services, reduce inequality, or create new forms of exclusion. If robots are primarily deployed in commercial sectors serving affluent consumers while public services remain under-resourced, perceptions of unfairness may emerge. Conversely, robotics technologies that enhance healthcare access, public transportation, or environmental sustainability may strengthen perceptions that innovation benefits society as a whole.

Another dimension of social cohesion relates to cultural acceptance. Technologies do not operate within a cultural vacuum. Different societies interpret and respond to machines according to historical experiences, social values, and symbolic meanings. In some cultures, robots are perceived positively as tools that enhance efficiency and social wellbeing. In others, they may evoke concerns about surveillance, labour displacement, or the erosion of human relationships. Understanding these cultural contexts is essential for designing governance frameworks that respect local values.

Cultural acceptance metrics may involve qualitative research methods such as community consultations, ethnographic studies, and focus groups that explore how individuals interpret the presence of robots in their environments. These insights can help policymakers identify potential sources of discomfort or misunderstanding before they escalate into public controversy. For example, communities may express concerns about robots operating in religious spaces, cultural heritage sites, or environments where human presence carries symbolic significance.

In addition to surveys and qualitative research, behavioural indicators can provide valuable insight into social acceptance. Observing how individuals interact with robots in public spaces—whether they approach them, avoid them, or express curiosity—can reveal attitudes that may not be captured through formal questionnaires. Pilot programmes deploying robots in controlled environments often include such observational studies to understand how human behaviour adapts to technological presence.

Governments and research institutions are increasingly developing technology acceptance indices that combine multiple indicators into comprehensive assessments of public sentiment. These indices may include measures of trust, perceived fairness, cultural compatibility, and willingness to interact with robotic systems. By tracking these metrics over time, policymakers can identify trends indicating whether public attitudes toward robotics are becoming more positive or more sceptical.

Social cohesion metrics are particularly valuable in identifying early signs of resistance or misunderstanding. If surveys reveal declining trust in certain categories of robotic systems, regulators can investigate the underlying causes before they develop into broader social conflicts. For example, concerns about privacy may arise if robots equipped with cameras operate in public spaces without clear safeguards. Addressing such concerns through transparent regulation and technological adjustments can prevent the erosion of public trust.

Another important application of these metrics involves evaluating the effectiveness of governance policies. Regulations designed to enhance safety, transparency, and fairness should ideally produce measurable improvements in public trust and acceptance. By analysing changes in social cohesion metrics following policy implementation, governments can assess whether regulatory interventions are achieving their intended outcomes.

International cooperation may also benefit from shared methodologies for measuring public attitudes toward robotics. As humanoid robots increasingly operate across borders, comparative research can help identify best practices for maintaining social trust in different cultural contexts. International organisations, academic institutions, and regulatory bodies may collaborate in developing standardised indicators that allow policymakers to compare experiences across countries.

However, it is important to recognise that public trust cannot be manufactured solely through measurement. Metrics provide valuable insights, but trust ultimately emerges from lived experiences with technology. If robotic systems consistently operate safely, respect privacy, and contribute positively to communities, public confidence will grow naturally. Conversely, if incidents occur that undermine safety or fairness, trust may erode regardless of how effectively it is measured.

For this reason, social cohesion metrics should be integrated into broader governance strategies that prioritise accountability and responsiveness. When communities express concerns about robotics deployment, authorities must demonstrate a willingness to listen and adapt policies accordingly. Participatory governance models—such as citizen advisory panels or public consultations—can help ensure that community voices remain part of the decision-making process.

The concept of social cohesion also highlights the importance of shared responsibility. Public acceptance of robotics technologies depends not only on government regulation but also on the behaviour of companies, engineers, and institutions deploying these systems. Responsible innovation practices—such as ethical design, transparent communication, and inclusive consultation—can significantly influence whether communities perceive robotics as beneficial or disruptive.

Ultimately, the success of human–robot coexistence will depend on whether societies view these technologies as legitimate participants in public life. Measuring trust, fairness perceptions, and cultural acceptance provides valuable tools for understanding this process. However, the deeper objective is not simply to track public attitudes but to cultivate a social environment in which technological innovation strengthens rather than weakens community bonds.

The final section of this chapter examines how societies may respond when coexistence encounters conflict. Even with careful governance and widespread trust, situations may arise in which robotic systems cause harm or disrupt social expectations. Addressing such incidents effectively requires legal mechanisms that emphasise repair, accountability, and dialogue between the humans and institutions involved. This leads to the emerging concept of restorative justice in the context of human–robot interactions.

Restorative Justice Between Humans and Machines

The growing presence of humanoid robots in human environments inevitably raises the possibility of conflict. Despite advances in safety engineering, behavioural calibration, and regulatory oversight, complex technological systems can never eliminate risk entirely. Robots may malfunction, misinterpret environmental conditions, or produce unintended consequences when interacting with unpredictable human environments. When such incidents occur, traditional legal responses have typically relied on adversarial litigation focused on assigning liability and awarding damages. While this approach remains necessary in many circumstances, it may not always provide the most constructive framework for addressing conflicts arising from human–robot interaction.

Restorative justice offers an alternative perspective that may prove particularly valuable in the emerging domain of robotics governance. Rather than focusing solely on punishment or financial liability, restorative justice emphasises dialogue, accountability, and the repair of harm. The approach seeks to bring together affected parties in a structured process that allows them to understand what occurred, identify the causes of the incident, and determine how harm can be addressed and prevented in the future. Although restorative justice has traditionally been applied in criminal and community contexts, its principles can also inform dispute resolution involving technological systems. Human–robot conflicts often involve complex chains of causation that extend beyond a single actor. A malfunctioning robot may reflect design decisions made by engineers, operational policies set by employers, maintenance practices carried out by technicians, or environmental conditions present in the workplace. Traditional litigation may struggle to capture this complexity, particularly when responsibility is distributed across multiple institutions. Restorative processes, by contrast, allow stakeholders to examine the broader ecosystem that contributed to the incident rather than focusing exclusively on legal blame.

Consider a hypothetical scenario in which a warehouse worker is injured when a humanoid logistics robot unexpectedly shifts direction while carrying a heavy load. Under conventional legal processes, the injured worker might pursue compensation through product liability claims against the robot manufacturer or through workplace safety claims against the employer operating the facility. These legal avenues remain essential, particularly where negligence or defective design can be demonstrated. However, they may not fully address the broader systemic issues that contributed to the incident.

A restorative justice framework could supplement these processes by bringing together the injured worker, representatives of the employer, engineers from the robot's manufacturer, and regulatory observers responsible for workplace safety oversight. In such a setting, participants could examine the sequence of events that led to the injury. Engineers might explain how the robot's navigation algorithms interpreted sensor data at the moment of the incident. The employer could discuss operational procedures governing human–robot collaboration in the warehouse environment. The injured worker could describe how the event affected their safety and confidence in workplace technology.

Through facilitated dialogue, the participants might identify contributing factors that would not emerge easily in adversarial proceedings. Perhaps the robot's sensors misinterpreted reflective surfaces in the warehouse environment. Perhaps human workers had not received sufficient training on safe distances from robotic units. Perhaps maintenance schedules failed to detect a calibration drift in the robot's navigation system. By examining these factors collaboratively, the restorative process can generate insights that lead to meaningful improvements in safety practices. Another advantage of restorative approaches lies in their capacity to address the psychological dimensions of technological incidents. Workers who experience injuries involving autonomous machines may develop anxiety or distrust toward robotics systems in their workplace. Simply receiving financial compensation may not fully resolve these concerns. Participating in a process where the causes of the incident are explained and corrective measures are implemented can help restore confidence in the technologies that workers must continue to interact with.

Restorative justice processes can also promote organisational accountability. Companies involved in robotics development or deployment may be more willing to participate in collaborative problem-solving environments than in adversarial litigation, where legal liability concerns often limit transparency. Structured restorative frameworks can create safe spaces for organisations to acknowledge mistakes and commit to improvements without immediately escalating legal disputes.

For such processes to function effectively, clear institutional frameworks must exist to support them. Regulatory agencies responsible for robotics safety could establish mediation panels specialising in technological disputes. These panels might include engineers, legal experts, labour representatives, and community stakeholders capable of evaluating incidents from multiple perspectives. Participation in restorative processes would not replace formal legal rights but would operate as a complementary mechanism aimed at resolving conflicts constructively.

Documentation and transparency are essential components of such systems. Outcomes of restorative proceedings should include publicly accessible reports describing the incident, the contributing factors identified, and the corrective measures implemented. These reports can serve as valuable learning tools for other organisations deploying similar technologies. Over time, the accumulation of such knowledge can contribute to safer and more responsible robotics ecosystems.

Restorative justice frameworks may also play a role in addressing broader societal concerns about robotics deployment. Communities may express fears about automation replacing human labour, increasing surveillance, or undermining social relationships. Structured dialogue processes involving citizens, policymakers, and technology developers can help address these concerns through open discussion and collaborative problem-solving. By engaging communities directly, institutions demonstrate that robotics governance is not imposed from above but shaped through democratic participation.

It is important to recognise that restorative justice is not suitable for all situations. Cases involving serious negligence, deliberate misconduct, or systemic regulatory violations may require formal legal sanctions. The purpose of restorative approaches is not to weaken accountability but to complement it by providing additional pathways for addressing harm and improving systems. In many cases, restorative processes may occur alongside formal legal proceedings rather than replacing them entirely.

The integration of restorative justice into robotics governance reflects a broader shift toward adaptive legal systems capable of responding to complex technological environments. As humans and machines increasingly coexist in workplaces, cities, and institutions, conflicts will inevitably arise. Addressing these conflicts effectively requires mechanisms that combine legal accountability with opportunities for dialogue, learning, and institutional improvement.

Human–robot coexistence is ultimately a social project as much as a technological one. Legal frameworks must not only regulate machines but also guide the relationships between people and the systems that shape their environments. Restorative justice offers one possible pathway for ensuring that when conflicts occur, they become opportunities for learning and progress rather than sources of enduring division.

With this final discussion, the chapter returns to its central theme: the governance of coexistence. Robots are no longer confined to laboratories or factories. They are becoming participants in the social spaces where human life unfolds. Law must therefore evolve not only to regulate their behaviour but also to maintain trust, fairness, and community cohesion in the shared environments that humans and machines increasingly inhabit together.

Between Us and Them - Episode Fourteen

The city was quiet at that hour. Most of the lights across the skyline had dimmed, leaving only the slow glow of infrastructure systems and the steady movement of machines that never slept. Autonomous freight units crossed the river bridge. Maintenance robots inspected the elevated rail line. Street sanitation units moved silently through empty avenues. From the observation deck of the Global Robotics Commission, Ògún watched the city as it operated through the night.

Ten years ago the same city would have looked very different after midnight. Now it never truly rested. Machines had quietly extended the rhythm of civilisation. Elias stood beside him. The humanoid's sensors traced the distant movement of robots across the skyline, mapping traffic flows, energy consumption, and logistics networks that stretched far beyond the city. "Night operations have increased by twenty-three percent this year," Elias said. Ògún nodded slowly. "Efficiency always wins." Below them, a fleet of logistics robots moved through a distribution terminal near the river. Their paths crossed and diverged with precise choreography. No collisions. No hesitation.

Thousands of decisions per minute. None of them human. For years the Commission had focused on governance frameworks. Licensing regimes. Safety standards. Ethical design rules. International treaties. Those frameworks had stabilised the transition. But now a different question was beginning to surface. It appeared quietly at first, buried in research reports and academic conferences. Then it began appearing in economic forecasts. Now it appeared in government briefings. Ògún tapped the console beside the window. A series of projections appeared in the air.

Labour automation rates. Industrial productivity curves. Energy efficiency forecasts. Global robotics deployment estimates. One graph in particular caught the eye. The curve showed something economists had once believed impossible. Productivity was rising while human labour demand was falling. "Post-scarcity indicators," Elias said quietly. Ògún frowned. "Not yet." "But approaching."

The projections suggested a world where machines performed an increasing share of productive activity. Factories. Agriculture. Logistics. Construction. Even parts of law and governance itself. Humans would still work. But increasingly by choice rather than necessity. A civilisation built for scarcity was beginning to encounter abundance. Ògún leaned back against the console. "For centuries," he said, "law was written for a world where human labour powered everything." Elias remained silent. "Property law," Ògún continued. "Contract law. Employment law. Taxation. Social welfare systems." He gestured toward the city. "All of it assumes the same thing." "That humans must work to produce value."

Elias looked back toward the skyline. "That assumption may no longer hold indefinitely." Ògún exhaled slowly. This was the conversation that governments had quietly begun having behind closed doors. What happens to law when machines produce most of the world's wealth? What happens to legal systems designed for industrial societies when automation changes the structure of economic life itself? Below them, the city continued operating with mechanical precision.

Maintenance robots replaced street lights. Autonomous buses prepared for the morning routes. A humanoid patrol unit paused at an intersection before continuing its route. None of these systems required wages. None of them voted. Yet their productivity increasingly sustained the city. "Do you think people are ready for this conversation?" Ògún asked. Elias considered the question. "People rarely become ready before change arrives." Ògún smiled faintly.

"History agrees with you." For a long moment they watched the skyline. Then a notification appeared on the Commission display. A new policy briefing had arrived. Ògún opened it. The title appeared in bold letters. Future Legal Architectures for Automated Civilisations. He read the opening sentence quietly. "Legal systems designed for the industrial age must evolve into adaptive governance frameworks capable of regulating societies shaped by artificial intelligence and robotics."

Ògún closed the file. For years the law had struggled simply to keep up with technological change. Now it faced something more profound. The possibility that the structure of civilisation itself was changing. "Elias," he said. "Yes." "Do you know what worries me?" The humanoid turned toward him. "That the laws we've spent centuries building might not be designed for the world we're entering." Elias did not respond immediately. Instead he looked again at the city. Thousands of machines were working. Quietly. Efficiently. Without complaint.

Finally he spoke. "Then perhaps," he said, "the law must learn to evolve as quickly as the systems it governs." Ògún nodded. "Yes." He turned back toward the console. "Which means the next chapter of robotics law is not about robots at all." Elias tilted his head slightly. "It is about law itself." Outside the window the first light of morning appeared over the horizon. Another day was beginning. And somewhere in the legal systems of the world, the last laws written exclusively for human societies were still quietly being enforced. But not for long.

XIII

13. Toward a Post-Human Legal Order

The preceding chapters of this book have examined the governance of humanoid robotics through the familiar institutions of modern law. Licensing regimes, safety standards, ethical design principles, liability frameworks, and international regulatory cooperation have all been explored as mechanisms for ensuring that robots operate safely within human environments. These legal structures represent the immediate tools through which societies can regulate emerging technologies and prevent harm during the transition toward increasingly automated systems. Yet as robotics and artificial intelligence continue to evolve, a deeper and more fundamental question begins to emerge. Are the legal institutions inherited from earlier eras sufficient for governing a world in which intelligent machines participate directly in economic, social, and infrastructural life? For centuries, legal systems have developed in response to technological transformations that reshaped human societies. The emergence of maritime commerce produced new doctrines of admiralty law that governed international trade across oceans. The industrial revolution required labour law, corporate law, and regulatory frameworks capable of organising large-scale economic activity. The digital revolution forced legal systems to confront questions of data protection, cybersecurity, and intellectual property in virtual environments. Each technological era required law to adapt to new patterns of human interaction and economic organisation.

The transformation now unfolding through robotics and artificial intelligence may prove even more consequential. Unlike earlier technologies that extended human capacity while remaining clearly subordinate to human control, advanced robotics systems increasingly perform tasks once considered exclusively human. Autonomous machines now manage logistics networks, maintain infrastructure, assist in medical procedures, monitor environmental systems, and perform complex analytical tasks in both public and private institutions. In doing so, they participate in processes that shape economic productivity, social welfare, and the functioning of modern states.

This shift challenges a foundational assumption embedded within most legal systems: that human labour remains the primary driver of economic value and institutional activity. From employment law to taxation, from welfare systems to intellectual property regimes, many legal doctrines presuppose that individuals and organisations create value primarily through human work. As automation expands across industries, that assumption becomes increasingly fragile. Machines may soon perform a growing share of productive tasks that once required human effort.

The consequences of this transition extend beyond labour markets. When robots operate infrastructure, coordinate logistics, and assist in governance processes, the structure of social organisation itself begins to evolve. Legal systems designed to regulate human interactions within industrial societies must now address relationships among humans, machines, and institutions operating within highly automated environments. The challenge is not merely to regulate technology but to reconsider the underlying architecture of law in a world where artificial systems participate in the creation and distribution of value.

This chapter explores the possibility that robotics and artificial intelligence may ultimately lead to what can be described as a post-human legal order. The term does not imply the disappearance or marginalisation of human beings. Rather, it recognises that the legal systems of the future may govern societies in which humans are no longer the sole actors shaping economic production, social organisation, and institutional decision-making. Law will still serve human welfare, dignity, and freedom, but it may need to operate within ecosystems that include artificial agents performing increasingly significant roles.

Such a transformation does not occur overnight. Legal systems tend to evolve gradually, responding to technological developments through incremental reforms rather than sweeping institutional redesign. Nevertheless, moments arise when incremental adaptation becomes insufficient and deeper structural change becomes necessary. The emergence of robotics and artificial intelligence may represent such a moment. If machines continue to expand their capabilities and influence across social systems, legal frameworks must evolve to remain effective instruments of governance.

One way to understand this challenge is to consider the analogy between law and engineering. Software engineers rarely design complex systems that remain static over long periods of time. Instead, software is updated continuously through modular improvements, testing cycles, and iterative development. Errors are identified through monitoring systems, corrected through patches, and refined through successive versions. In contrast, legal systems often rely on statutes and regulations that may remain unchanged for decades despite rapid technological change.

This discrepancy creates what scholars sometimes describe as regulatory lag—the gap between the pace of technological innovation and the speed at which legal systems adapt. When this gap becomes too wide, governance frameworks struggle to maintain effective oversight. Emerging technologies may operate in legal grey areas where existing rules provide insufficient guidance. Policymakers may find themselves responding reactively to crises rather than proactively shaping technological development.

Robotics governance illustrates this dynamic clearly. Many regulatory frameworks addressing artificial intelligence and robotics have been developed only after technologies were already deployed widely in commercial and social contexts. Legislators often struggle to keep pace with innovations introduced by private sector developers or research institutions. As a result, legal systems may appear perpetually behind the technological curve. The challenge therefore extends beyond individual regulations. It concerns the methodology through which law adapts to technological change. If robotics and artificial intelligence are expected to evolve rapidly over the coming decades, legal systems must develop mechanisms capable of evolving alongside them. Static legal frameworks designed for slower technological cycles may prove inadequate in environments where innovations emerge continuously.

This chapter proposes that law governing advanced technologies must become more adaptive, modular, and interdisciplinary. Instead of relying exclusively on traditional legislative processes, future governance frameworks may incorporate mechanisms resembling those used in complex engineering systems. Laws could be designed as modular regulatory architectures capable of evolving through iterative updates. Monitoring systems could track emerging technological capabilities and identify regulatory gaps in real time. Collaborative networks involving lawyers, engineers, ethicists, and policymakers could develop responsive governance solutions that reflect the technical realities of emerging technologies.

Such an approach does not abandon the principles of democratic governance or the rule of law. On the contrary, it seeks to preserve those principles by ensuring that legal institutions remain capable of guiding societies undergoing rapid technological transformation. Law must remain stable enough to protect fundamental rights and provide predictable rules for social interaction. At the same time, it must remain flexible enough to respond effectively to technological innovation. Another dimension of the post-human legal order concerns the economic implications of automation. If robotics technologies continue to increase productivity while reducing reliance on human labour, societies may confront new questions about the distribution of wealth and opportunity. Economic systems historically built on wage labour may require new legal frameworks capable of ensuring social stability in environments where machines generate a substantial portion of productive output.

Some scholars have proposed concepts such as universal basic income or universal high-income systems as mechanisms for distributing the economic benefits of automation. Whether such proposals prove feasible remains a matter of debate. What is clear, however, is that legal systems must consider how economic governance evolves when technological abundance becomes possible. Questions of taxation, ownership of automated production systems, and the allocation of technological benefits will increasingly occupy policymakers in the coming decades. Addressing these challenges requires collaboration across disciplines. Traditional legal education has often focused primarily on statutory interpretation, case analysis, and doctrinal reasoning. While these skills remain essential, the governance of emerging technologies also demands familiarity with engineering principles, data science, ethics, and systems design. Lawyers working in robotics governance must be capable of understanding how algorithms function, how hardware interacts with software, and how technological systems behave within complex environments. Legal innovation therefore depends not only on new regulations but also on new forms of professional expertise. Law schools and professional training programmes may need to incorporate interdisciplinary curricula that prepare future lawyers to engage effectively with technological systems. Lawyers who understand both legal doctrine and technical architecture will be better equipped to design governance frameworks that align legal principles with technological realities.

Ultimately, the concept of a post-human legal order is not about diminishing the role of humanity within legal systems. Rather, it recognises that human societies increasingly coexist with artificial systems that influence economic production, institutional decision-making, and social interaction. The challenge for law is to maintain its commitment to human dignity, justice, and accountability while adapting to a world in which humans are no longer the only actors shaping the systems that sustain civilisation.

The sections that follow explore how such adaptation might occur. They examine how legal frameworks can be redesigned as adaptive governance systems capable of evolving alongside technological innovation. They introduce tools for tracking regulatory gaps and coordinating responses across institutions. They explore interdisciplinary models of legal education and policymaking that integrate technological understanding into legal practice. They also consider the economic implications of automation and the responsibilities that legal systems may bear in ensuring that technological abundance benefits society as a whole. The future of robotics governance ultimately depends on the willingness of legal institutions to evolve alongside the technologies they regulate. The law has always been a human creation, shaped by the needs and values of the societies it serves. As those societies enter an era shaped by intelligent machines and automated systems, the law must once again reinvent itself—not by abandoning its principles, but by extending them into a world where human and artificial actors coexist within shared systems of governance.

Law as Adaptive Engineering

The rapid evolution of robotics and artificial intelligence exposes a structural tension within modern legal systems. Technologies develop through cycles of experimentation, iteration, and refinement, while laws often evolve through slower processes of legislation, judicial interpretation, and administrative rulemaking. In many areas of governance this difference in pace is manageable. In the domain of robotics and artificial intelligence, however, the gap between technological change and legal adaptation risks becoming so wide that regulatory frameworks struggle to remain effective. Bridging this gap requires rethinking not only the substance of law but also the methodology through which legal systems are designed and updated.

One useful conceptual shift is to view law as a form of adaptive engineering. In engineering disciplines, complex systems are rarely built as static structures. Instead, they are designed to evolve through modular updates, testing cycles, and feedback loops that respond to changing conditions. Software development provides a particularly clear illustration of this approach. Engineers do not write a single final version of a programme and leave it unchanged indefinitely. Instead, they create modular code structures that can be revised, patched, and expanded as new requirements emerge. Bugs are detected through monitoring systems, corrected through updates, and refined through iterative development.

Legal systems historically have not operated in this manner. Statutes are often drafted as comprehensive documents intended to remain stable for extended periods. Amendments occur only after lengthy legislative processes. Judicial interpretation may gradually adapt rules to new circumstances, but this process can take years or decades. While such stability is essential for protecting legal certainty and democratic legitimacy, it may become problematic when governing technologies that evolve on timescales measured in months rather than generations.

Robotics governance therefore invites a new way of thinking about how law can evolve alongside technology. Instead of treating legislation as a fixed architecture that must be periodically replaced, policymakers might design legal frameworks as modular governance systems capable of continuous refinement. In this model, statutory provisions would establish core principles and oversight mechanisms, while technical standards and regulatory guidelines could be updated more rapidly in response to emerging developments. This concept does not undermine the rule of law. On the contrary, it seeks to preserve legal effectiveness in environments where rigid frameworks risk becoming obsolete before they are fully implemented. By designing law with adaptability in mind, policymakers can ensure that governance structures remain aligned with the technologies they regulate.

The analogy between law and software engineering becomes particularly powerful when considering how robotics systems operate in practice. Robots function through layers of software, firmware, and hardware architecture that guide behaviour in real time. Safety protocols, navigation rules, and interaction guidelines are often encoded directly into machine instructions. These instructions determine how a robot responds to environmental conditions, human commands, and unexpected events.

If legal requirements remain external to these operational systems, enforcement may rely primarily on after-the-fact compliance checks and liability mechanisms. While such approaches remain necessary, they do not fully leverage the capacity of modern technology to integrate governance directly into operational processes. A more proactive approach would involve embedding legal constraints and ethical guidelines into the design architecture of robotic systems themselves.

This concept can be described as law as proactive engineering. Instead of relying solely on legal enforcement after violations occur, policymakers and engineers collaborate to translate legal requirements into technical protocols that guide robotic behaviour automatically. For example, safety standards governing robotic mobility could be encoded into navigation software, ensuring that machines automatically maintain safe distances from pedestrians. Data protection rules could be implemented through privacy-by-design architectures that restrict data collection and storage according to regulatory standards.

Embedding legal norms within technological systems does not eliminate the need for human oversight or legal accountability. Rather, it creates an additional layer of governance that reduces the likelihood of violations occurring in the first place. By integrating legal requirements into system design, regulators can ensure that compliance becomes an operational feature of technology rather than merely an external obligation. Developing such systems requires close collaboration between legal scholars, engineers, and policymakers. Translating statutory language into technical specifications is not a straightforward task. Legal rules often rely on flexible standards and contextual interpretation, while software systems require precise instructions that can be executed consistently. Bridging this gap requires interdisciplinary expertise capable of interpreting legal principles in ways that can be operationalised within technological architectures.

One possible roadmap for implementing proactive legal engineering involves several stages. The first stage is normative translation, in which legal principles are analysed to identify operational requirements that can be expressed in technical terms. For example, a legal requirement that robots must operate safely in public spaces could be translated into measurable parameters such as speed limits, collision avoidance thresholds, and sensor redundancy standards.

The second stage involves technical integration, where these operational requirements are embedded within software and firmware controlling robotic behaviour. Engineers develop algorithms that enforce compliance with regulatory constraints while allowing systems to function efficiently in dynamic environments. These algorithms may incorporate real-time monitoring and automatic adjustments when conditions approach regulatory limits.

The third stage is verification and testing. Just as software engineers test new code before deployment, regulatory systems must ensure that embedded legal protocols function reliably under diverse conditions. Simulation environments and real-world pilot programmes allow developers and regulators to evaluate whether robotic systems behave in accordance with legal expectations.

The fourth stage involves continuous monitoring and iteration. As robotics technologies evolve and new use cases emerge, embedded legal frameworks may require updates. Modular governance architectures allow regulators to update technical standards without rewriting entire legislative frameworks. This iterative process mirrors the continuous improvement cycles common in engineering disciplines.

Implementing such adaptive governance systems also requires institutional innovation. Regulatory agencies may need technical units capable of analysing robotics architectures and verifying compliance with embedded legal standards. Governments may establish collaborative platforms where engineers and legal experts jointly develop technical guidelines for robotics systems. International organisations could play a role in coordinating standardised governance protocols that ensure consistency across jurisdictions.

The European Union's regulatory approach to artificial intelligence provides an early example of this direction. The EU AI Act introduces risk-based classification systems and technical requirements for high-risk AI systems, encouraging developers to incorporate compliance mechanisms directly into system design. While the regulation focuses primarily on artificial intelligence, similar principles can guide robotics governance more broadly.

Another promising development involves the emergence of regulatory sandboxes—controlled environments in which new technologies can be tested under regulatory supervision. These sandboxes allow policymakers to observe how robotic systems behave in real-world conditions and refine legal frameworks accordingly. By integrating experimentation into governance processes, regulators can develop more responsive and evidence-based policies.

Despite its potential advantages, the concept of law as adaptive engineering also raises important questions about democratic oversight and legal legitimacy. Laws derive their authority from political processes that reflect societal values and public deliberation. If legal rules are embedded directly into technological systems, care must be taken to ensure that these systems remain subject to transparent governance and democratic accountability.

For this reason, adaptive legal frameworks must preserve clear lines of authority between legislators, regulators, engineers, and the public. Legislatures remain responsible for establishing the fundamental principles and objectives guiding robotics governance. Regulatory agencies translate these principles into operational standards. Engineers implement technical protocols consistent with regulatory guidance. Public oversight ensures that these processes remain transparent and accountable. The ultimate goal of this approach is not to transform law into software but to enable legal systems to function effectively in technological environments where traditional regulatory methods may prove insufficient. By adopting adaptive engineering principles, law can maintain its capacity to guide innovation while protecting human interests.

Robotics governance therefore represents an opportunity to rethink how legal institutions interact with technological systems. Instead of responding reactively to technological developments, legal frameworks can become proactive partners in shaping the design and operation of emerging technologies. This shift requires a cultural transformation within both legal and engineering communities, encouraging collaboration that transcends disciplinary boundaries.

As robotics technologies continue to evolve, the ability of legal systems to adapt will determine whether societies can harness their benefits while mitigating associated risks. Viewing law as an adaptive engineering discipline provides a conceptual foundation for this transformation, ensuring that governance structures remain capable of guiding technological progress in ways that promote safety, fairness, and human wellbeing.

The next section builds on this perspective by introducing tools for identifying regulatory gaps in real time. If law is to evolve alongside rapidly advancing technologies, policymakers must be able to detect when existing frameworks no longer align with technological capabilities. Developing mechanisms for tracking such gaps will be essential for maintaining effective governance in an era of accelerating innovation.

Dynamic Legal Gap Tracking

One of the most persistent challenges in technology governance is the phenomenon commonly referred to as regulatory lag. Innovations in robotics and artificial intelligence often advance at a pace that far exceeds the speed at which laws can be written, debated, enacted, and interpreted. By the time legislators address a technological development through statutory reform, the underlying technology may already have evolved into new forms that raise different risks and opportunities. This structural delay is not merely an inconvenience. It can produce governance vacuums in which emerging technologies operate without clear legal boundaries, creating uncertainty for innovators and risk for society.

In the context of humanoid robotics, regulatory lag is particularly visible. Autonomous machines are rapidly acquiring new capabilities in mobility, perception, manipulation, and decision-making. Each advancement creates new questions for law: how should robots operating in public spaces be licensed, what safety standards must govern their interaction with humans, who bears responsibility when autonomous behaviour produces harm, and how should cross-border robotic operations be regulated? Legislators often respond only after these technologies have already begun to shape economic and social systems. The result is a recurring cycle in which law trails behind innovation.

To address this challenge, legal systems must develop tools capable of identifying emerging regulatory gaps before they become systemic problems. One promising approach involves the concept of dynamic legal gap tracking. Instead of waiting for regulatory deficiencies to appear through crises or litigation, policymakers could monitor technological developments continuously and evaluate whether existing legal frameworks remain aligned with current capabilities. In effect, governance systems would operate more like real-time monitoring infrastructures than static regulatory architectures.

Dynamic legal gap tracking begins with a simple premise: technological change can be measured, and so can the responsiveness of law. By analysing the relationship between technological capability and regulatory adaptation, policymakers can identify areas where legal frameworks are falling behind. The objective is not merely to measure delay but to create mechanisms that trigger timely policy responses when the gap between innovation and regulation becomes too wide.

A practical way to implement this concept is through the development of gap-tracker dashboards—data-driven monitoring systems that visualise the relationship between technological progress and legal frameworks. Such dashboards would integrate information from multiple sources, including technological research, industry deployment statistics, regulatory updates, and judicial decisions. By analysing these inputs, the system could generate indicators showing where governance frameworks may require revision.

For example, consider a scenario in which humanoid robots capable of autonomous navigation become widely deployed in urban logistics. Researchers and companies may report rapid improvements in navigation algorithms, sensor accuracy, and operational reliability. At the same time, municipal regulatory frameworks governing sidewalk mobility may remain limited to earlier forms of robotic delivery devices with narrower capabilities. A gap-tracker system analysing both technological and legal developments could detect this divergence. It might show that robot capabilities have advanced significantly while legal standards governing their operation have not been updated accordingly.

Such insights would allow policymakers to respond proactively rather than reactively. Regulatory agencies could initiate consultations with industry experts, urban planners, and safety authorities to assess whether existing rules remain adequate. If necessary, they could propose updated guidelines or pilot programmes designed to test new governance approaches. By intervening early, regulators reduce the likelihood that technological deployment will outpace legal oversight.

The design of gap-tracker dashboards requires careful consideration of what metrics should be monitored. Several categories of indicators are particularly relevant for robotics governance. One category involves technological capability metrics. These indicators measure advances in areas such as autonomous decision-making, sensor accuracy, mobility speed, manipulation dexterity, and system scalability. Tracking these metrics provides insight into how rapidly robotics technologies are evolving.

A second category involves deployment metrics. Even highly capable technologies may pose limited governance challenges if they remain confined to research laboratories. Regulatory attention becomes more urgent when technologies move from experimental environments into widespread commercial or public use. Deployment indicators therefore track the number of robotic systems operating in specific sectors, geographic regions, or operational contexts.

A third category involves legal adaptation metrics. These indicators measure how frequently laws, regulations, and standards are updated to reflect technological change. They may include the number of legislative amendments addressing robotics, the issuance of new regulatory guidelines, or the development of technical standards by international organisations. By comparing these metrics with technological and deployment indicators, policymakers can estimate the degree of alignment between law and innovation.

A fourth category involves risk and incident indicators. Reports of accidents, system failures, or ethical controversies provide valuable information about where existing governance frameworks may be insufficient. Integrating incident data into gap-tracker systems allows regulators to detect patterns indicating systemic weaknesses in current rules. When these data streams are combined, policymakers can generate visual representations showing the lag between technological capability and legal adaptation. For instance, a dashboard might display timelines comparing major robotics innovations with the dates on which relevant regulatory frameworks were introduced or revised. If technological developments consistently appear years before corresponding legal updates, the system highlights areas where governance processes require acceleration.

Dynamic gap tracking also enables comparative analysis across jurisdictions. Different countries often adopt varying approaches to regulating emerging technologies. Some governments implement proactive regulatory frameworks designed to encourage innovation while maintaining oversight. Others take a more cautious approach, introducing regulations only after technologies become widely deployed. Gap-tracker systems can reveal these differences by comparing technological progress and legal adaptation across regions.

Such comparative insights may encourage international coordination. If one jurisdiction develops effective governance models for specific robotics applications, other countries can learn from these experiences. International organisations may use gap-tracker data to identify areas where harmonised standards would improve safety and interoperability across borders.

The development of dynamic legal gap tracking systems also raises important institutional questions. Which organisations should operate these monitoring infrastructures? Government regulatory agencies may possess the legal authority to analyse technological trends and propose policy responses. However, effective monitoring often requires collaboration with academic researchers, technology developers, and civil society organisations that possess specialised expertise.

A multi-stakeholder governance model may therefore be appropriate. Universities and research institutes could contribute data on technological progress. Industry participants could provide information about deployment patterns and operational experiences. Regulatory agencies would analyse this information and coordinate policy responses. Civil society groups could ensure that governance frameworks reflect broader societal concerns about safety, privacy, and fairness.

Technological infrastructure itself may play a role in supporting these processes. Advances in data analytics and artificial intelligence could enable automated systems to analyse large volumes of information related to technological development and legal frameworks. Such systems might identify emerging trends or correlations that human analysts might overlook. Nevertheless, human oversight remains essential to interpret these findings within broader legal and ethical contexts.

While dynamic gap tracking offers significant benefits, it must be implemented carefully to avoid unintended consequences. Excessively reactive governance may create regulatory instability that discourages innovation. If laws change too frequently, developers and investors may struggle to plan long-term projects. Policymakers must therefore balance responsiveness with stability, ensuring that regulatory updates occur in a predictable and transparent manner.

Another challenge involves ensuring that monitoring systems themselves remain transparent and accountable. Data used to assess technological capabilities and regulatory performance must be accessible and verifiable. If gap-tracker dashboards rely on proprietary or incomplete information, they may produce misleading conclusions. Public access to relevant data sources can help maintain trust in the governance process.

Despite these challenges, dynamic legal gap tracking represents an important step toward modernising technology governance. By transforming regulatory oversight from a reactive process into a proactive monitoring system, policymakers can maintain closer alignment between law and innovation. This alignment reduces uncertainty for developers, improves safety for users, and strengthens public confidence in emerging technologies.

In the broader context of a post-human legal order, gap-tracking systems illustrate how law itself can evolve into a responsive infrastructure that adapts continuously to technological change. Instead of relying solely on periodic legislative reform, governance frameworks can incorporate feedback mechanisms that identify emerging challenges and guide timely policy responses.

The next section expands on this perspective by exploring the concept of modular governance systems. If law is to operate effectively in rapidly evolving technological environments, it may need to adopt layered regulatory architectures that allow different actors—governments, international organisations, industry groups, and civil society—to collaborate in shaping adaptive governance frameworks capable of guiding future innovations.

Modular Governance Systems

If law is to remain effective in an age of accelerating technological transformation, it cannot rely exclusively on rigid, monolithic regulatory structures. Traditional legal frameworks are often designed as comprehensive regimes that attempt to address an entire domain of activity through detailed statutory provisions. While such approaches provide clarity and authority, they may struggle to adapt quickly enough when the technologies being regulated evolve continuously. Robotics and artificial intelligence illustrate this tension clearly. By the time a comprehensive regulatory framework is enacted, the technologies it was designed to govern may already have developed new capabilities that fall outside the scope of the original legislation.

One response to this challenge is the development of modular governance systems. In engineering and software design, modularity refers to the practice of constructing complex systems from smaller, interchangeable components that can be updated independently without disrupting the entire architecture. When one component becomes obsolete or requires improvement, it can be replaced without dismantling the entire system. This approach allows technological systems to evolve gradually while maintaining overall stability.

Applying this principle to law suggests a regulatory architecture composed of layered governance modules rather than single comprehensive statutes. At the highest level, legal frameworks establish foundational principles and institutional responsibilities that remain relatively stable over time. Beneath these foundational layers, more specialised regulatory modules address particular technological domains or operational contexts. These modules can be updated, refined, or expanded as technologies evolve, ensuring that governance structures remain responsive without requiring constant legislative overhaul.

In the context of robotics governance, a modular approach may begin with core normative frameworks that define the overarching objectives of regulation. These frameworks articulate fundamental principles such as human safety, transparency, accountability, fairness, and respect for human dignity. They also establish institutional responsibilities, specifying which regulatory bodies oversee robotics deployment and how coordination occurs among agencies responsible for technology, labour, infrastructure, and public safety. Because these principles reflect enduring societal values, they are likely to remain stable even as specific technologies evolve. The next layer of governance consists of sector-specific regulatory modules that address the unique characteristics of robotics applications in particular domains. Humanoid robots operating in healthcare environments, for example, raise different legal concerns from robots used in industrial logistics or urban maintenance. Sector-specific modules allow regulators to tailor rules to the operational realities of each domain. Healthcare robotics regulations may focus on patient safety, clinical accountability, and medical device certification. Industrial robotics frameworks may emphasise workplace safety and labour law. Urban robotics modules may address mobility regulations, zoning laws, and public infrastructure integration.

A further layer may involve technical standards and certification frameworks developed in collaboration with industry and international standards organisations. These standards translate legal principles into detailed operational requirements governing system design, testing procedures, and performance thresholds. Because technical standards must evolve alongside technological capabilities, they are particularly well suited to modular governance structures. Standards bodies can update technical specifications periodically without requiring full legislative revisions, allowing regulatory frameworks to remain aligned with technological progress.

An essential feature of modular governance is its reliance on multi-stakeholder collaboration. Governing complex technologies requires expertise that extends beyond traditional legal institutions. Engineers understand how robotics systems function at a technical level. Industry practitioners possess knowledge about real-world deployment conditions. Academic researchers contribute insights into emerging technological trends. Civil society organisations provide perspectives on ethical implications and societal concerns. Modular governance frameworks create institutional pathways for these diverse actors to contribute to regulatory design.

International cooperation also becomes easier within modular regulatory architectures. Robotics technologies increasingly operate across borders, particularly in sectors such as logistics, manufacturing, and digital infrastructure. Comprehensive global treaties addressing every aspect of robotics governance may be difficult to negotiate and maintain. Modular governance allows international organisations to coordinate specific components of regulation—such as safety standards, interoperability protocols, or cross-border liability frameworks—without requiring complete harmonisation of national legal systems. The aviation sector offers an instructive precedent. International aviation safety is governed through a layered architecture involving national regulators, regional regulatory bodies, and global institutions such as the International Civil Aviation Organization. Core principles of aviation safety remain consistent worldwide, while technical standards and operational procedures are updated continuously through specialised committees and expert groups. This system allows aviation governance to evolve alongside technological innovation while maintaining a high level of global coordination.

Robotics governance may follow a similar trajectory. National governments establish fundamental legal frameworks defining the responsibilities of developers, operators, and regulators. Regional bodies such as the European Union develop harmonised standards for member states. International organisations coordinate technical standards and ethical guidelines that support interoperability across jurisdictions. Industry groups contribute expertise in developing operational best practices. Civil society organisations ensure that public values remain integrated within governance processes.

Modular governance systems also support experimental regulatory approaches. Technologies often develop through iterative experimentation before stabilising into mature forms. Rigid regulatory frameworks may discourage such experimentation by imposing strict compliance requirements that are difficult to meet during early development stages. Modular governance architectures allow policymakers to create regulatory sandboxes or pilot programmes where new technologies can be tested under controlled conditions. Insights from these experiments inform the evolution of regulatory modules, ensuring that governance frameworks remain grounded in real-world experience.

Another advantage of modular governance lies in its capacity to support future technological convergence. Robotics does not exist in isolation from other technological domains. Artificial intelligence, biotechnology, advanced materials, and quantum computing may increasingly intersect in ways that produce entirely new categories of technological systems. Regulatory frameworks designed exclusively for robotics may struggle to address such convergence. Modular governance systems allow regulators to integrate new regulatory modules addressing emerging technological intersections without dismantling existing frameworks. Nevertheless, modular governance also presents important challenges. One concern involves maintaining coherence across different regulatory modules. If sector-specific frameworks evolve independently without adequate coordination, inconsistencies may emerge. For example, safety standards developed for industrial robotics may conflict with mobility rules governing robots operating in urban environments. Effective modular governance therefore requires strong coordinating institutions capable of ensuring that different regulatory components remain aligned with overarching legal principles. Another challenge involves democratic accountability. When regulatory authority becomes distributed across multiple institutions and technical standards bodies, the connection between governance decisions and democratic oversight may become less visible. Legislatures must therefore retain the authority to establish the core normative frameworks guiding technological governance. Regulatory agencies and standards organisations operate within these frameworks but remain accountable to democratic institutions.

Transparency is equally important. Stakeholders must understand how different components of the governance system interact and how decisions affecting robotics deployment are made. Public reporting mechanisms, open consultation processes, and accessible regulatory documentation can help maintain trust in modular governance systems. Despite these challenges, modular governance represents a promising pathway for managing technologies that evolve more rapidly than traditional legal frameworks can accommodate. By structuring regulation as a layered system of principles, sectoral rules, technical standards, and collaborative oversight mechanisms, policymakers can create governance architectures capable of evolving alongside technological progress.

The emergence of robotics and artificial intelligence therefore invites legal systems to adopt governance models that resemble the complex systems they regulate. Just as engineers design modular technologies capable of continuous improvement, legal institutions may need to design regulatory architectures that remain flexible, responsive, and inclusive. The next section explores another dimension of this transformation: the need for cross-disciplinary legal innovation. If law is to operate effectively within technological societies, future legal professionals must possess knowledge that extends beyond traditional doctrinal analysis. Integrating insights from engineering, ethics, and design into legal education will be essential for developing governance frameworks capable of guiding the technologies of the future.

Cross-Disciplinary Legal Innovation

The emergence of robotics and artificial intelligence as central forces in economic and social transformation requires more than new laws. It requires new kinds of legal thinkers. For centuries, the legal profession has been trained primarily within the intellectual boundaries of doctrinal reasoning, case law analysis, and statutory interpretation. These skills remain essential foundations of legal practice, but they are no longer sufficient for governing technological systems whose architecture is shaped as much by code and engineering decisions as by legal rules.

Humanoid robotics, autonomous systems, and advanced artificial intelligence operate through complex technological infrastructures that many traditional legal frameworks struggle to interpret. Decisions affecting human safety, privacy, labour, and social interaction may now emerge from algorithmic processes embedded in software and hardware systems. If legal professionals lack the ability to understand these systems at a conceptual level, governance risks becoming reactive rather than proactive. Law may end up responding to technological consequences after they occur rather than shaping the design of technologies before they are deployed.

For this reason, the future of robotics governance demands a new intellectual approach: cross-disciplinary legal innovation. Lawyers must increasingly operate at the intersection of law, engineering, ethics, design, and public policy. Legal education must therefore evolve to equip future practitioners with the conceptual tools necessary to understand and influence the technological systems that shape modern societies.

At the heart of this transformation lies a simple insight. Technology is not neutral. The design choices embedded within technological systems determine how those systems behave and how they affect human life. Engineers decide how algorithms interpret data, how robots navigate physical environments, and how systems prioritise competing objectives. These design decisions often carry profound legal and ethical implications. For example, a robot's collision-avoidance algorithm may determine how it balances the safety of different individuals in crowded environments. Data-processing architectures may determine whether personal information is stored, anonymised, or discarded. Governance of such systems cannot occur effectively if legal professionals remain distant from the design process. Cross-disciplinary legal innovation therefore calls for a new model of collaboration between lawyers and technologists.

Rather than engaging only after technologies have been developed, legal expertise should be integrated into the design and development stages of technological systems. This approach allows legal principles—such as accountability, transparency, fairness, and privacy—to influence the architecture of systems before they are deployed widely. Design thinking provides a useful framework for this integration. Originating in engineering and product development, design thinking emphasises iterative problem-solving, user-centred perspectives, and collaborative experimentation. When applied to legal innovation, design thinking encourages lawyers to approach regulatory challenges not only through abstract legal reasoning but also through practical engagement with how technologies function in real environments. Lawyers working alongside engineers and designers can identify potential risks early and help ensure that systems are developed in ways that respect legal norms.

This collaborative model also reflects a broader transformation occurring across professional disciplines. Complex global challenges—from climate change to cybersecurity—often require interdisciplinary solutions that integrate knowledge from multiple fields. Robotics governance is no different. Effective regulatory frameworks must combine legal reasoning with technical understanding, ethical reflection, and policy analysis. Lawyers capable of navigating these intersections will become essential contributors to the governance of emerging technologies.

Legal education must therefore evolve accordingly. Traditional law school curricula often focus heavily on doctrinal subjects such as contract law, property law, tort law, and constitutional law. While these subjects remain indispensable, students preparing to work in technology governance may also require exposure to subjects that have historically been considered outside the legal domain. Courses in software architecture, data science fundamentals, robotics engineering, and human–computer interaction can provide valuable insight into how technological systems operate. Equally important is the integration of ethics and social impact analysis within legal training. Technologies do not exist in isolation from social contexts. Decisions about automation, data collection, and artificial intelligence deployment influence employment patterns, social inequality, cultural norms, and democratic governance. Lawyers responsible for regulating these technologies must therefore be capable of evaluating not only legal compliance but also broader societal consequences.

One promising development in this direction is the emergence of technology law laboratories within universities and research institutions. These laboratories bring together students and researchers from law, computer science, engineering, and social sciences to collaborate on governance challenges related to emerging technologies. Participants work on practical projects such as designing regulatory frameworks for autonomous vehicles, developing ethical guidelines for artificial intelligence systems, or analysing legal implications of robotic healthcare technologies. Through such collaborative environments, students gain experience working across disciplinary boundaries while developing solutions to real-world governance problems.

Another important area of innovation involves the development of legal engineering as a professional discipline. Legal engineering refers to the process of translating legal rules and regulatory requirements into operational systems that can be implemented within technological platforms. For example, a legal engineer might collaborate with software developers to embed compliance requirements directly into algorithms governing automated decision-making systems. This role requires both legal expertise and technical literacy, bridging the gap between abstract legal principles and practical technological implementation.

Legal engineering is particularly relevant in the context of robotics governance. As discussed earlier in this chapter, embedding legal constraints within technological systems can enhance compliance and reduce the likelihood of regulatory violations. Achieving this integration requires professionals capable of interpreting legal norms and translating them into technical specifications. Future legal professionals trained in both law and engineering will be uniquely positioned to perform this function.

Professional institutions also have a role to play in promoting cross-disciplinary innovation. Bar associations, judicial training programmes, and regulatory agencies may need to develop continuing education initiatives that familiarise legal practitioners with technological concepts relevant to their work. Judges adjudicating disputes involving robotics systems may benefit from specialised training that explains how autonomous technologies function. Policymakers responsible for drafting legislation governing artificial intelligence may require technical advisors capable of explaining the implications of specific regulatory choices. The integration of cross-disciplinary knowledge also strengthens democratic governance. When legal professionals understand technological systems more deeply, they can communicate regulatory challenges more effectively to legislators and the public. Clear explanations of how technologies function and what risks they present help ensure that public debates about technology policy remain informed and constructive rather than driven by speculation or misunderstanding.

At the same time, cross-disciplinary collaboration must preserve the independence and critical perspective that legal analysis brings to governance processes. Engineers and technologists often focus primarily on solving technical problems, while lawyers are trained to consider broader questions about rights, accountability, and institutional responsibility. Effective collaboration requires respecting the strengths of each discipline rather than subsuming one within the other. The goal is not to transform lawyers into engineers or engineers into lawyers, but to create professionals capable of working together in shaping responsible technological futures. The need for cross-disciplinary legal innovation will only intensify as robotics technologies continue to evolve. Emerging developments such as human–robot collaboration, machine learning autonomy, and cyber-physical infrastructure systems blur the boundaries between technological design and social governance. Legal professionals who understand both domains will play a critical role in ensuring that technological progress remains aligned with human values.

Ultimately, the evolution of legal education and professional practice reflects a broader transformation in how societies govern innovation. Law can no longer operate solely as an external regulator responding to technological change after the fact. Instead, it must become an active participant in shaping the systems that structure modern life. Cross-disciplinary legal innovation represents one pathway toward achieving this transformation. The next section turns to the economic implications of widespread automation. As robotics technologies generate increasing productivity, societies may encounter a future in which abundance replaces scarcity as the defining condition of economic life. Law must therefore consider how governance frameworks evolve when machines produce wealth on a scale that challenges traditional assumptions about work, income, and social stability.

Law for Abundant and Universal High-Income Futures

For most of recorded history, law has developed within the constraints of scarcity. Economic systems were organised around the assumption that human labour, land, and capital were limited resources that must be allocated carefully through markets, property rights, and contractual exchange. Legal institutions evolved to govern this reality. Property law determined ownership of scarce resources. Labour law regulated the relationship between employers and workers whose time and effort created economic value. Taxation systems redistributed portions of that value to support public institutions and social welfare. These frameworks functioned effectively within societies where productivity depended primarily on human work and where economic output was constrained by the physical limits of human labour.

The rise of robotics and artificial intelligence introduces a potential shift in this historical pattern. Autonomous systems increasingly perform tasks that once required human effort across manufacturing, logistics, agriculture, infrastructure management, and even aspects of professional services. Machines capable of operating continuously without fatigue or wages may dramatically expand productive capacity. If such technologies continue to advance, societies may encounter an economic environment where the production of goods and services becomes less constrained by human labour than at any previous point in history.

This possibility raises profound questions for legal systems designed around assumptions of scarcity. If machines generate substantial economic surplus, the challenge of governance may shift from managing limited resources to distributing abundant output fairly and sustainably. Law must therefore begin to consider how institutional frameworks evolve in environments where automation reshapes the foundations of work, wealth creation, and social stability.

The concept of an abundant economy does not imply that scarcity disappears entirely. Natural resources, environmental constraints, and human attention will always impose limits on economic activity. However, automation may dramatically reduce the cost of producing many goods and services that currently require significant human labour. Autonomous factories, robotic agriculture, and intelligent logistics networks may produce large quantities of essential products with minimal human intervention. In such circumstances, economic value may increasingly derive from the ownership and governance of technological systems rather than from the direct labour of individuals.

This transformation challenges one of the central pillars of modern economic law: the relationship between labour and income. In industrial societies, wages represent the primary mechanism through which individuals obtain purchasing power and economic security. Employment contracts link labour contribution to income generation, while welfare systems provide support to individuals who cannot participate in the labour market. If robotics technologies gradually reduce the demand for human labour in certain sectors, these traditional relationships may weaken.

Legal systems must therefore confront the possibility that future societies require new mechanisms for distributing economic value. The emergence of universal income models represents one possible response. Under such frameworks, individuals receive a baseline level of economic support independent of traditional employment. Advocates argue that universal income systems could ensure social stability in economies where automation reduces reliance on human labour while maintaining consumer purchasing power necessary for economic activity.

Whether universal income becomes politically feasible remains uncertain. Nevertheless, the discussion illustrates a broader point. Robotics and artificial intelligence may require legal systems to reconsider how economic rights and obligations are structured. If machines generate substantial economic surplus, societies must decide how the benefits of that surplus are distributed among citizens, corporations, and public institutions.

Taxation frameworks will play a central role in this debate. Historically, governments have relied heavily on taxes derived from wages, corporate profits, and consumption. Automation may alter this balance. If robotic systems replace human labour in certain industries, payroll tax revenues could decline even as productivity increases. Policymakers may therefore explore alternative taxation models that capture value generated by automated production systems.

Some economists have proposed forms of automation taxation or robot dividends, where revenues derived from highly automated industries contribute to public funds supporting social programmes. Such proposals remain controversial, particularly among technology developers concerned that excessive taxation could discourage innovation. Yet the underlying question remains unavoidable. Legal systems must determine how the economic value generated by autonomous technologies contributes to the broader social contract.

Property law may also evolve in response to automated production. Ownership of robotic infrastructure, artificial intelligence systems, and automated factories could become a central determinant of economic power. If ownership becomes concentrated within a small number of corporations or individuals, inequalities may widen despite overall increases in productivity. Legal frameworks governing corporate governance, antitrust policy, and intellectual property rights will therefore influence how the benefits of automation are distributed across society.

Another dimension of the abundant economy concerns public goods and infrastructure. Automated technologies may enable governments to provide certain services more efficiently than in the past. Autonomous systems could assist in maintaining transportation networks, managing energy grids, monitoring environmental conditions, and delivering essential services in remote regions. Legal frameworks governing public procurement and infrastructure investment must adapt to incorporate these technological capabilities while ensuring transparency and accountability.

The transition toward automation also raises questions about the future of human work itself. Even if machines perform many tasks more efficiently than humans, work remains a central source of meaning, identity, and social participation for many individuals. Legal frameworks must therefore address not only the economic consequences of automation but also its social implications. Policies supporting education, lifelong learning, and creative industries may help ensure that individuals continue to find meaningful roles within societies increasingly shaped by technological systems.

One potential outcome is the emergence of pluralistic labour ecosystems in which human work shifts toward areas where human creativity, empathy, and cultural understanding remain particularly valuable. Fields such as education, arts, research, and community services may expand as automation reduces the necessity of labour in more repetitive sectors. Legal frameworks supporting these sectors—through cultural policy, intellectual property rights, and public funding mechanisms—may play a significant role in shaping the social fabric of automated societies. Social stability represents another crucial consideration. Economic transformations that disrupt traditional employment patterns can generate political and social tensions if they occur without adequate institutional support. Historical examples such as the industrial revolution illustrate how technological change can produce both extraordinary prosperity and profound social dislocation. Law therefore has a responsibility to guide technological transitions in ways that preserve social cohesion and protect vulnerable populations.

International dimensions also emerge in discussions of automated abundance. Countries with advanced robotics industries may experience productivity gains that alter global economic balances. Developing economies that previously relied on labour-intensive manufacturing may encounter challenges if automation reduces the comparative advantage of low-cost labour. International trade law and development policy must therefore consider how global economic relationships evolve in response to widespread automation.

Environmental considerations further complicate the picture. Automation may enable more efficient use of resources through precision agriculture, intelligent energy management, and optimised logistics networks. At the same time, increased production capacity could lead to greater consumption of materials and energy if sustainability considerations are not integrated into technological development. Legal frameworks governing environmental protection and sustainable development must therefore remain central to the governance of automated economies.

The concept of universal high-income futures reflects the possibility that technological productivity could enable societies to achieve higher overall standards of living while reducing the economic pressures historically associated with scarcity. If managed responsibly, robotics and artificial intelligence could contribute to a world in which essential goods and services become more accessible, working hours decline, and individuals have greater freedom to pursue education, creativity, and community engagement.

Realising such a future requires thoughtful legal design. Laws must ensure that technological wealth does not concentrate excessively in ways that undermine democratic institutions or social stability. Governance frameworks must promote fair distribution of benefits while preserving incentives for innovation and entrepreneurship. Achieving this balance will require experimentation, public dialogue, and international cooperation as societies navigate the economic transformations associated with automation.

Ultimately, the legal challenges posed by abundant automated economies illustrate a broader theme running throughout this book. Robotics governance is not merely about preventing technological harm. It is also about shaping the societal conditions under which technological progress unfolds. Law must guide innovation in ways that align technological capabilities with human values, ensuring that the benefits of robotics and artificial intelligence contribute to collective wellbeing rather than exacerbating existing inequalities.

The final section of this chapter returns to the role of the reader and the broader legal community in shaping this future. Technological transformation does not occur independently of human choices. Lawyers, policymakers, engineers, scholars, and citizens all participate in designing the institutions that govern emerging technologies. The future of robotics law therefore depends on collective engagement with the questions explored throughout this book.

Call to Action

Every generation of lawyers inherits a legal system shaped by the technological and social realities of its time. The jurists of the maritime age developed rules for governing trade across oceans. The architects of industrial law constructed frameworks for labour relations, corporations, and mass production. The late twentieth century witnessed the emergence of digital law, privacy regulation, and cybersecurity doctrines in response to the expansion of the internet and information technologies. Each transformation required legal thinkers willing to reimagine how law should function in a changing world.

The emergence of robotics and artificial intelligence represents another such moment. Yet the challenge facing today's legal community may be more profound than those encountered in previous technological transitions. Machines are no longer merely tools extending human capability. They are increasingly autonomous systems capable of interacting with physical environments, making decisions, and generating economic value at scales that reshape entire industries. Law must therefore evolve not only to regulate machines but to govern societies in which humans and intelligent systems coexist within shared infrastructures.

Throughout this book, we have explored the foundations of such governance. We have examined the regulatory frameworks necessary to ensure safety and accountability when robots operate in human environments. We have considered how ethical design principles, licensing regimes, and liability rules can guide the development of responsible robotic technologies. We have also explored deeper philosophical questions concerning the legal status of intelligent machines and the evolving relationship between human and artificial actors within legal systems.

The final chapters have pushed this discussion further by considering how legal institutions themselves must adapt to the technological transformations now unfolding. Law must become more adaptive, more interdisciplinary, and more capable of responding to rapid innovation. Governance frameworks must incorporate mechanisms for monitoring technological change, updating regulatory standards, and coordinating responses across jurisdictions. Legal professionals must develop new forms of expertise that bridge traditional doctrinal analysis with technological understanding.

Yet none of these transformations will occur automatically. Legal systems do not evolve simply because technology advances. They evolve because individuals and institutions recognise the need for change and act collectively to shape the future of governance. Lawyers, judges, policymakers, engineers, scholars, and citizens all play roles in determining how emerging technologies are integrated into society.

The purpose of this final section is to issue a challenge. The governance of robotics and artificial intelligence will define many of the most important legal questions of the twenty-first century. Decisions made in the coming decades will influence how technological power is distributed, how human dignity is protected, and how societies navigate the transition toward increasingly automated economic systems.

The legal community must therefore take an active role in shaping these developments. Waiting for technological disruption to occur before responding through reactive legislation is no longer sufficient. Lawyers must engage directly with technologists, participate in interdisciplinary research, and contribute to the design of governance frameworks capable of guiding innovation responsibly. Legal scholarship must expand beyond traditional doctrinal analysis to explore the institutional architectures required for governing complex technological systems.

One practical step toward achieving this goal is the creation of living legal frameworks that evolve alongside technological development. Static regulatory models may struggle to keep pace with rapidly advancing robotics technologies. Instead, governance systems should incorporate mechanisms for continuous monitoring, iterative policy updates, and collaborative problem-solving among stakeholders. Legal institutions must become more dynamic while preserving the stability and legitimacy that remain essential to the rule of law.

To support this vision, this book proposes the development of an open living annex—a collaborative, continuously updated resource documenting global developments in robotics and artificial intelligence governance. Such an annex would track new legislation, regulatory initiatives, court decisions, and policy experiments from around the world. By maintaining an evolving repository of governance knowledge, the legal community can respond more effectively to technological change and learn from experiences across jurisdictions.

The living annex would not belong to a single institution or government. Instead, it would function as a collaborative platform where scholars, policymakers, engineers, and practitioners contribute insights and updates. Universities, research institutes, international organisations, and professional associations could all participate in maintaining and expanding this knowledge infrastructure. In doing so, they would help create a shared global understanding of how robotics governance is evolving.

Such collaborative initiatives reflect a broader principle underlying this book: the governance of emerging technologies must be a collective endeavour. Robotics and artificial intelligence do not respect national borders. Autonomous systems developed in one country may operate globally through supply chains, digital infrastructure, and international commerce. Effective governance therefore requires cooperation among jurisdictions, institutions, and disciplines.

At the same time, technological progress must remain guided by enduring human values. The purpose of law is not merely to facilitate innovation but to ensure that innovation serves human flourishing. As machines become more capable, the responsibility of legal systems to protect dignity, justice, and social stability becomes even more important. Robotics governance must therefore remain anchored in principles that prioritise human wellbeing even as societies integrate increasingly autonomous technologies.

The transition toward a post-human legal order does not diminish the importance of humanity within legal systems. On the contrary, it reinforces the need for human wisdom in shaping the institutions that govern technological societies. Machines may perform tasks, analyse data, and optimise processes, but the fundamental questions of justice, fairness, and collective responsibility remain human questions. Law provides the institutional framework through which societies address those questions.

Readers of this book are therefore invited to participate in the continuing development of robotics law. Whether as lawyers, engineers, policymakers, educators, business leaders, technologists, students, or informed citizens, individuals can contribute to shaping governance frameworks that align technological progress with human values. The ideas presented in these chapters represent only a starting point for an ongoing global conversation about how law should evolve in response to increasingly intelligent and autonomous systems.

Readers are also encouraged to engage directly with the sixteen questions presented in the Notes section. These questions have been designed not merely as a summary of the book, but as a framework for critical reflection on the legal, ethical, technological, and societal challenges explored throughout these pages. The ability to answer them thoughtfully is, in many respects, a measure of one's understanding of the foundations, complexities, and future directions of robotics law.

The invitation extends further. Throughout the Between Us and Them narrative, readers are encouraged to apply the four recurring questions introduced in Episode One to the unfolding story of Ògún and Elias. What does each episode reveal about what it means to be human? Which human quality, belief, fear, or aspiration is being reflected through the relationship between humans and humanoids? What responsibilities emerge when intelligence becomes embodied and shares our world? Which legal, ethical, or governance assumptions are being challenged? The answers may evolve as the story progresses, just as society's understanding of intelligent machines will evolve in the years ahead.

The future of robotics governance will not be written solely in legislative chambers, courtrooms, laboratories, boardrooms, or academic journals. It will be shaped through the combined efforts of individuals and institutions committed to building legal systems capable of guiding technological transformation responsibly. By engaging with the questions explored in this book, reflecting on the journey of Ògún and Elias, and contributing to collaborative governance initiatives, readers can play a role in ensuring that the rise of intelligent machines strengthens rather than undermines the foundations of human society.

Law has always been humanity's way of organising itself in the face of change. The emergence of artificial intelligence, embodied intelligence, and physical intelligence presents one of the most profound transformations our species has ever encountered. The challenge before us is not simply to regulate machines, but to govern wisely the future we are building alongside them. Whether that future is defined by trust or fear, dignity or exploitation, shared prosperity or division, will depend on the choices we make today. The conversation has already begun. The next chapter is ours to write.

Notes

Legislation & Regulatory Instruments

Council of Europe, Framework Convention on Artificial Intelligence, Human Rights, Democracy and the Rule of Law (2024).

Council Regulation (EC) No 428/2009 setting up a Community regime for the control of exports, transfer, brokering and transit of dual-use items [2009] OJ L 134/1.

Directive 85/374/EEC on liability for defective products [1985] OJ L 210/29.

Directive 2001/95/EC on general product safety [2002] OJ L 11/4.

Directive 2006/42/EC on machinery [2006] OJ L 157/24.

Directive 2014/30/EU on electromagnetic compatibility [2014] OJ L 96/79.

Directive 2014/53/EU on radio equipment [2014] OJ L 153/62.

Directive (EU) 2020/1828 on representative actions for the protection of the collective interests of consumers [2020] OJ L 409/1.

Directive (EU) 2024/2853 on liability for defective products (recast) [2024] OJ L.

European Commission, Proposal for a Directive on adapting non-contractual civil liability rules to artificial intelligence (AI Liability Directive) COM (2022) 496 final.

European Commission, Proposal for a Regulation on Machinery Products (recast) COM (2021) 202 final.

European Parliament, Resolution of 16 February 2017 with recommendations to the Commission on Civil Law Rules on Robotics (2015/2103(INL)).

Federal Aviation Administration, Remote Identification of Unmanned Aircraft 86 Fed Reg 4390 (15 January 2021).

ISO 10218-1:2011 Robots and robotic devices — Safety requirements for industrial robots — Part 1: Robots.

ISO 13482:2014 Robots and robotic devices — Safety requirements for personal care robots.

ISO/IEC 42001:2023 Artificial intelligence management systems — Requirements.

IEEE 7000-2021 IEEE Standard Model Process for Addressing Ethical Concerns during System Design.

Japan, Act on the Protection of Personal Information (APPI), Act No 57 of 2003 (as amended).

Japan Cabinet Secretariat, Basic Act on Artificial Intelligence (AI Basic Act) (2023).

National Traffic and Motor Vehicle Safety Act, 49 USC ch 301 (2018).

Outer Space Treaty, Treaty on Principles Governing the Activities of States in the Exploration and Use of Outer Space (1967).

Regulation (EU) 2016/679 General Data Protection Regulation [2016] OJ L 119/1.

Regulation (EU) 2019/1020 on market surveillance and compliance of products [2019] OJ L 169/1.

Regulation (EU) 2021/821 setting up a Union regime for the control of exports, brokering, technical assistance, transit and transfer of dual-use items [2021] OJ L 206/1.

Regulation (EU) 2023/1230 on machinery products [2023] OJ L 165/1.

Regulation (EU) 2024/1689 of the European Parliament and of the Council laying down harmonised rules on artificial intelligence (Artificial Intelligence Act) [2024] OJ L.

Restatement (Third) of Torts: Products Liability (American Law

Institute 1998).

United Kingdom Automated and Electric Vehicles Act 2018.

United Kingdom Consumer Protection Act 1987.

United Kingdom Data Protection Act 2018.

United Nations Convention on Certain Conventional Weapons (1980).

United States Congress, Algorithmic Accountability Act of 2022 HR 6580, 117th Congress.

US Export Administration Regulations 15 CFR §§ 730–774.

US International Traffic in Arms Regulations 22 CFR §§ 120–130.

International Treaties & Instruments

Additional Protocol I to the Geneva Conventions relating to the Protection of Victims of International Armed Conflicts (adopted 8 June 1977, entered into force 7 December 1978) 1125 UNTS 3.

Articles on Responsibility of States for Internationally Wrongful Acts, UNGA Res 56/83 (12 December 2001).

Charter of the United Nations (adopted 26 June 1945, entered into force 24 October 1945) 1 UNTS XVI.

Convention on Certain Conventional Weapons (Convention on Prohibitions or Restrictions on the Use of Certain Conventional Weapons Which May Be Deemed to Be Excessively Injurious or to Have Indiscriminate Effects) (adopted 10 October 1980, entered into force 2 December 1983) 1342 UNTS 137.

Convention on the Prohibition of Anti-Personnel Mines (Ottawa Convention) (adopted 18 September 1997, entered into force 1 March 1999) 2056 UNTS 211.

Geneva Convention (I) for the Amelioration of the Condition of the Wounded and Sick in Armed Forces in the Field (adopted 12 August 1949, entered into force 21 October 1950) 75 UNTS 31.

Outer Space Treaty (Treaty on Principles Governing the Activities of States in the Exploration and Use of Outer Space, including the Moon and Other Celestial Bodies) (adopted 27 January 1967, entered into force 10 October 1967) 610 UNTS 205.

Vienna Convention on the Law of Treaties (adopted 23 May 1969, entered into force 27 January 1980) 1155 UNTS 331.

Wassenaar Arrangement on Export Controls for Conventional Arms and Dual-Use Goods and Technologies (adopted 12 July 1996, as amended).

Books

Abbott R, The Reasonable Robot: Artificial Intelligence and the Law (Cambridge University Press 2020).

Acemoglu D and Johnson S, Power and Progress: Our Thousand-Year Struggle over Technology and Prosperity (PublicAffairs 2023).

Asimov I, I, Robot (Gnome Press 1950).

Asimov I, The Rest of the Robots (Doubleday 1964).

Autor D, Mindell D and Reynolds E, The Work of the Future: Building Better Jobs in an Age of Intelligent Machines (MIT Press 2022).

Bostrom N, Super intelligence: Paths, Dangers, Strategies (Oxford University Press 2014).

Brynjolfsson E and McAfee A, The Second Machine Age: Work, Progress, and Prosperity in a Time of Brilliant Technologies (W W Norton 2014).

Brynjolfsson E and McAfee A, Machine, Platform, Crowd: Harnessing Our Digital Future (W W Norton 2017).

Buolamwini J, Unmasking AI: My Mission to Protect What Is Human in a World of Machines (Random House 2023).

Calo R, Froomkin AM and Kerr IM (eds), Robot Law (Edward Elgar 2016).

Cath C and Floridi L (eds), The Ethics of Artificial Intelligence (Oxford University Press 2023).

Crawford K, Atlas of AI: Power, Politics, and the Planetary Costs of Artificial Intelligence (Yale University Press 2021).

Eitel-Porter R, Dongha P and Vogel M, Governing the Machine (Kogan Page 2024).

Endicott T, Administrative Law (1st edn, Oxford University Press 2009).

Ford M, Rise of the Robots: Technology and the Threat of a Jobless Future (Basic Books 2015).

Gurney JK, Driving Towards Driverless: A Roadmap for Autonomous Vehicles (Carolina Academic Press 2021).

Harari YN, Homo Deus: A Brief History of Tomorrow (Harvill Secker 2016).

Harari YN, 21 Lessons for the 21st Century (Jonathan Cape 2018).

Kaplan J, Artificial Intelligence: What Everyone Needs to Know (Oxford University Press 2016).

Kerr I, Millar J and Turton K (eds), Robot Law (Edward Elgar 2016).

Lee KF, AI Superpowers: China, Silicon Valley, and the New World Order (Houghton Mifflin Harcourt 2018).

Marcus G, Taming Silicon Valley (MIT Press 2024).

Marcus G and Davis E, Rebooting AI: Building Artificial Intelligence We Can Trust (Pantheon 2019).

Pagallo U, The Laws of Robots: Crimes, Contracts, and Torts (Springer 2013).

Pasquale F, The Black Box Society: The Secret Algorithms That Control Money and Information (Harvard University Press 2015).

Pasquale F, New Laws of Robotics: Defending Human Expertise in the Age of AI (Harvard University Press 2020).

Rashidi S, Scaling AI: The Governance and Security Playbook for Executives (Wiley 2024).

Rashidi S, Your AI Survival Guide (Wiley 2023).

Rogers J, The Liability of Autonomous Systems (Oxford University Press 2022).

Russell S, Human Compatible: Artificial Intelligence and the Problem of Control (Penguin 2019).

Schwab K, The Fourth Industrial Revolution (World Economic Forum 2016).

Schwab K and Malleret T, The Great Reset (Forum Publishing 2020).

Schmitt MN and Heintschel von Heinegg W (eds), The Tallinn Manual 2.0 on the International Law Applicable to Cyber Operations (Cambridge University Press 2017).

Soares S, AI Governance (Technics Publications 2022).

Soares S, The IBM Data Governance Unified Process (Technics Publications 2010).

Susskind R, The Future of the Professions (Oxford University Press 2015).

Susskind R, Tomorrow's Lawyers (Oxford University Press 2017).

Susskind R and Susskind D, The Future of the Professions: How Technology Will Transform the Work of Human Experts (Oxford University Press 2015).

Teubner G, Law as an Autopoietic System (Blackwell 1993).

Turner J, Robot Rules: Regulating Artificial Intelligence (Palgrave Macmillan 2019).

Varian HR, Shapiro C and others, Information Rules: A Strategic Guide to the Network Economy (Harvard Business School Press 1999).

Zuboff S, The Age of Surveillance Capitalism (Profile Books 2019).

Policy Documents & Communications

European Commission, 'Artificial Intelligence for Europe' COM (2018) 237 final.

European Commission, Ethics Guidelines for Trustworthy AI (High-Level Expert Group on Artificial Intelligence 2019).

European Commission, 'Proposal for a Directive on adapting non-contractual civil liability rules to artificial intelligence (AI Liability Directive)' COM (2022) 496 final.

European Commission, 'Proposal for a Regulation laying down harmonised rules on artificial intelligence (Artificial Intelligence Act)' COM (2021) 206 final.

European Commission, 'White Paper on Artificial Intelligence: A European Approach to Excellence and Trust' COM (2020) 65 final.

European Commission, 'Guidance on the Application of the Artificial

Intelligence Act to General-Purpose AI Models' (2024).

European Parliament, Resolution of 16 February 2017 with recommendations to the Commission on Civil Law Rules on Robotics [2017] OJ C 252/239.

European Parliament, Resolution on Artificial Intelligence in Criminal Law and its Use by the Police and Judicial Authorities in Criminal Matters (2020/2016(INI)).

G7, Hiroshima AI Process Comprehensive Policy Framework (2023).

IEEE, Ethically Aligned Design: A Vision for Prioritizing Human Wellbeing with Autonomous and Intelligent Systems (1st edn, IEEE 2019).

International Committee of the Red Cross, Autonomous Weapon Systems: Technical, Military, Legal and Humanitarian Aspects (ICRC 2014).

International Committee of the Red Cross, Position on Autonomous Weapon Systems (ICRC 2023).

International Organization for Standardisation, ISO 13482:2014 Robots and Robotic Devices — Safety Requirements for Personal Care Robots.

International Organization for Standardisation, ISO/IEC 42001:2023 Artificial Intelligence Management Systems — Requirements.

OECD, Recommendation of the Council on Artificial Intelligence OECD/LEGAL/0449 (2019).

OECD, OECD Principles on Artificial Intelligence (2019).

UK Department for Science, Innovation and Technology, A Pro-Innovation Approach to AI Regulation (March 2023).

United Nations Educational, Scientific and Cultural Organization (UNESCO), Recommendation on the Ethics of Artificial Intelligence (2021).

United Nations Group of Governmental Experts on Emerging Technologies in the Area of Lethal Autonomous Weapons Systems, Reports (2017–2023).

United Nations Secretary-General, Our Common Agenda: Policy Brief on Artificial Intelligence (2023).

United States National Institute of Standards and Technology, Artificial Intelligence Risk Management Framework 1.0 (NIST 2023).

United States, Executive Order 14110 on the Safe, Secure, and Trustworthy Development and Use of Artificial Intelligence (30 October 2023).

World Economic Forum, Global Governance of Artificial Intelligence for Humanity (WEF 2022).

Academic Journals

Abbott R, 'I Think, Therefore I Invent: Creative Computers and the Future of Patent Law' (2016) 57 Boston College Law Review 1079.

Balkin JM, 'The Path of Robotics Law' (2015) 6 California Law Review Circuit 45.

Bryson JJ, 'Robots Should Be Slaves' in Yorick Wilks (ed), Close Engagements with Artificial Companions (John Benjamins 2010).

Burri T, 'The International Regulation of Autonomous Military Robots' (2017) 7 Journal of International Humanitarian Legal Studies 139.

Calo R, 'Artificial Intelligence Policy: A Primer and Roadmap' (2017) 51 UC Davis Law Review 399.

Calo R, 'Robotics and the Lessons of Cyberlaw' (2015) 103 California Law Review 513.

Crootof R, 'The Killer Robots Are Here: Legal and Policy Implications' (2015) 36 Cardozo Law Review 1837.

Crootof R, 'War Torts: Accountability for Autonomous Weapons' (2016) 164 University of Pennsylvania Law Review 1347.

Dignum V, 'Responsible Artificial Intelligence: Developing and Using AI in a Responsible Way' (2019) 24 AI & Society 1.

European Parliamentary Research Service, Civil Law Rules on Robotics (PE 571.379, 2016).

Floridi L and Cowls J, 'A Unified Framework of Five Principles for AI in Society' (2019) 1 Harvard Data Science Review.

Gless S, Silverman E and Weigend T, 'If Robots Cause Harm, Who Is to Blame?' (2016) 19 New Criminal Law Review 412.

Goodall N, 'Ethical Decision Making During Automated Vehicle Crashes' (2014) 12 Transportation Research Record 58.

Hildebrandt M, 'Law as Information in the Era of Data-Driven Agency' (2016) 79 Modern Law Review 1.

Marchant GE and Lindor RA, 'The Coming Collision Between Autonomous Vehicles and the Liability System' (2012) 52 Santa Clara Law Review 1321.

Martin K, 'Ethical Implications and Accountability of Algorithms' (2019) 160 Journal of Business Ethics 835.

McKinsey Global Institute, The Future of Work After COVID-19 (2021).

Mittelstadt BD, Allo P, Taddeo M, Wachter S and Floridi L, 'The Ethics of Algorithms: Mapping the Debate' (2016) 3 Big Data & Society.

OECD, AI, Robotics and the Future of Work (OECD Publishing 2021).

Pagallo U, 'The Legal Challenges of Big Data' (2017) 16 International Journal of Communications Law and Policy 1.

Rahwan I, 'Society-in-the-Loop: Programming the Algorithmic Social Contract' (2018) 2 Ethics and Information Technology.

Sparrow R, 'Killer Robots' (2007) 24 Journal of Applied Philosophy 62.

Turner J, 'Liability and Artificial Intelligence in the EU' (2020) 21 ERA Forum 143.

Vallor S, 'The Moral Deskilling of Robots' (2015) 29 Philosophy & Technology 107.

Wachter S, Mittelstadt B and Russell C, 'Counterfactual Explanations without Opening the Black Box: Automated Decisions and the GDPR' (2018) 31 Harvard Journal of Law & Technology 841.

Reports & Technical Standards

Alan Turing Institute, Understanding Artificial Intelligence Ethics and Safety (2019).

Boston Dynamics, Atlas Technical Overview (company white paper, 2023).

Brookings Institution, Governing AI: Principles and Policies (2023).

European Union Agency for Fundamental Rights, Getting the Future Right: Artificial Intelligence and Fundamental Rights (Publications Office of the European Union 2020).

Figure AI, Helix System Overview (company materials, 2024).

High-Level Expert Group on Artificial Intelligence, Ethics Guidelines for Trustworthy AI (European Commission 2019).

High-Level Expert Group on Artificial Intelligence, Policy and

Investment Recommendations for Trustworthy AI (European Commission 2019).

IEEE, IEEE 7000-2021 Standard Model Process for Addressing Ethical Concerns During System Design (2021).

IEEE, IEEE 7010-2020 Recommended Practice for Assessing the Impact of Autonomous and Intelligent Systems on Human Well-Being (2020).

International Civil Aviation Organization, Annex 19: Safety Management (ICAO 2016).

International Federation of Robotics, World Robotics Report (2023).

International Organization for Standardisation, ISO 13482:2014 Robots and Robotic Devices — Safety Requirements for Personal Care Robots (ISO 2014).

International Organization for Standardisation and International Electrotechnical Commission, ISO/IEC 23894:2023 Artificial Intelligence — Risk Management (ISO/IEC 2023).

MIT Technology Review, 'Humanoid Robots Are Having a Moment' (2023).

NATO, Autonomy in Defence Systems (2022).

Royal Society, Machine Learning: The Power and Promise of Computers that Learn by Example (2017).

Stanford Institute for Human-Centered Artificial Intelligence, AI Index Report 2024 (Stanford University 2024).

Tesla Inc, AI Day Presentations (2022–2024).

World Customs Organization, Harmonized Commodity Description and Coding System (HS Code 8479) (WCO).

World Economic Forum, A Blueprint for Equity and Inclusion in Artificial Intelligence (2022).

World Economic Forum, Navigating the Regulatory Landscape for Artificial Intelligence (2023).

National AI & Robotic Strategies

Cabinet Office, Government of Japan, Society 5.0 (Cabinet Office 2016).

China State Council, New Generation Artificial Intelligence Development Plan (State Council 2017).

European Commission, Coordinated Plan on Artificial Intelligence COM (2018) 795 final.

European Commission, Coordinated Plan on Artificial Intelligence 2021 Review COM (2021) 205 final.

Government of Canada, Pan-Canadian Artificial Intelligence Strategy (CIFAR 2017, renewed 2022).

Government of Singapore, National Artificial Intelligence Strategy (Smart Nation and Digital Government Office 2019).

Government of Singapore, National AI Strategy 2.0: AI for the Public Good, for Singapore and the World (Smart Nation Group 2023).

Ministry of Economy, Trade and Industry, Japan, Robot Strategy: Japan's Robot Strategy, Vision, Strategy, Action Plan (METI 2015).

Ministry of Industry and Information Technology, China, Guiding Opinions on the Innovative Development of Humanoid Robots (MIIT 2023).

Republic of Korea, National Strategy for Artificial Intelligence (Government of the Republic of Korea 2019).

UAE Government, UAE Strategy for Artificial Intelligence 2031 (UAE Government 2017).

UK Department for Science, Innovation and Technology, National AI Strategy (HM Government 2021).

United States National Artificial Intelligence Initiative Office, The National Artificial Intelligence Research and Development Strategic Plan: 2023 Update (NSTC 2023).

United States National Robotics Initiative, National Robotics Initiative 3.0: Innovations in Integration of Robotics (National Science Foundation 2021).

Robotics Engineering & Safety Frameworks

American National Standards Institute and Robotic Industries Association, ANSI/RIA R15.06-2012 Industrial Robots and Robot Systems: Safety Requirements (ANSI/RIA 2012).

ASTM International, ASTM F45 Standards on Robotics, Automation and Autonomous Systems (ASTM International).

British Standards Institution, BS 8611:2016 Robots and Robotic Devices: Guide to the Ethical Design and Application of Robots and Robotic Systems (BSI 2016).

IEC, IEC 61508 Functional Safety of Electrical/Electronic/Programmable Electronic Safety-Related Systems (IEC 2010).

IEEE, IEEE 7000-2021 Standard Model Process for Addressing Ethical Concerns During System Design (IEEE 2021).

IEEE, IEEE 7010-2020 Recommended Practice for Assessing the Impact of Autonomous and Intelligent Systems on Human Well-Being (IEEE 2020).

International Organization for Standardisation, ISO 10218-1:2011 Robots and Robotic Devices: Safety Requirements for Industrial Robots, Part 1: Robots (ISO 2011).

International Organization for Standardisation, ISO 10218-2:2011

Robots and Robotic Devices: Safety Requirements for Industrial Robots, Part 2: Robot Systems and Integration (ISO 2011).

International Organization for Standardisation, ISO 13482:2014 Robots and Robotic Devices: Safety Requirements for Personal Care Robots (ISO 2014).

International Organization for Standardisation, ISO/TS 15066:2016 Robots and Robotic Devices: Collaborative Robots (ISO 2016).

International Organization for Standardisation, ISO/IEC 23894:2023 Artificial Intelligence: Guidance on Risk Management (ISO/IEC 2023).

International Organization for Standardisation, ISO/IEC 42001:2023 Artificial Intelligence Management System (ISO/IEC 2023).

SAE International, J3016 Taxonomy and Definitions for Terms Related to Driving Automation Systems for On-Road Motor Vehicles (SAE International 2021).

Underwriters Laboratories, UL 4600 Standard for Safety for the Evaluation of Autonomous Products (UL 2020).

Corporate Technical Architecture Materials

Agility Robotics, Digit Product and Deployment Materials (Agility Robotics 2023-2025).

Boston Dynamics, Atlas Technical Overview and Product Materials (Boston Dynamics 2024).

Boston Dynamics, Spot Product Documentation and Safety Materials (Boston Dynamics 2023).

Figure AI, Helix: A Vision-Language-Action Model for Generalist Humanoid Control (Figure AI 2025).

Fourier Intelligence, GR-1 and GR-2 Humanoid Robot Product Materials (Fourier Intelligence 2023-2025).

Hanson Robotics, Sophia Technical and Company Materials (Hanson Robotics).

NVIDIA, Project GR00T: Generalist Robot Foundation Model Materials (NVIDIA 2024).

Sanctuary AI, Phoenix Humanoid Robot and Cognitive Architecture Materials (Sanctuary AI 2023-2025).

Tesla, AI Day 2022: Optimus and Tesla Bot Presentation Materials (Tesla 2022).

Tesla, We, Robot and Optimus Demonstration Materials (Tesla 2024).

Unitree Robotics, H1 Humanoid Robot Product Materials (Unitree Robotics 2023-2025).

XPENG Robotics, IRON Humanoid Robot Product Materials (XPENG Robotics 2024).

Zhiyuan Robotics, AgiBot Humanoid Robot Product and Company Materials (Zhiyuan Robotics 2023-2025).

Labour Transition & Infrastructure Governance Reports

Brookings Institution, Automation and Artificial Intelligence: How Machines Are Affecting People and Places (Brookings Institution 2019).

International Labour Organization, World Employment and Social Outlook: The Role of Digital Labour Platforms in Transforming the World of Work (ILO 2021).

International Labour Organization, Generative AI and Jobs: A Global Analysis of Potential Effects on Job Quantity and Quality (ILO 2023).

International Monetary Fund, Gen-AI: Artificial Intelligence and the Future of Work (IMF 2024).

McKinsey Global Institute, A Future That Works: Automation, Employment, and Productivity (McKinsey & Company 2017).

McKinsey Global Institute, The Future of Work After COVID-19 (McKinsey & Company 2021).

OECD, Employment Outlook 2023: Artificial Intelligence and the Labour Market (OECD Publishing 2023).

OECD, The Future of Work: OECD Employment Outlook 2019 (OECD Publishing 2019).

United Nations, World Social Report 2023: Leaving No One Behind in an Ageing World (United Nations 2023).

UN-Habitat, World Cities Report 2022: Envisaging the Future of Cities (UN-Habitat 2022).

World Bank, World Development Report 2019: The Changing Nature of Work (World Bank 2019).

World Economic Forum, The Future of Jobs Report 2023 (WEF 2023).

World Economic Forum, The Future of Jobs Report 2025 (WEF 2025).

Regulatory Enforcement & Guidance

Competition and Markets Authority, AI Foundation Models: Initial Report (CMA 2023).

Competition and Markets Authority, AI Foundation Models: Update Paper (CMA 2024).

European Data Protection Board, Guidelines 05/2020 on Consent under Regulation 2016/679 (EDPB 2020).

European Data Protection Board, Guidelines 04/2022 on the Calculation of Administrative Fines under the GDPR (EDPB 2022).

European Data Protection Supervisor, Opinion on the Proposal for an Artificial Intelligence Act (EDPS 2021).

Federal Trade Commission, Aiming for Truth, Fairness, and Equity in Your Company's Use of AI (FTC 2021).

Federal Trade Commission, Keep Your AI Claims in Check (FTC 2023).

Food and Drug Administration, Artificial Intelligence and Machine Learning Software as a Medical Device Action Plan (FDA 2021).

Food and Drug Administration, Marketing Submission Recommendations for a Predetermined Change Control Plan for Artificial Intelligence/Machine Learning-Enabled Device Software Functions (FDA 2024).

Information Commissioner's Office, Guidance on AI and Data Protection (ICO 2023).

National Highway Traffic Safety Administration, Automated Driving Systems 2.0: A Vision for Safety (NHTSA 2017).

National Highway Traffic Safety Administration, Standing General Order on Crash Reporting for Automated Driving Systems and Level 2 Advanced Driver Assistance Systems (NHTSA 2021).

US Equal Employment Opportunity Commission, Assessing Adverse Impact in Software, Algorithms, and Artificial Intelligence Used in Employment Selection Procedures (EEOC 2023).

Robotics Governance, Infrastructure & Strategy

Alan Turing Institute, Understanding Artificial Intelligence Ethics and Safety (Alan Turing Institute 2019).

Andy Clark, Natural-Born Cyborgs: Minds, Technologies and the Future of Human Intelligence (Oxford University Press 2003).

Brookings Institution, Automation and Artificial Intelligence: How Machines Are Affecting People and Places (Brookings Institution 2019).

Center for Strategic and International Studies, The Geopolitics of AI Chips and Compute (CSIS 2024).

Council of Europe, Framework Convention on Artificial Intelligence, Human Rights, Democracy and the Rule of Law (Council of Europe 2024).

European Commission, Regulation (EU) 2023/1781 Establishing a Framework of Measures for Strengthening Europe's Semiconductor Ecosystem (European Chips Act) [2023] OJ L229/1.

European Data Protection Supervisor, TechDispatch: Emotional AI (EDPS 2021).

European Union Agency for Fundamental Rights, Facial Recognition Technology: Fundamental Rights Considerations in the Context of Law Enforcement (FRA 2019).

International Committee of the Red Cross, Autonomous Weapon Systems: Technical, Military, Legal and Humanitarian Aspects (ICRC 2021).

International Labour Organization, Generative AI and Jobs: A Global Analysis of Potential Effects on Job Quantity and Quality (ILO 2023).

Karl Friston and others, 'The Free-Energy Principle: A Unified Brain

Theory?' (2009) 11 Nature Reviews Neuroscience 127.

Martin Ford, Rise of the Robots: Technology and the Threat of a Jobless Future (Basic Books 2015).

NASA, Autonomous Systems and Robotics Roadmap (NASA Technology Area 4, 2015).

NATO, NATO Artificial Intelligence Strategy (NATO 2021).

NIST, Face Recognition Vendor Test (FRVT) Ongoing Reports (National Institute of Standards and Technology).

OECD, Job Creation and Local Economic Development in the Changing World of Work (OECD Publishing 2024).

OECD, The Semiconductor Supply Chain and Policy Implications (OECD Publishing 2023).

Rolf Pfeifer and Josh Bongard, How the Body Shapes the Way We Think: A New View of Intelligence (MIT Press 2006).

Rodney Brooks, Flesh and Machines: How Robots Will Change Us (Pantheon Books 2002).

Stockholm International Peace Research Institute, Governing Artificial Intelligence in Military Systems (SIPRI 2024).

Daniel Susskind, A World Without Work: Technology, Automation and How We Should Respond (Allen Lane 2020).

UK Information Commissioner's Office, Biometric Recognition Guidance (ICO 2023).

UN-Habitat, World Cities Report 2022: Envisaging the Future of Cities (UN-Habitat 2022).

UNESCO, Recommendation on the Ethics of Artificial Intelligence (UNESCO 2021).

United Nations Convention on Certain Conventional Weapons,

Report of the 2023 Session of the Group of Governmental Experts on Emerging Technologies in the Area of Lethal Autonomous Weapons Systems (UN CCW 2023).

United Nations Human Rights Council, Human Rights and Technical Standard-Setting Processes for New and Emerging Digital Technologies UN Doc A/HRC/53/42 (2023).

United Nations Human Rights Council, The Right to Privacy in the Digital Age UN Doc A/HRC/RES/48/4 (2021).

United States Department of Commerce, Bureau of Industry and Security, Export Controls on Advanced Computing and Semiconductor Manufacturing Items (US Department of Commerce 2023).

United States Department of Defense, Directive 3000.09: Autonomy in Weapon Systems (DoD 2023).

World Economic Forum, The Future of Jobs Report 2025 (WEF 2025).

World Health Organization, Ethics and Governance of Artificial Intelligence for Health (WHO 2021).

World Intellectual Property Organization, Technology Trends 2025: The Future of Transportation (WIPO 2025).

World Bank, World Development Report 2019: The Changing Nature of Work (World Bank 2019).

Advanced Governance, Embodiment, Semiconductor & Societal Transition.

Andy Clark, Natural-Born Cyborgs: Minds, Technologies and the Future of Human Intelligence (Oxford University Press 2003).

Council of Europe, Framework Convention on Artificial Intelligence, Human Rights, Democracy and the Rule of Law (Council of Europe 2024).

Daniel Susskind, A World Without Work: Technology, Automation and How We Should Respond (Allen Lane 2020).

European Commission, Regulation (EU) 2023/1781 Establishing a Framework of Measures for Strengthening Europe's Semiconductor Ecosystem (European Chips Act) [2023] OJ L229/1.

European Data Protection Supervisor, TechDispatch: Emotional AI (EDPS 2021).

European Union Agency for Fundamental Rights, Facial Recognition Technology: Fundamental Rights Considerations in the Context of Law Enforcement (FRA 2019).

International Labour Organization, Generative AI and Jobs: A Global Analysis of Potential Effects on Job Quantity and Quality (ILO 2023).

Karl Friston and others, 'The Free-Energy Principle: A Unified Brain Theory?' (2009) 11 Nature Reviews Neuroscience 127.

Martin Ford, Rise of the Robots: Technology and the Threat of a Jobless Future (Basic Books 2015).

NIST, Face Recognition Vendor Test (FRVT) Ongoing Reports (National Institute of Standards and Technology).

OECD, Job Creation and Local Economic Development in the Changing World of Work (OECD Publishing 2024).

OECD, The Semiconductor Supply Chain and Policy Implications (OECD Publishing 2023).

Rolf Pfeifer and Josh Bongard, How the Body Shapes the Way We Think: A New View of Intelligence (MIT Press 2006).

Rodney Brooks, Flesh and Machines: How Robots Will Change Us (Pantheon Books 2002).

Stockholm International Peace Research Institute, Governing Artificial Intelligence in Military Systems (SIPRI 2024).

UK Information Commissioner's Office, Biometric Recognition Guidance (ICO 2023).

UNESCO, Recommendation on the Ethics of Artificial Intelligence (UNESCO 2021).

United Nations Convention on Certain Conventional Weapons, Report of the 2023 Session of the Group of Governmental Experts on Emerging Technologies in the Area of Lethal Autonomous Weapons Systems (UN CCW 2023).

United States Department of Commerce, Bureau of Industry and Security, Export Controls on Advanced Computing and Semiconductor Manufacturing Items (US Department of Commerce 2023).

United States Department of Defense, Directive 3000.09: Autonomy in Weapon Systems (DoD 2023).

World Economic Forum, Future of Jobs Report 2025 (WEF 2025).

World Health Organization, Ethics and Governance of Artificial Intelligence for Health (WHO 2021).

World Bank, World Development Report 2019: The Changing Nature of Work (World Bank 2019).

Questions

Question One: What makes Robotics Law 3.0 different from earlier legal approaches to industrial automation, artificial intelligence, and autonomous systems?

Question Two: Why does embodiment change the legal meaning of action, risk, responsibility, and harm?

Question Three: How should law distinguish between automation, autonomy, adaptive learning, and physically intelligent decision-making?

Question Four: Why do humanoid robots challenge traditional legal concepts such as intention, consent, foreseeability, causation, and liability?

Question Five: When a humanoid robot causes harm, how should responsibility be distributed among manufacturers, software developers, deployers, operators, data providers, and institutions?

Question Six: Why is humanlikeness legally significant, and how does anthropomorphism affect trust, emotional reliance, consent, and public perception?

Question Seven: What legal duties arise when humanoid robots operate in vulnerable environments such as homes, hospitals, schools, care facilities, workplaces, and public spaces?

Question Eight: How should human rights law respond to humanoid robots capable of surveillance, emotional influence, behavioural prediction, and physical intervention?

Question Nine: Why is legal personhood for robots not simply a question of rights, but also a question of duties, accountability, utility, and institutional design?

Question Ten: What role should international law play in governing cross-border robotic harm, export controls, military use, trade, intellectual property, and global deployment?

Question Eleven: Why does Robotics Law 3.0 require a Global Robotics Charter, and what problems would such a charter seek to solve?

Question Twelve: What are the Twelve Pillars of Humanoid Robotics Governance, and how do they translate ethical principles into operational governance?

Question Thirteen: Why must robotics governance move from static compliance toward continuous monitoring, behavioural calibration, legal gap tracking, and adaptive regulation?

Question Fourteen: How should law govern shared agency between humans and humanoids when decisions emerge from collaboration rather than unilateral human command?

Question Fifteen: What economic, labour, taxation, and social-stability questions arise if humanoid robots become a major source of productivity and abundance?

Question Sixteen: What kind of legal mindset is required to ensure that humans and humanoid robots coexist without undermining dignity, trust, accountability, and the foundations of law itself?

Index

A

Accountability, robotic systems, 64–69, 118–124, 181–187, 238–244, 296–302, 337–341, 402–409

algorithmic accountability, 182–187, institutional accountability structures, 296–302, shared accountability between humans and robots, 402–405. see also Distributed liability

Adaptive governance, robotics regulation in rapidly evolving technological environments, 48–52, 118–120, 292–296, 468–472

Adaptive learning systems, legal implications of machine learning robots, 203–210, 256–262, 317–320

Administrative oversight, robotics regulators and supervisory authorities, 118–124, 292–298

Agency, human–robot shared agency, 320–327, 409–414

Agricultural robotics, automation in farming systems, 24–28, 56–60, 82–86, 173–176

AI governance, relationship between artificial intelligence and robotics regulation, 92–101, 164–171, 276–284

AI risk management, regulatory frameworks for AI safety, 276–284

AI safety, governance mechanisms for preventing harm from autonomous systems, 182–187, 276–284

AI standards, international standards for artificial intelligence systems, 284–290

Algorithmic bias, ethical and legal implications in robotic decision-making, 181–187

Algorithmic governance, automated decision-making systems within public institutions, 164–171

Algorithmic transparency, regulatory requirements for explainability, 182–187, 256–260

Artificial agents, legal recognition of autonomous entities, 350–356

Artificial intelligence, technological foundations of robotic systems, 89–101, 164–171, 276–284

AI-driven robotics systems, 164–171, machine learning in robotics, 203–210, ethical implications, 256–262

Artificial personhood, legal debates on whether robots may possess legal status, 344–359

corporate personhood comparisons, 347–350, philosophical arguments, 350–356

Artificial systems, governance of non-human decision-making entities, 164–171, 350–356

Automation, societal consequences of robotics-driven automation, 168–176, 437–452

automation and labour markets, 168–

176, automation and economic restructuring, 437–452

Autonomous behaviour, robotic decision-making without direct human control, 203–210

Autonomous decision-making, legal implications for liability and governance, 181–187, 203–210

Autonomous robotics, governance challenges posed by self-operating machines, 164–171, 181–187

Autonomous weapons systems, military robotics and international humanitarian law, 298–309

arms-control debates, 300–305

legal vacuum concerns, 305–309

Autonomy certification, regulatory classification of robotics autonomy levels, 142–149

Autonomy levels, tiered regulatory models for robotic capabilities, 142–149

Autopoietic law, legal theory applied to adaptive regulatory systems, 463–468

B

Behavioural calibration, governance of robotic conduct through programmed behavioural constraints, 214–221, 256–260, 301–304

ethical calibration of decision-making systems, 256–260, regulatory certification of behavioural parameters, 214–221, relationship to safety engineering, 301–304

Behavioural governance, legal regulation of robotic behavioural outputs, 212–218, 301–304

Behavioural learning, adaptive robotic learning and regulatory oversight, 221–229, 256–262, 304–309

continuous machine learning systems, 221–229, alignment monitoring, mechanisms, 304–309

Behavioural transparency, monitoring and interpretability of robotic actions, 258–262

Bias in artificial intelligence, discrimination risks in robotic decision-making, 184–190, 259–262

algorithmic bias mitigation strategies, 259–262, regulatory oversight of biased systems, 184–190

Black box recording, robotic data logging for incident investigation, 245–252

ethical black-box systems, 245–252, post-incident analysis of robotic behaviour, 247–250

Black box requirements, regulatory obligations for robotic system logging, 245–252

Boston Dynamics, humanoid robotics development, 34–37, 72–75

Built environments, robotics integration into human physical spaces, 381–388

robot navigation in urban environments, 381–388

Business governance of robotics, corporate

oversight responsibilities, 296–301

Business liability, corporate responsibility for autonomous systems, 337–341

Behavioural risk management, regulatory frameworks addressing unpredictable robotic actions, 214–221, 304–309

Bio-inspired robotics, engineering design based on biological systems, 72–75

Bounded autonomy, legal constraints placed on autonomous systems, 146–149

Boundary setting in robotic missions, see Mission boundaries

Built-in legal safeguards, embedding legal constraints into system architecture, 468–474

Behavioural audit, auditing robotic actions and decisions, 301–304

Behavioural compliance, alignment of robotic behaviour with regulatory standards, 214–221

Behavioural drift, deviation of machine learning systems from intended operational parameters, 221–229, 304–309

C

Calibration of robotic behaviour, see Behavioural calibration

Certification of autonomy levels, regulatory classification of robotic intelligence and decision-making capacity, 138–147, 214–218

tiered autonomy certification systems, 140–147, regulatory implications of advanced autonomy, 214–218

Certification regimes, robotics safety certification and approval systems, 134–142

Civil liability, liability frameworks for autonomous systems, 324–337

Codes of ethics, voluntary ethical frameworks for robotics design, 250–256

Collective robotics governance, international coordination of robotics regulation, 309–316

Commercial robotics, industrial and service-sector applications of robotics systems, 30–36, 70–76

Component accountability, responsibility across robotics supply chains, 300–306

Compliance-by-design, embedding legal and ethical safeguards into robotic systems architecture, 256–263, 468–472

Compliance monitoring, oversight of robotic system deployment and operation, 296–302

Computational decision-making, algorithmic processes in robotic systems, 181–187

Conflict law, application of international humanitarian law to robotics systems, 298–309

Connected robotics systems, networked

robots operating across digital infrastructures, 164–171

Continuous monitoring, oversight of deployed robotics systems, 292–298

Corporate governance, oversight of robotics technologies within corporate structures, 296–301

Corporate liability, responsibility of companies deploying robotics systems, 324–337

Cross-border robotics regulation, jurisdictional challenges in international robotics deployment, 309–316

Cross-border robotic mobility, movement of robotics systems across national jurisdictions, 146–149

Cyber-physical systems, integrated physical and computational robotics architectures, 164–171

Cultural design sensitivity, adapting robotics systems to diverse cultural and legal environments, 262–268

Customs regulation, trade classification of robotics technologies, 312–316

Cybersecurity in robotics, protection of robotic systems from digital threats, 300–306

Cyberlaw, legal precedents influencing robotics regulation, 92–96

Cognitive robotics, robots capable of advanced decision-making processes, 203–210

Coexistence between humans and robots, legal frameworks governing shared environments, 372–401

urban coexistence, 372–381, workplace coexistence, 381–388, public-space coexistence, 388–395, psychological implications of human–robot interaction, 395–401

Collaborative robots (cobots), robotics designed for direct human interaction, 70–75, 381–388

Compensation mechanisms, remedies for harm caused by robotic systems, 334–337

Consumer robotics, robotics used in domestic and personal environments, 32–36, 70–75

Control systems, technical architecture governing robotic actions, 203–210

Convention on Certain Conventional Weapons, international legal framework addressing autonomous weapons, 298–304

Cross-disciplinary legal innovation, integration of law, ethics, engineering, and design in robotics governance, 470–476

D

Data governance, management of data used in robotic systems, 165–173, 239–245, 283–290

data provenance and integrity, 168–171, data protection and privacy implications, 171–173, data lifecycle governance in robotics systems, 283–290

Data logging, recording robotic system behaviour for safety and accountability, 245–252

Data protection, regulatory obligations governing personal data used by robotic systems, 170–173, 284–288

Decision autonomy, robotic capacity for independent decision-making, 203–210

Decision transparency, visibility of robotic reasoning processes, 182–187, 258–262

Decision traceability, recording decision pathways within autonomous systems, 245–252

Defective robotics systems, legal liability for malfunctioning robotic technologies, 324–332

Delegated decision-making, transfer of decision authority from humans to autonomous systems, 181–187

Design ethics, ethical considerations embedded in robotics development, 250–256

Design for compliance, see Compliance-by-design

Design governance, oversight of robotics design processes and engineering choices, 250–256

Digital infrastructure, network systems supporting robotics deployment, 164–171

Digital twins, virtual models used to simulate robotic system performance, 204–208

Distributed liability, allocation of responsibility across robotics ecosystems, 334–341

- manufacturers' liability, 334–336,
- operator responsibility, 336–338,
- shared liability models, 338–341

Distributed robotics systems, networks of coordinated robotic units operating collaboratively, 164–171

Documentation requirements, regulatory documentation for robotics systems, 239–245

Due diligence, corporate responsibility in robotics development and deployment, 296–301

Due process in automated systems, procedural fairness in algorithmic decision-making, 182–187

Dynamic governance, regulatory systems capable of adapting to evolving technologies, 468–472

Dynamic legal gap tracking, mechanisms for identifying regulatory lag in emerging technologies, 472–477

Defence robotics, military applications of robotics technologies, 298–309

Dual-use technologies, robotics technologies capable of civilian and military application, 309–316

Data-driven robotics, machine-learning systems trained on large datasets, 203–210

Digital accountability systems,

regulatory infrastructure for monitoring autonomous technologies, 292–298

E

Economic transformation, societal impact of robotics-driven productivity, 432–441

Embodied artificial intelligence, integration of cognition and physical robotics systems, 198–205

Embodied autonomy, robotics systems capable of acting independently in physical environments, 203–210

Embodied ethics, ethical principles embedded in humanoid robotics systems, 266–274

design-stage ethical constraints, 266–269, human dignity considerations, 269–272, alignment with global ethical frameworks, 272–274

Embodied intelligence, physical instantiation of artificial intelligence systems in robotic bodies, 198–205

Embodied robotics, robotics systems operating in real-world physical environments, 198–205

Emergency override mechanisms, safety mechanisms allowing humans to interrupt robotic actions, 230–235

Engineering ethics, professional ethical obligations in robotics development, 250–256

Ethical alignment, ensuring robotic systems operate within human ethical norms, 256–263

Ethical black-box systems, recording robotic behaviour for post-incident investigation, 245–252

Ethical design, integration of ethical principles into robotics architecture, 250–256

Ethical engineering, embedding legal and ethical constraints in robotics systems, 250–256

Ethical governance, institutional oversight of robotics technologies, 276–284

Ethical impact assessments, regulatory evaluation of societal risks posed by robotics systems, 263–268

Ethical machine behaviour, programming robots to operate within acceptable moral parameters, 256–263

Ethical risk management, governance strategies addressing harms from autonomous systems, 276–284

Ethics-by-design, incorporation of ethical safeguards into system development, 250–256

Explainability, requirement that robotic decisions be understandable to humans, 182–187

Explainable artificial intelligence (XAI), transparency mechanisms for AI decision-making, 182–187

Explainable embodiment, interpretability of robotic actions in physical environments, 274–280 behavioural transparency, 274–276, interpretability of robotic decisions, 276–280

Export controls, regulation of robotics technologies with strategic or military applications, 309–316

Export regulation of autonomous systems, national security controls on robotics technologies, 309–316

Ethics guidelines for trustworthy AI, international governance frameworks influencing robotics regulation, 276–284

Ethical oversight bodies, regulatory and advisory institutions supervising robotics governance, 292–298

Environmental implications of robotics, sustainability and energy impacts of robotic systems, 441–446

Extraterritorial regulation, application of national robotics laws across jurisdictions, 312–316

F

Fairness in artificial intelligence, prevention of discriminatory outcomes in robotic decision-making, 182–190, 259–262

bias detection and mitigation strategies, 184–188, fairness auditing of autonomous systems, 259–262

Federal regulation of robotics, national regulatory oversight structures governing robotics technologies, 118–124, 309–316

Feedback loops in machine learning, iterative learning mechanisms affecting robotic behaviour, 203–210

Firmware governance, embedding legal constraints directly into robotic system software, 466–471

Flexible regulatory frameworks, adaptive governance approaches for emerging robotics technologies, 48–52, 468–472

Formal verification, mathematical validation of robotic safety and reliability, 230–235

Foundations of robotics law, emergence of robotics as a distinct legal discipline, 18–24

Frameworks for robotics governance, institutional structures guiding the regulation of robotic systems, 276–284, 292–298

Free movement of robotics technologies, cross-border trade and regulatory harmonisation, 309–316

Future of work, transformation of labour markets through robotics and automation, 168–176, 432–441

automation and employment displacement, 168–172, robotics-driven productivity expansion, 432–441

Future regulatory institutions, emerging bodies responsible for robotics governance, 292–298

Fundamental rights, protection of human rights in robotics deployment, 170–173,

256–263, 364–372

privacy and surveillance concerns, 170–173, human dignity protections, 256–263, constitutional implications of robotics governance, 364–372

Functional autonomy, operational independence of robotic systems, 203–210

Functional safety, engineering approaches to preventing robotic harm, 230–235

Future economic models, robotics-enabled abundance and social redistribution, 437–452

Future legal systems, transformation of law in highly automated societies, 463–476

G

General-purpose AI models, regulatory treatment under emerging AI governance frameworks, 276–283, 296–300

Global governance of robotics, international coordination of robotics regulation, 292–301, 309–316

Global Robotics Charter, proposed multilateral framework governing AI and robotics systems across jurisdictions, 456–464

principles of international robotics governance, 458–460, verification and monitoring mechanisms, 460–462, dispute-resolution mechanisms, 462–464

Global robotics registry, proposed international system for identifying and tracking humanoid robots, 144–149

Global safety standards, international technical standards governing robotics systems, 284–290

Global technology governance, institutional frameworks managing emerging technologies, 292–301

Government oversight, supervisory authority over robotics development and deployment, 118–124, 292–298

Governance dashboards, real-time monitoring systems for deployed robotic fleets, 452–456

Governance frameworks, institutional and regulatory systems guiding robotics development, 276–284

Governance-by-design, embedding governance mechanisms into technological architecture, 256–263

Governance ecosystems, interaction between regulators, industry, and international organisations, 292–301

Governance principles for robotics, ethical and legal foundations guiding robotics regulation, 266–274

Governance transparency, disclosure requirements for robotics operations, 258–262

Governance of autonomous weapons, international humanitarian law approaches to military robotics, 298–309

Governance of AI safety, regulatory

approaches to managing risks from advanced artificial intelligence systems, 276–284

Governance of human-robot interaction, institutional frameworks guiding coexistence between humans and robots, 372–381

Governance of robotic supply chains, oversight of component production and distribution, 300–306

Governance innovation, evolving legal and institutional approaches for emerging technologies, 468–476

Governance metrics, measurement systems used to evaluate robotics governance effectiveness, 292–298

Governance legitimacy, public trust in robotics regulatory systems, 396–401

Governance institutions, international organisations involved in robotics regulation, 309–316

International Telecommunication Union (ITU), 310–311, International Civil Aviation Organization (ICAO), 311–312, World Intellectual Property Organization (WIPO), 312–314

H

Harm caused by robots, legal responsibility and compensation mechanisms, 324–337

civil liability regimes, 324–332, compensation frameworks, 332–337

Health and safety regulation, robotics deployment in public and industrial environments, 134–142, 230–235

High-risk AI systems, classification within regulatory frameworks, 276–283

Human autonomy, preservation of human decision authority in automated systems, 181–187, 364–370

Human control, maintaining meaningful oversight of autonomous systems, 230–235

Human dignity, ethical foundation of robotics governance, 266–272, 364–372

Human enhancement technologies, relationship between robotics and augmentation of human capabilities, 198–205

Human oversight, supervisory control over robotic systems, 230–235, 320–327

oversight-by-design architectures, 230–235, regulatory requirements for human supervision, 320–327

Human-in-the-loop systems, governance model requiring human review of autonomous decisions, 230–235

Human-on-the-loop supervision, supervisory monitoring of autonomous robotics operations, 230–235

Human–robot coexistence, integration of humanoids into social environments, 372–401

urban coexistence, 372–381, workplace integration, 381–388, public-space interaction, 388–395, social acceptance of robotics systems, 396–401

Human–robot interaction, legal and ethical considerations in shared environments, 372–381

Human–robot trust, public confidence in robotics systems, 396–401

Humanoid robotics, advanced robotics systems designed to resemble human bodies and behaviours, 34–37, 72–75, 134–142, 372–381

industrial humanoids, 72–75, service-sector humanoids, 372–381

Humanoid robot registries, national and international registration systems, 144–149

Humanoid robotics governance, regulatory frameworks governing embodied AI systems, 266–274, 292–301

Humanoid robotics safety standards, international technical standards for humanoid systems, 284–290

Human rights, application of fundamental rights frameworks to robotics governance, 170–173, 364–372

privacy rights, 170–173, non-discrimination protections, 182–190, freedom of expression implications, 364–372

Hybrid decision-making systems, collaborative human-machine governance models, 320–327

Hybrid governance architectures, regulatory models combining human oversight and automated monitoring, 468–472

I

Impact assessments, regulatory evaluation of robotics systems before and after deployment, 263–268, 292–298

ethical impact assessments, 263–268, regulatory impact assessments for high-risk systems, 292–298, social impact evaluation of humanoid deployment, 396–401

Incident reporting, mandatory disclosure of robotics failures and harmful events, 245–252, 292–298

Inclusive design, robotics development accommodating diverse users and environments, 267–268, 388–395

Industrial automation, robotics-driven transformation of production systems, 168–176, 432–441

Industrial robotics, legal regulation of robots in manufacturing and logistics, 30–36, 70–76, 134–142

industrial safety obligations, 230–235, collaborative robots in workplaces, 381–388

Industrial safety standards, technical and legal rules governing robotics in workplaces, 230–235, 284–290

Information asymmetry, imbalance of knowledge between robotics developers, operators, and users, 171–173, 396–401

Infrastructure law, regulation of robotics systems operating in public and critical infrastructure, 372–381, 441–446

Innovation governance, balancing robotics innovation with public safety and accountability, 276–284, 468–476

Inspection robots, autonomous systems used for infrastructure monitoring and maintenance, 372–381

Institutional accountability, organisational responsibility for robotics governance failures, 296–302, 337–341

Institutional design, architecture of regulatory bodies and governance systems for robotics, 292–301

Intellectual property, ownership and control of robotics innovation, 312–316, 433–438

  patents and AI-generated inventions, 433–436

  software ownership in robotic systems, 436–438

Interdisciplinary legal education, integration of law, engineering, ethics, and design in legal training, 476–482

International Civil Aviation Organization (ICAO), role in robotics-related standardisation and cross-border coordination, 311–312

International Committee of the Red Cross (ICRC), role in debates on autonomous weapons and humanitarian law, 298–304

International coordination, multilateral efforts to harmonise robotics governance, 292–301, 309–316

International criminal law, possible implications of autonomous military systems, 304–309

International humanitarian law, application to autonomous weapons and military robotics, 298–309

  distinction, proportionality, and precaution, 300–305

International law of robotics, emerging legal frameworks governing robotics across borders, 292–316

International Organisation for Standardisation (ISO), role in robotics standards and safety requirements, 284–290

International registries, cross-border tracking systems for autonomous and humanoid robots, 144–149, 309–316

International responsibility, state responsibility for robotics-related harm across borders, 317–324

International Telecommunication Union (ITU), communications and infrastructure aspects of robotics governance, 310–311

International trade law, regulation of robotics goods, components, and dual-use technologies, 309–316

International weapons law, treatment of autonomous military systems, 298–309

Interoperability, capacity of robotic systems and regulatory systems to function across jurisdictions, 144–149, 309–316

Interpretability, requirement that robotic systems remain understandable to human operators and regulators, 182–187, 274–280

Investment in robotics, public and private funding structures shaping robotics deployment, 432–438

Iterative regulation, legal systems updated through continuous refinement rather than static reform, 468–472

Iterative governance, adaptive policy approaches responding to technological change, 468–472

J

Judicial oversight, role of courts in reviewing robotics regulation and governance decisions, 118–124, 324–329

Judicial review, challenges to robotics regulation before national and international courts, 118–124, 309–316

Judicial interpretation of AI law, evolving case law interpreting robotics and artificial intelligence regulations, 324–329

Justice in robotics governance, ensuring fairness and accountability in automated decision systems, 182–190, 396–401

Justice systems, use of artificial intelligence and robotics within courts and legal institutions, 170–173

Jurisprudence of artificial personhood, legal theory surrounding recognition of non-human entities as legal subjects, 344–359

Jurisprudence of robotics law, development of robotics law as a distinct field of legal scholarship, 18–24, 463–468

Jurisdiction, legal authority over robotics systems operating across borders, 309–316

cross-border jurisdictional conflicts, 309–316, extraterritorial application of robotics law, 312–316

Jurisdictional harmonisation, international efforts to coordinate robotics regulation between legal systems, 292–301

Jurisdictional responsibility, allocation of regulatory authority among national and international institutions, 292–301

K

Killer robots, public and policy debates surrounding lethal autonomous weapons systems, 298–304

Knowledge asymmetry, imbalance of technical knowledge between robotics developers, regulators, and users, 171–173, 396–401

Knowledge governance, regulatory frameworks ensuring responsible management of technical knowledge and data in robotics systems, 283–290

Knowledge transparency, accessibility of technical information necessary for robotics oversight and accountability, 182–187

Knowledge systems, institutional and technical structures used to monitor robotics governance and compliance, 292–298

Knowledge transfer, dissemination of robotics technology and expertise across institutions and jurisdictions, 309–316

Knowledge-based robotics, robotics systems relying on structured knowledge models and reasoning systems, 203–210

L

Law as adaptive engineering, concept of law evolving in response to technological change, 466–472

Law as proactive engineering, embedding legal requirements directly into robotic systems architecture, 468–472

Law and robotics, emergence of robotics law as a specialised legal discipline, 18–24

Law beyond the human subject, transformation of legal systems in the presence of intelligent machines, 362–371, 463–468

Legal accountability, responsibility mechanisms governing robotics systems and their operators, 324–337

Legal architecture of robotics governance, structural organisation of regulatory systems governing robotics technologies, 292–301

Legal engineering, integration of legal constraints into robotics design and development processes, 256–263, 468–472

Legal frameworks for robotics, institutional and statutory systems regulating robotic technologies, 118–124, 276–284

Legal gap tracking, mechanisms identifying regulatory lag in emerging technologies, 472–477

Legal innovation, development of new legal tools and institutions to govern robotics, 468–476

Legal oversight, supervisory roles of regulators and courts in robotics governance, 118–124, 292–301

Legal personhood, recognition of non-human entities as legal subjects, 344–359

corporate personhood analogies, 347–350, philosophical foundations, 350–356

Legal pluralism in robotics governance, interaction of national, regional, and international regulatory systems, 292–301

Legal responsibility, allocation of accountability for robotics-related harm, 324–337

Legal status of robots, debates on rights, duties, and personhood of autonomous systems, 344–359

Liability for autonomous systems, legal responsibility for harm caused by robots, 324–337

manufacturer liability, 324–329, operator liability, 329–332, shared liability models, 332–337

Liability regimes, frameworks governing compensation and accountability in robotics incidents, 324–337

Licensing of robotics systems, regulatory

approval mechanisms for robotic technologies, 134–142

hardware licensing regimes, 134–138, operational licensing requirements, 138–142

Licensing authorities, regulatory bodies responsible for approving robotics deployment, 118–124

Licensing regimes for humanoid robots, proposed regulatory systems for controlling robot deployment, 134–142

Lifelong learning, continuous adaptation of robotic systems through ongoing learning processes, 221–229

governance of adaptive robotics behaviour, 221–229

Living law, concept of continuously evolving legal frameworks responding to technological innovation, 468–476

Living regulatory systems, adaptive governance systems capable of responding to robotics innovation, 468–476

Logistics robotics, use of robotics in transportation and supply chain systems, 70–76, 173–176

M

Machine autonomy, operational independence of advanced robotic systems, 203–210

Machine ethics, programming robots to comply with ethical principles, 250–256

Machine learning, computational techniques enabling robots to learn from data, 203–210, 221–229

supervised learning systems, 203–206, unsupervised learning architectures, 206–208, reinforcement learning systems, 208–210

Machine learning governance, regulatory oversight of adaptive learning algorithms, 221–229, 276–284

Machine reasoning, algorithmic systems capable of structured decision-making, 203–210

Market regulation of robotics, legal oversight of robotics markets and commercial deployment, 276–284, 432–438

Market concentration in AI industries, economic implications of large-scale robotics platforms, 432–438

Meaningful human control, requirement that humans retain authority over autonomous systems, 230–235, 320–327

Medical robotics, robotics used in healthcare and surgical systems, 32–36, 372–381

Military robotics, robotic systems deployed in defence and armed conflict, 298–309

autonomous weapons systems, 298–304, international humanitarian law, implications, 300–305

Mission boundaries, constraints defining the permissible scope of robotic actions, 214–221

operational limitations, 214–218,

safety restrictions in autonomous systems, 218–221

Modular governance systems, layered regulatory structures adaptable to technological change, 472–476

Monitoring frameworks, oversight systems tracking robotic performance and safety, 292–298

Monitoring technologies, tools used to observe and regulate robotics operations, 292–298

Moral agency of robots, philosophical debate on whether robots can possess ethical responsibility, 350–356

Moral responsibility, attribution of ethical responsibility in human–robot interactions, 350–356

Multilateral robotics governance, international cooperation on robotics regulation, 292–301, 309–316

Multidisciplinary governance, integration of law, engineering, ethics, and policy in robotics regulation, 468–476

Manufacturing robotics, industrial robots used in production environments, 30–36, 70–76

Mobility robotics, robots operating within transportation systems and urban infrastructure, 372–381

Model transparency, interpretability of machine learning models used in robotics, 182–187

Monitoring of autonomous systems, regulatory observation of robotic behaviour after deployment, 292–298

N

National humanoid robot registries, proposed systems for registering and identifying humanoid robots within national jurisdictions, 144–149

regulatory oversight through registration systems, 146–149

National regulatory authorities, government institutions responsible for robotics oversight, 118–124, 292–298

National security considerations, strategic implications of robotics and artificial intelligence technologies, 309–316

Networked robotics systems, interconnected robotic platforms operating across digital infrastructure, 164–171

Network governance, coordination of robotics regulation across distributed institutional actors, 292–301

Non-human legal subjects, entities recognised by law without human status, 344–359

corporations as legal persons, 347–350, legal recognition of rivers and ecosystems, 350–353, implications for robotic legal status, 353–359

Normative frameworks for robotics governance, ethical and legal principles guiding robotics regulation, 266–274, 276–284

Normative alignment, ensuring robotic behaviour corresponds with human ethical and legal norms, 256–263

Normative regulation, use of principles and ethical guidelines alongside statutory rules in robotics governance, 276–284

Normative pluralism, coexistence of multiple regulatory approaches across jurisdictions, 292–301

Non-discrimination in automated decision-making, safeguards against biased outcomes in robotic systems, 182–190

Notification requirements, obligations to report robotics incidents or safety failures to regulatory authorities, 245–252

Non-state actors in robotics governance, role of corporations, research institutions, and international organisations in shaping robotics regulation, 292–301

Network transparency, visibility of communications and coordination among distributed robotics systems, 182–187

O

Object recognition systems, visual perception technologies enabling robots to identify environmental elements, 203–210

Obstacle detection, sensor-based systems used by robots to detect physical barriers, 203–210

Obstacle avoidance, autonomous navigation capability preventing collisions with humans or infrastructure, 203–210, 372–381

Operational autonomy, degree of independent functioning permitted to robotic systems, 138–147, 203–210

Operational boundaries, defined limits governing permissible robotic actions, see Mission boundaries

Operational licensing, regulatory authorisation for robotics deployment, 134–142

Operational monitoring, ongoing observation of robotic systems after deployment, 292–298

Operational oversight, supervisory mechanisms ensuring robots operate within regulatory constraints, 118–124, 292–298

Operational safety standards, technical and regulatory requirements ensuring safe robotics operation, 230–235, 284–290

Operational transparency, disclosure of robotic system behaviour and decision processes, 182–187, 258–262

Operator responsibility, accountability of human operators supervising robotic systems, 329–332

Organisational accountability, institutional responsibility for robotics governance failures, 296–301

Oversight authorities, regulatory institutions supervising robotics deployment and safety compliance, 118–124

Oversight-by-design, system architectures enabling continuous human supervision of autonomous systems, 230–235

Open governance frameworks, participatory regulatory models involving governments, industry, and civil society, 292–301

Open standards, publicly available technical standards guiding robotics development, 284–290

Open regulatory ecosystems, collaborative governance structures coordinating robotics oversight across institutions, 292–301

Organisational governance, internal corporate oversight mechanisms governing robotics systems, 296–301

Operational risk management, identification and mitigation of risks associated with robotics deployment, 276–284

Operational compliance, ensuring robotics systems adhere to regulatory requirements during deployment, 292–298

P

Patent law and robotics, intellectual property rights relating to robotic inventions and AI-generated innovations, 433–438

Personhood, legal recognition of non-human entities within legal systems, 344–359

- artificial personhood debates, 346–356, comparisons with corporate personhood, 347–350, implications for robotic rights and responsibilities, 350–359

Physical autonomy, capacity of robots to act independently within physical environments, 203–210

Physical embodiment, manifestation of artificial intelligence within robotic bodies, 198–205

Physical security, safeguarding robotic systems and environments from physical harm or manipulation, 214–221, 300–306

- infrastructure protection, 214–218, security protocols for robotics deployment, 300–306

Platform robotics ecosystems, interconnected robotics systems operating across digital and physical infrastructures, 164–171

Policy frameworks for robotics governance, national and international policy structures guiding robotics regulation, 276–284

Post-human legal order, transformation of legal systems in societies where humans and intelligent machines coexist, 463–476

Post-market monitoring, regulatory observation of robotics systems after deployment, 292–298

Predictive algorithms, data-driven systems forecasting future events or behaviours, 203–210

Privacy, protection of personal data collected by robotics systems, 170–173

Procedural fairness, ensuring automated decisions respect legal due process, 182–187

Product liability, responsibility for harm caused by defective robotics systems, 324–332

Product safety regulation, legal standards governing robotics manufacturing and deployment, 134–142

Professional responsibility, ethical obligations of robotics developers and engineers, 250–256

Public infrastructure robotics, robots deployed in transportation, utilities, and urban management systems, 372–381

Public policy, governmental strategies shaping robotics innovation and regulation, 276–284

Public safety, protection of individuals and communities from robotics-related risks, 230–235

Public trust in robotics, societal confidence in autonomous technologies and governance institutions, 396–401

Purpose limitation, principle restricting robotics systems to clearly defined operational goals, 214–221

Q

Quality assurance in robotics, verification processes ensuring robotics systems meet safety and regulatory standards, 134–142, 230–235

Quality control in robotics manufacturing, technical inspection and compliance procedures for robotic hardware and components, 134–142

Quantified governance metrics, numerical indicators used to evaluate robotics governance effectiveness, 292–298

Quantitative risk assessment, statistical evaluation of risks posed by robotics systems, 276–284

Quantum computing, potential impact of quantum technologies on artificial intelligence and robotics systems, 444–448

Quasi-autonomous systems, systems operating with partial autonomy under human supervision, 138–147, 230–235

Quorum-based decision systems, distributed robotics systems requiring consensus among multiple agents before executing actions, 164–171

Quality standards for robotics, technical standards governing design, safety, and reliability of robotics systems, 284–290

R

Real-time governance, continuous oversight of robotics systems during operation, 292–298, 452–456

Real-time monitoring, live supervision of humanoid and autonomous systems, 292–298, 452–456

Reasonable robot standard, legal benchmark for evaluating robotic conduct and foreseeability of harm, 324–332

Reciprocal Responsibility Index, proposed test for recognition of artificial legal standing based on proportional duties and rights, 356–362

Redress, mechanisms for remedying harm caused by robots and autonomous systems, 332–337, 408–412

Redress and accountability mechanisms, 408–412

compensation pathways, 332–337, administrative redress, 408–410, judicial remedies, 410–412

Registration of humanoid robots, national systems for identifying, recording, and supervising deployed robots, 142–149

Registries, regulatory databases for robotic identification, ownership, and compliance tracking, 142–149, 309–316

national humanoid robot registries, 144–149, international registries, 309–316

RegTech for robotics, use of digital compliance tools in robotics governance, 452–456, 472–477

Regulation of artificial intelligence, statutory and policy frameworks governing AI systems used in robotics, 276–284

Regulation of robotics, legal frameworks governing design, deployment, and oversight of robotic systems, 118–124, 276–284

Regulatory adaptability, ability of legal systems to evolve alongside robotics innovation, 468–472

Regulatory agencies, institutions responsible for enforcing robotics governance, 118–124, 292–298

Regulatory architecture, structure of legal and institutional systems governing robotics, 292–301, 463–468

Regulatory certification, approval processes confirming robotic systems meet legal and safety standards, 134–142

Regulatory coherence, consistency across robotics laws, standards, and oversight regimes, 292–301

Regulatory convergence, harmonisation of robotics rules across jurisdictions, 309–316

Regulatory gap analysis, identification of areas where law lags behind technological capability, 472–477

Regulatory lag, delay between technological development and legal adaptation, 472–477

Regulatory sandboxes, controlled legal environments for testing emerging robotics technologies, 276–284, 468–472

Regulatory standards, technical and legal benchmarks governing robotics systems, 284–290

Regulatory supervision, oversight of robotics deployment by public authorities, 118–124, 292–298

Regulatory transparency, openness of regulatory processes and requirements for robotics governance, 258–262, 292–298

Reimagining law beyond the human subject, transformation of legal theory in the presence of intelligent non-human systems, 362–371

Reinforcement learning, machine learning technique allowing robots to optimise through feedback, 208–210, 221–229

Relational autonomy, human autonomy shaped by interaction with robotic systems and institutions, 364–370, 395–401

Relational ethics, ethical obligations arising in ongoing human–robot relationships, 256–263, 395–401

Remote identification, digital identification of autonomous systems for safety and traceability, 144–149, 309–316

Repair and maintenance obligations, duties relating to the upkeep and safe operation of robotics systems, 300–306, 329–332

Representation of artificial entities, legal mechanisms through which robots or robotic systems might appear in legal processes, 347–356

Reputational risk, organisational exposure associated with robotics failures or unethical deployment, 296–301, 396–401

Reskilling, transition measures for workers affected by robotics and automation, 168–176, 437–441

Resilience of governance systems, ability of legal institutions to respond to robotics-related disruptions, 468–476

Responsibility, attribution of legal, institutional, and moral responsibility within robotics ecosystems, 320–341, 409–414

- human responsibility, 320–327,
- institutional responsibility, 337–341,
- moral responsibility, 350–356

Responsible AI, governance approach prioritising safety, accountability, transparency, and fairness, 276–284

Responsible robotics, legal and ethical design, deployment, and oversight of robots, 250–256, 266–274, 276–284

Restorative justice between humans and machines, frameworks for mediated repair after robotics-related harm, 401–408

- mediation in workplace robotics incidents, 404–408

Rights of robots, philosophical and legal debates on whether robots should hold rights, 344–356

Rights-based governance, legal frameworks rooted in dignity, fairness, and non-discrimination, 364–372

Risk assessment, evaluation of technical, ethical, and legal risks in robotics systems, 276–284

Risk-based controls, governance obligations calibrated to the level of technological risk, 276–284

Risk-based regulation, proportional regulation according to system capability and potential harm, 276–284

Risk categorisation, classification of robotics systems by safety and rights impact, 138–147, 276–284

Risk governance, institutional management of uncertainty and harm in robotics deployment, 276–284

Risk mitigation, technical and legal measures designed to reduce robotics-related harm, 230–235, 276–284

Risk monitoring, ongoing observation of deployed systems to detect emerging dangers, 292–298

Robotic agency, operational role of robots in decision environments, 350–356

Robotic autonomy, degree of independent functioning in robots, 138–147, 203–210

Robotic behaviour, observable actions of robots in physical and institutional settings, 214–229, 256–262

Robotic coexistence, see Human–robot coexistence

Robotic decision-making, algorithmic and embodied decision processes in autonomous systems, 181–187, 203–210

Robotic dignity, debate on whether dignity concepts can apply to machines or only to humans interacting with them, 350–356, 364–370

Robotic ecosystems, interconnected environments of robots, humans, institutions, and infrastructure, 292–301, 372–381

Robotic harm, injury, loss, or social disruption caused by robotic systems, 324–337, 401–408

Robotic legal status, debates on whether robots are objects, agents, or legal subjects, 344–359

Robotic mobility, movement of robots through urban and institutional environments, 372–381

Robotic rights, see Rights of robots

Robotic supply chains, international production and distribution of robotics components and systems, 300–306, 309–316

Robotic subjecthood, debates about inclusion of robots within legal subject categories, 344–359

Robotic systems architecture, layered technical structure of embodied AI systems, 198–205, 203–210

Robotics, 18–24, 30–37, 70–76, 89–101, 118–124, 198–205, 276–284, 372–381, 463–476

definition of, 18–21, industrial development of, 30–37, relationship with artificial intelligence, 89–101, societal integration of, 372–381

Robotics governance, 118–124, 266–274, 276–284, 292–301, 463–476

global governance, 292–301, institutional governance, 118–124, ethical governance, 266–274, adaptive governance, 463–476

Robotics law, emergence as a specialised field of legal inquiry, 18–24, 463–468

Robotics literacy, need for legal and public understanding of robotics systems, 396–

401, 476–482

Robotics policy, governmental strategies shaping the development and oversight of robotics systems, 276–284

Robots, personhood, and legal subjecthood, 344–371

Rule of law in automated societies, preservation of legality and accountability in highly automated environments, 463–476

Rules-based governance, use of formal legal norms to constrain robotics systems and institutions, 118–124, 276–284

S

Safety certification, regulatory approval confirming robotic systems meet safety requirements, 134–142

Safety engineering, technical discipline focused on preventing robotics-related harm, 230–235

Safety standards, internationally recognised technical benchmarks governing robotics design and operation, 284–290

Safety-by-design, integration of safety considerations into robotics architecture from the earliest development stages, 230–235

Scenario planning for robotics governance, anticipatory governance techniques used to prepare for emerging robotics risks, 468–472

Scientific governance, incorporation of scientific expertise into robotics regulatory frameworks, 292–298

Security of robotic infrastructure, protection of robotics systems and supporting infrastructure against sabotage or interference, 214–221, 300–306

Self-learning systems, robotics systems capable of improving performance through continuous learning, 221–229

Sensorial integrity, ensuring the accuracy, reliability, and security of robotic sensing systems, 214–221

- sensor data reliability, 214–218, environmental perception integrity, 218–221

Sensors, hardware components enabling robots to perceive and interpret environmental information, 198–205

Shared agency, distributed decision-making authority between humans and robots, 338–343

- collaborative autonomy, 338–341, human–robot coordination, 341–343

Shared environments, physical spaces where humans and robots interact regularly, 372–381

Social acceptance of robotics, public attitudes toward robots and autonomous systems, 396–401

Social cohesion, societal stability and trust in the presence of robotics technologies, 396–401

Social contract in automated societies, evolving relationship between citizens, governments, and technological systems, 437–441, 463–468

Societal co-design, participatory governance model involving public stakeholders in robotics development and regulation, 250–256

public participation mechanisms, 250–253, democratic governance of robotics technologies, 253–256

Societal governance of robotics, institutional mechanisms ensuring robotics development aligns with public values, 292–301

Socio-technical systems, interconnected networks of technology, institutions, and social practices, 266–274

Software governance, legal oversight of robotics software and algorithmic decision systems, 276–284

Software liability, legal responsibility for harm caused by robotics software defects, 324–332

Sovereignty and robotics, implications of robotics technologies for national authority and geopolitical stability, 309–316

Space robotics, autonomous systems used in space exploration and orbital infrastructure, 444–448

Standards development organisations, institutions responsible for establishing technical robotics standards, 284–290

State responsibility, attribution of international legal responsibility for robotics actions attributable to states, 309–316

State regulation of robotics, governmental authority over robotics manufacturing, deployment, and use, 118–124

Strategic robotics industries, national economic sectors focused on robotics innovation and production, 432–438

Strategic stability, implications of robotics technologies for geopolitical balance and military deterrence, 298–304

Supply chain governance, oversight of global robotics production and component sourcing networks, 300–306

Supply chain security, protection of robotics supply chains from vulnerabilities and interference, 300–306

Sustainability and robotics, environmental implications of robotics technologies and automated production systems, 444–448

System architecture of robots, structural design integrating hardware, software, and decision systems, 198–205

System transparency, ability of regulators and stakeholders to understand robotics system behaviour, 182–187

T

Technological governance, regulatory oversight of emerging robotics and artificial intelligence systems, 276–284, 292–301

Technological sovereignty, national control over critical robotics and AI technologies, 309–316, 432–438

Technological unemployment, labour displacement caused by automation and robotics systems, 168–176, 437–441

Technological risk, potential harms arising from robotics deployment and autonomous decision-making, 276–284

Technology regulation, legal frameworks governing the development and use of advanced technologies, 118–124, 276–284

Technical certification, approval processes verifying robotics systems comply with regulatory and safety standards, 134–142

Technical standards, international benchmarks guiding robotics safety, design, and interoperability, 284–290

Testing regimes, evaluation procedures used to verify robotics system safety and reliability, 134–142, 230–235

Testing environments, controlled settings used for robotics experimentation and regulatory validation, 276–284

Traceability, ability to track the origin, operation, and behaviour of robotic systems, 142–149, 182–187

Transparency, principle requiring disclosure of robotic system design, operation, and decision-making processes, 182–187, 258–262

- algorithmic transparency, 182–187,
- operational transparency, 258–262

Transparency-by-design, integration of explainability and traceability into robotics architecture, 258–262

Treaties governing robotics, international legal instruments addressing robotics technologies and autonomous systems, 298–309

Treaty obligations, binding commitments imposed by international agreements governing robotics use, 309–316

Treaty verification mechanisms, international oversight procedures ensuring compliance with robotics-related treaties, 309–316

Trust in robotics, societal confidence in robotics technologies and governance systems, 396–401

Trustworthy AI, governance principles ensuring AI systems operate safely, fairly, and transparently, 266–274, 276–284

Transnational robotics governance, international cooperation and regulatory harmonisation across jurisdictions, 292–301, 309–316

Transnational liability, legal responsibility for robotics-related harm occurring across borders, 324–332, 309–316

Transportation robotics, robots used in logistics, mobility systems, and autonomous transport networks, 372–381

Treaty-based governance models, international frameworks coordinating robotics oversight among states, 309–316

Technological ecosystems, interconnected networks of robotics, data systems, institutions, and users, 292–301

Technological ethics, moral frameworks guiding robotics design and deployment, 250–256

Technological accountability, mechanisms ensuring developers and operators remain responsible for robotics systems, 320–341

U

Universal Basic Equity (UBE), proposal for distributing ownership stakes in automated production systems to citizens, 438–444

Universal Basic Income (UBI), guaranteed income model proposed as a response to automation-driven labour displacement, 437–441

Universal high-income futures, economic model where robotics productivity significantly increases societal wealth and living standards, 437–445

Urban robotics governance, legal frameworks regulating robotics operation in cities and shared urban environments, 372–381

Urban planning for robotic coexistence, infrastructure design enabling safe integration of humanoid robots into cities, 372–381

Urban robotics infrastructure, public infrastructure supporting robotics mobility and services in urban spaces, 372–381

User responsibility, obligations of individuals or organisations operating robotics systems, 329–332

User safety protections, legal safeguards protecting individuals interacting with robots in public or private environments, 372–381

Utility-based legal reasoning, evaluation of robotics governance based on societal benefits and harms, 356–362

Use-of-force principles, international legal rules governing deployment of autonomous systems in military contexts, 298–304

Use of autonomous weapons, legal debates surrounding lethal autonomous weapon systems, 298–304

Unpredictable AI behaviour, emergent behaviour in adaptive robotics systems that may deviate from expected outcomes, 181–187, 221–229

Unsupervised learning systems, machine learning models identifying patterns without labelled datasets, 206–208

Universal access to automated services, legal and policy considerations ensuring robotics benefits are broadly distributed across society, 437–445

Universal service robotics, robotics systems providing essential public services such as healthcare, logistics, and infrastructure maintenance, 372–381

V

Validation of robotics systems, processes

ensuring robotic technologies perform as intended in operational environments, 134–142, 230–235

Verification regimes, regulatory and technical procedures confirming robotics systems meet safety and legal requirements, 134–142, 284–290

Verification of autonomous decision systems, evaluation of algorithmic decision-making processes in robots, 182–187

Values in robotics governance, ethical principles guiding the development and deployment of robotics technologies, 250–256, 266–274

Value alignment, ensuring robotic behaviour remains consistent with human ethical norms and societal expectations, 256–262

Value-sensitive design, design methodology integrating ethical and social values into technology development, 250–256

Vicarious liability, legal responsibility assigned to organisations or individuals for actions performed by robotic systems under their control, 324–332

Victim compensation mechanisms, legal frameworks for compensating individuals harmed by robotics systems, 332–337, 408–412

Virtual testing environments, simulated environments used to test robotics safety and performance before deployment, 230–235

Virtual governance environments, digital platforms supporting oversight, monitoring, and regulatory coordination of robotics systems, 452–456

Virtual infrastructure for robotics regulation, digital systems supporting licensing, monitoring, and compliance processes, 452–456

Virtual robotics ecosystems, simulated environments supporting robotics development, training, and regulatory experimentation, 230–235, 468–472

Vision systems, perception technologies enabling robots to interpret visual data, 198–205

Verification of safety standards, evaluation of compliance with technical robotics safety frameworks, 284–290

Value-driven governance, governance frameworks prioritising ethical and societal considerations in robotics regulation, 266–274

W

Warfare and autonomous systems, use of robotic technologies in armed conflict and military operations, 298–309

Weapons systems, robotics technologies designed for defence and military purposes, 298–304

Weapons autonomy, degree of independent decision-making permitted in lethal autonomous weapons systems, 298–304

Weapons regulation, international legal frameworks governing military robotics and autonomous weapons, 309–316

Well-being metrics, measures used to evaluate the societal impact of robotics technologies on human welfare, 396–401

Whanganui River legal personhood, recognition of a natural entity as a legal subject and its implications for artificial personhood debates, 347–350

Workplace robotics, deployment of robots within industrial and service-sector workplaces, 168–176, 372–381

Worker displacement, labour market disruption caused by robotics and automation technologies, 168–176, 437–441

Worker protection frameworks, legal safeguards protecting workers interacting with robotic systems, 168–176, 372–381

Workforce transition, policies supporting reskilling and adaptation to robotics-driven economic change, 168–176, 437–441

World governance of robotics, international coordination of robotics regulation across national jurisdictions, 292–301, 309–316

World robotics markets, global economic ecosystems surrounding robotics development and deployment, 432–438

World Robotics Report, global statistical analysis of robotics adoption and deployment trends, 432–438

Worldwide robotics standards, international standards governing robotics design, safety, and interoperability, 284–290

World trade and robotics, international trade implications of robotics technologies and automated production, 309–316, 432–438

Welfare state transformation, evolution of social welfare institutions in automated economies, 437–445

Work–automation balance, emerging relationship between human labour and robotics productivity, 437–445

X

X-risk (existential risk), catastrophic global risks arising from uncontrolled artificial intelligence or autonomous systems, 456–463

Explainability in autonomous systems, see Explainable embodiment

Explainability requirements, regulatory obligations requiring robotics systems to provide understandable explanations of decisions, 182–187, 258–262

Explainable artificial intelligence (XAI), technologies enabling interpretation of machine-learning decisions in robotics systems, 182–187

Explainable embodiment, governance pillar requiring robotic behaviour to be interpretable and accountable, 221–229

Experimental governance environments, regulatory sandboxes and testing environments used to evaluate emerging robotics technologies, 276–284, 468–472

Experimental robotics regulation, adaptive regulatory approaches allowing controlled experimentation with robotics deployment, 468–472

Y

Yield from automated economies, economic surplus generated by robotics and artificial intelligence systems, 437–445

Youth workforce transition, impact of robotics and automation on emerging labour markets and early-career employment, 168–176, 437–441

Youth education in robotics governance, integration of robotics ethics, law, and technology education into academic curricula, 468–472

Human–machine symbiosis, cooperative relationships between humans and intelligent machines in social and economic systems, 372–381, 463–468

Human–machine collaboration, shared operational environments where humans and robots perform complementary roles, 372–381

Human–robot interaction, behavioural, psychological, and legal dynamics of human engagement with robots, 372–381, 395–401

Human–robot trust relationships, development of societal trust between humans and autonomous systems, 396–401

Z

Zero-risk aspiration in robotics governance, regulatory goal of minimising harm through robust design, testing, and oversight frameworks, 230–235, 276–284

Zero-trust security architecture, cybersecurity framework requiring continuous verification of robotics systems and network interactions, 214–221, 300–306

Zoning for robotic deployment, legal regulation of geographic areas where robotics systems may operate, 372–381

Zoning laws for robotics integration, urban planning regulations governing the safe coexistence of humans and humanoid robots in cities, 372–381

Zones of robotic coexistence, designated environments where humans and robots interact regularly within regulated frameworks, 372–381

Zones of restricted robotics operation, regulatory areas where robotics use is limited or prohibited due to safety or security concerns, 118–124, 372–381

About BabajideAI

BabajideAI is an AI-Native Governance, Compliance, Publication and Knowledge Platform designed to help organisations responsibly adopt, govern, and scale artificial intelligence and emerging technologies. Founded on the belief that trust will become one of the most valuable forms of infrastructure in the AI economy, BabajideAI develops practical governance frameworks, compliance systems, educational resources, and operational tools that help organisations navigate the rapidly evolving landscape of artificial intelligence, robotics, autonomous systems, and digital transformation. As intelligent systems become increasingly embedded within business operations, public institutions, critical infrastructure, healthcare, education, finance, mobility, and everyday life, organisations face a growing challenge. The question is no longer whether artificial intelligence will transform society. The question is how organisations can adopt these technologies responsibly, safely, legally, and at scale.

BabajideAI was founded to help answer that question. The platform operates at the intersection of AI, governance, ethics, policy, and business strategy. Rather than treating governance as a static checklist, BabajideAI approaches governance as a living system that must continuously evolve alongside technological capability. This philosophy reflects a broader shift occurring globally as regulators, companies, investors, and governments move beyond theoretical discussions about AI and begin implementing operational governance structures capable of managing real-world deployment. At its core, BabajideAI seeks to transform governance from a barrier into an enabler. Effective governance should not slow innovation. It should accelerate adoption by creating the trust necessary for organisations, regulators, employees, customers, and societies to embrace intelligent systems with confidence.

BabajideAI's vision is to shape a future where organisations worldwide harness artificial intelligence with confidence, guided by trusted governance frameworks, playbooks, and knowledge systems that drive ethical, adaptive, and scalable AI adoption. Its mission is to equip organisations with cutting-edge AI governance frameworks, compliance tools, and adaptive knowledge systems that enable responsible innovation, regulatory readiness, and sustainable growth in the age of intelligent systems. The Governance Challenge is real. AI is transforming nearly every sector of the global economy. Yet governance systems often struggle to keep pace with technological acceleration. Organisations face increasing complexity arising from evolving regulations, emerging risks, changing stakeholder expectations, and growing public scrutiny. Questions surrounding accountability, transparency, explainability, human oversight, cybersecurity, algorithmic bias, data governance, intellectual property, and cross-border compliance have become strategic business concerns rather than purely legal issues. BabajideAI was developed to address these challenges through an integrated governance ecosystem that combines practical implementation with strategic foresight. The platform helps organisations navigate major frameworks and standards including the European Union Artificial Intelligence Act, the General Data Protection Regulation, ISO/IEC 42001 Artificial Intelligence Management Systems, the NIST Artificial Intelligence Risk Management Framework, and other international governance requirements shaping the future of AI deployment.

Beyond compliance, BabajideAI believes the future belongs to organisations that move beyond compliance alone. The next generation of governance requires continuous monitoring, adaptive oversight, behavioural auditing, runtime accountability, and governance systems capable of responding dynamically to evolving technologies. This philosophy is reflected throughout the platform's educational resources, governance tools, research initiatives, policy development work, and advisory services. The goal is not merely to help organisations satisfy regulatory obligations. The goal is to help them build trustworthy systems that create long-term value while protecting human dignity, institutional integrity, and public confidence.

Through research, publication, and knowledge Infrastructure, BabajideAI also serves as a growing knowledge ecosystem dedicated to advancing global conversations surrounding artificial intelligence governance, robotics governance, compliance, ethics, and emerging technologies. Through book publications, practical frameworks, governance guides, industry analysis, educational content, and strategic advisory work, the platform contributes to the development of governance infrastructure for the age of intelligent systems. The long-term vision extends beyond compliance. BabajideAI aims to help shape the institutional foundations required for a future in which humans and intelligent systems coexist responsibly, productively, and sustainably. As AI evolves into embodied intelligence and physical intelligence, governance must evolve alongside it. That belief ultimately inspired the publication of Robotics Law 3.0.

About the Author

Saheed Babajide Okuboyejo is an international technology lawyer, author, and entrepreneur specialising in artificial intelligence, robotics governance, and sustainable innovation. A Solicitor and Advocate of the Supreme Court of Nigeria and a member of the Nigerian Bar Association, he holds an LL.B. from the University of Lagos, an MBA from the Gordon Institute of Business Science at the University of Pretoria, and an LL.M. in European Private Law from the University of Amsterdam. His work spans AI governance, robotics governance, renewable energy, and emerging technologies, with a focus on building practical, globally relevant frameworks for responsible innovation. He is the founder of BabajideAI, an AI-native platform for governance, compliance, and the ethical adoption of intelligent systems. Robotics Law 3.0 builds on his earlier work, Drone Law 3.0, extending his legal and governance vision into the age of embodied and physical intelligence. Across his writing, he combines legal depth, strategic foresight, and a strong commitment to shaping frameworks that help law respond meaningfully to technological change. Saheed is based in Amsterdam, the Netherlands, where he currently advises on AI and robotics governance and continues to research and publish future titles in the Law 3.0 Series.

www.ingramcontent.com/pod-product-compliance
Lightning Source LLC
LaVergne TN
LVHW090254210926
848480LV00001B/19
9798902435501